P9-DWG-698

CALIFORNIA NATIVE TREES & SHRUBS

For Garden & Environmental Use
In Southern California
And Adjacent Areas

LEE W. LENZ
Director,
Rancho Santa Ana Botanic Garden
Professor of Botany,
Claremont Graduate School

JOHN DOURLEY
Superintendent,
Rancho Santa Ana Botanic Garden
Advisor,
California Department Water Resources

Published by
Rancho Santa Ana Botanic Garden
Claremont, California 91711

The Cover: The photograph on the cover is of *Fremontodendron mexicanum,* one of California's most handsome plants, known only from a few populations in southern San Diego Co., it is now listed as an endangered species. Flowers shown approximately natural size.

Library of Congress Catalog Card Number 81-50257.

LIBRARY OF CONGRESS CATALOGING IN PUBLICATION DATA

Lenz, Lee W.

California native trees & shrubs
Bibliography: p.
Includes indexes

1. Ornamental trees — California. 2. Ornamental shrubs — California. 3. Trees — California. 4. Shrubs — California. I. Dourley, John. II. Title.

SB435.52.C2L46 635.9'77'09794 81-50257
AACR2

ISBN 0-905808-1-6
ISBN 0-905808-0-8 (pbk.)

Retrieval terms: gardening, horticulture, landscaping, environmental landscaping, California, trees, shrubs, water conservation, botany, plant climates, plant communities.

Manufactured in the United States of America.

The winter of 1977-78 brought desperately needed rain to California in record amounts. It filled the rivers, and lakes and reservoirs. It deposited tons of welcome snow in the mountains. And it lulled us into a false sense of security. The drought was over. . . . And we forgot rather quickly just how dry California can be.

California's weather patterns are anything but predictable. Perhaps winter rains will come again. Perhaps not. But the experts tell us that even normal rainfall may be incapable of filling our needs in years to come. But the state's history of lower than average rainfall and drought virtually guarantees a limited supply for future generations.

How To Have A Green Garden In A Dry State. Metropolitan Water District of Southern California.

Water shortages will become more serious, primarily in places where they are already a problem. The cost of developing new water supplies will increase.

Global 2000. Report of the Council on Environmental Quality and the United States Department of State.

Dedicated to the memory of

RALPH DALTON CORNELL

gentleman, and southern California's first, and one of its most distinguished landscape architects, who as a youth more than sixty-five years ago advocated the planting of native trees and shrubs, pre-adapted by Nature to the harsh environment of semi-arid southern California.

Southern Oak Woodland

Contents

IN A TIMELY ARTICLE on the present state of ornamental horticulture (*Pacif. Hort.* 39 (3): 15-23), Professor Russell Beatty of the University of California, Berkeley, noted that we are living in an age of re-evaluation and that horticulture and its related professions are not exempt from this period of introspection. According to Beatty, one important consideration is the ecological implication of horticulture, and he writes that the horticulturist has up to the present dealt with city plants while the ecologist has remained camped in the mountains. According to him, ecology needs horticultural expertise, and horticulture must employ ecological principles. Although we may not think of it as such, a city is, as Beatty points out, an ecosystem, and plants must be considered as an integral part of the environment rather than just decorative greenery. In suburban or rural landscape the selection of plants, according to Beatty, should relate to the ecosystem in which they are planted and he emphasizes the fact that this does not necessarily mean using only plants native to a particular area. In planting where the native vegetation is dominant, care should be taken to complement native vegetation, and wherever possible the use of plants of species already growing in a plant community should be encouraged. He concludes his essay by saying that in the West, a quiet revolution is under way which will change the face of horticulture.

From Beatty's essay one gains the impression that many horticulturists are only now becoming aware of how important an understanding of environmental biology and ecosystems is to the practise of ornamental horticulture, as well as of the value of native plants in today's planting schemes. History however discloses the fact that in the past there have been pioneers who not only advocated the use of native plants but also the application of sound ecological principles to the practise of horticulture. Unfortunately those pioneers were for the most part ignored.

As early as 1912, Ralph D. Cornell, southern California's first

Preface

and one of its most gifted and respected landscape architects, but then a twenty-two-year-old student at Pomona College, was promoting the use of native trees in southern California landscaping. Writing in the *Pomona College Journal of Economic Botany* (2: 301-314) he said, "Why not plant dry hillsides in such a manner that they will produce maximum results at minimum expenditure? Use native plants, plants already accustomed to the semi-arid conditions of our soil, and climate. Such a park would be decidedly typical and distinctive of California; it would be a garden spot of nature. . . ."

Another pioneer was Theodore Payne. No man has ever done more to promote the use of California plants in horticulture than did Payne who in 1915 planted two-and-a-half acres in Exposition Park in Los Angeles to native plants and then spent countless hours on the grounds explaining to visitors the horticultural virtues of the plants. The first nurseryman in southern California to specialize in natives, Payne continued until his death in 1963 to champion their cause.

Still another pioneer was Guy Fleming, later District Superintendent of the California State Parks Southern District, who in 1916 was the first to suggest that landscape plantings should be arranged in ecological associations or life zones. A bold concept then, Fleming's ideas were many years ahead of their time.

Also, more than fifty years ago there were established in southern California three botanic gardens whose goals included the promoting of California native plants for use in horticulture. But the time was not ripe for horticulturists to listen to these pioneers. Californians were out en masse to create a tropical paradise in the desert, a dream that has been shattered by the events of the past few years. If a revolution is now taking place in California horticulture the seeds for it were sown many years ago, and it is the critical shortage of water in the West, along with the recent surge in environmental awareness on the part of the public that are now fueling the flames.

THIS VOLUME is directed to several audiences: to landscape architects, park superintendents, nurserymen, highway planners and commercial developers; to college and university students in ornamental horticulture and environmental science and to concerned citizens.

It is based upon over 50 years of experience acquired by the Rancho Santa Ana Botanic Garden in the growing and use of native plants and although intended primarily for southern California, many of the recommendations apply equally well to areas as far north as San Luis Obispo Co., east to Fresno Co., southern Nevada, western Arizona and northern Baja California. It is our belief that an understanding of the environmental conditions under which the native flora evolved will better enable the horticulturist to match plants with the environment into which they are being placed. For that reason we have included sections on southern California climates, soils and natural plant communities. California possesses one of the most interesting and distinctive floras of the world; however, we are aware that some of the most beautiful California natives do not in general make satisfactory garden subjects and are best enjoyed in their native settings or at botanic gardens which attempt to meet their special requirements. Under cultivation some natives may be attacked by insect pests and plant disease or suffer from air pollution and these facts are noted.

We would like to support a recent author (Hildreth, 1977) in attempting to dispel the idea that native plants require no care after they are planted, as well as the impression that all natives are drought tolerant and water conserving. These misconceptions must be corrected if native plants are to assume their proper role in the horticulture of the future. Neither do we advocate the use of only native plants. Other areas of the world with a mediterranean climate have evolved distinctive and interesting floras, and representative species from these other regions may be grown together with the California natives to produce esthetically satisfying landscapes.

Introduction

In recommending plants for specific regions if we have erred it is probably on the conservative side and some plants may have a much wider range of tolerance than here indicated.

It is not possible for us to name all those who have contributed to this work and deserve our thanks. Without the highly original and successful record keeping system established by Carl Wolf and Percy Everett, pioneers in the growing of California natives, we would not have been able to trace the history of many of the plantings and the attendant changes in methods of cultivation which have taken place over the years. We are also much indebted to Warren Sullivan who for nearly 30 years was propagator at the garden, and to Nick Lolonis for his observations on insect pests and plant diseases. We wish to thank the many unnamed plant collectors who have called our attention to new and untried materials and have often provided us with seeds or cuttings.

We also wish to thank the people at our sister institution, the Santa Barbara Botanic Garden, particularly Ralph Philbrick and Dara Emery for information on the growing and handling of native trees and shrubs in coastal areas. We thank Bea Beck who was very helpful in bibliographic matters. We particularly wish to acknowledge our debt to Dick (C. W.) Tilforth for his many helpful suggestions made throughout the preparation of this work. We express our appreciation to the Board of Trustees of the Rancho Santa Ana Botanic Garden for their interest and support.

And finally we thank the students and teachers in the many southern California colleges and universities who have inspired us and have provided us with the stimulus to complete the work.

Claremont, California
November, 1980

Chapter 1

ENVIRONMENTAL CONSIDERATIONS

As HERE CIRCUMSCRIBED, southern California and northern Baja California is defined as that portion of California south of a line extending along the crest of the Santa Ynez Mts., north through Sespe Canyon, east through Lockwood Valley, crossing I-5 near Lebec, around the south base of the Tehachapi Mts. and then eastward where it follows the Garlock Fault-Leach Line to the Nevada border. The southern boundary of the area is a line that extends from near El Rosario in Baja California Norte, east to the Gulf of California. The western portion of this line marks the natural southern limit of the Californian Botanical Province, whereas in the east the line arbitarily cuts across the southern extension of the San Felipe Desert.

Southern California is a region of great topographic and geological diversity. Included within its limits are high mountain ranges, valleys whose floors lie below sea-level, precipitous canyons and broad basins at many elevations and an assemblage of other physiographic features that reflect a complex geological history (Jahns, 1954). Elevations vary from about 235 ft. below sea-level at the Salton Sea to 11,502 ft. on Mt. San Gorgonio. Precipitation averages from slightly over two inches a year at El Centro to more than 50 inches in the higher mountains. Temperatures vary from an extreme summer high of 125 F at Indio to winter lows of —25 F in the higher mountains.

Geologists recognize in southern California eight natural provinces (Jahns, 1954), four of which are confined to the area here defined as southern California; they are the Mojave and Colorado deserts and the Transverse and Peninsular ranges (Fig. 1).

Within southern California the Mojave Desert Province is the most extensive and is described as being in large part a gigantic fault-bounded wedge that points westward, bounded between the Garlock Fault on the north and the San Andreas Fault on the southwest and in California bounded in the northeast by the Nevada state line. Included within the Mojave Desert are numerous desert ranges with the Providence and New York mts. having the greatest elevations. The only major drainage system is that of the Mojave River.

In southern California the Colorado Desert Province is an elongate low-lying depression north of, and separated from, the Gulf of California by the delta of the Colorado River. Much of the area lies below sea-level and a portion of it is occupied by the Salton Sea. To the north it lies between the San Jacinto Mts. on the west and the Little San Bernardino Mts. to the east with its northernmost limit at about White River. To the south it extends from the base of the Laguna Mts. east to the Colorado River.

The Transverse Range Province is made up of a series of east-west trending mountain ranges that include (from west to east) the Santa Ynez, Santa Monica, San Gabriel, San Bernardino, Little San Bernardino and Eagle mts. Mt. San Gorgonio (11,502 ft.) in the San Bernardino Mts. is the highest mountain in southern California. Mt. San Antonio (Baldy) in the San Gabriel Mts. has an elevation of 10,064 ft. and there are other peaks over 9000 ft.

The Peninsular Range Province lies to the south of the Transverse ranges and consists of a series of more-or-less north-south trending mountain ranges most of

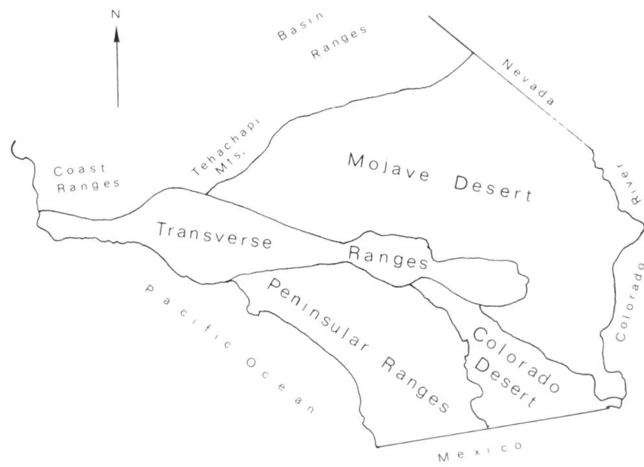

Fig. 1. The geological provinces of southern California (Jahns, 1954).

them of moderate elevations; they include, in California, the Santa Ana, Palomar and San Jacinto mts.; in Baja California, the Sierra Juárez and the Sierra San Pedro Mártir ranges.

For information on the geology of southern California the reader is referred to Jahns (1954); more recent information on plate tectonics and related matters may be found in Anderson (1971), Dewey (1972) and Hallan (1972). The origin and relationships of the California flora has been admirably covered by Raven and Axelrod (1978). For information on the geology of Baja California the reader should consult Durham and Allison (1960), and Gastel, Philips and Allison (1975). For the history and relationships of the land flora of Baja California, Wiggins (1960) is recommended.

Living within southern California (approximately 42,540 square miles) are 13.25 million people, or about 58% of the state's population, and in Baja California Norte there are about 1.3 million people or over 99% of the population of the peninsula.

The Climates of Southern California

Numerous classifications of the world's climates have been proposed. One of them, that of the Austrian climatologist, Wladimir Köppen, has been widely accepted in this country by geographers and climatologists alike (Trewartha, 1954). His classification is based upon annual and monthly means of temperature and precipitation. Native vegetation is looked upon as the best expression of the totality of a climate so that many climatic boundaries are selected with vegetation in mind. A unique and distinctive feature of the Köppen system is the employment of an ingenious symbolic nomenclature in designating climates (Trewartha, 1954).

Trewartha in presenting his classification, which is based upon that of Köppen, and the one that we will follow, says that "for those, such as geographers, biologists, or agriculturists, who need to understand and use the climatic environment for their own purposes, the facts of climate must be presented realistically. . . . First and foremost one must be guided by the observed facts." Following Köppen, Trewartha recognizes five 'Groups of climates' (*A*) *Tropical*; (*B*) *Dry*; (*C*) *Humid mesothermal*; (*D*) *Humid microthermal*; and (*E*) *Polar*. Of these groups, two (*B*) and (*C*) are found in southern California. Groups (*A*), (*C*) and (*D*) are subdivided into three categories depending upon the timing of the dry period: no dry season, *f*; dry season in the summer, *s*; and dry season in the winter, *w* (Fig. 2). Trewartha divides the *Humid mesothermal* (*C*) climate (where *C* indicates the average temperature of the coldest month to be between 18 C (64.4 F) and 0 C (32 F) into six subcategories only one of which is found in southern California, the *Cs*, dry summer subtropical, better known as *mediterranean*-type climate.

The *Cs* climate is then divided into *a*, hot summers

Köppen's Climatic Groups and Types

Climatic group	*Symbol*	*Dry period*	*Degrees of dryness of cold*	
Tropical rainy climates	*A*	*f* (*s*) *w*		
Dry climates	*B*	. . . . .	*S*	*W*
Warm temperate rainy climates	*C*	*f s w*		
Cold snowy forest climates	*D*	*f* (*s*) *w*		
Polar climates	*E*	. . . . .	*T*	*F*

Fig. 2. Köppen's world climatic groups and types (Trewartha, 1954).

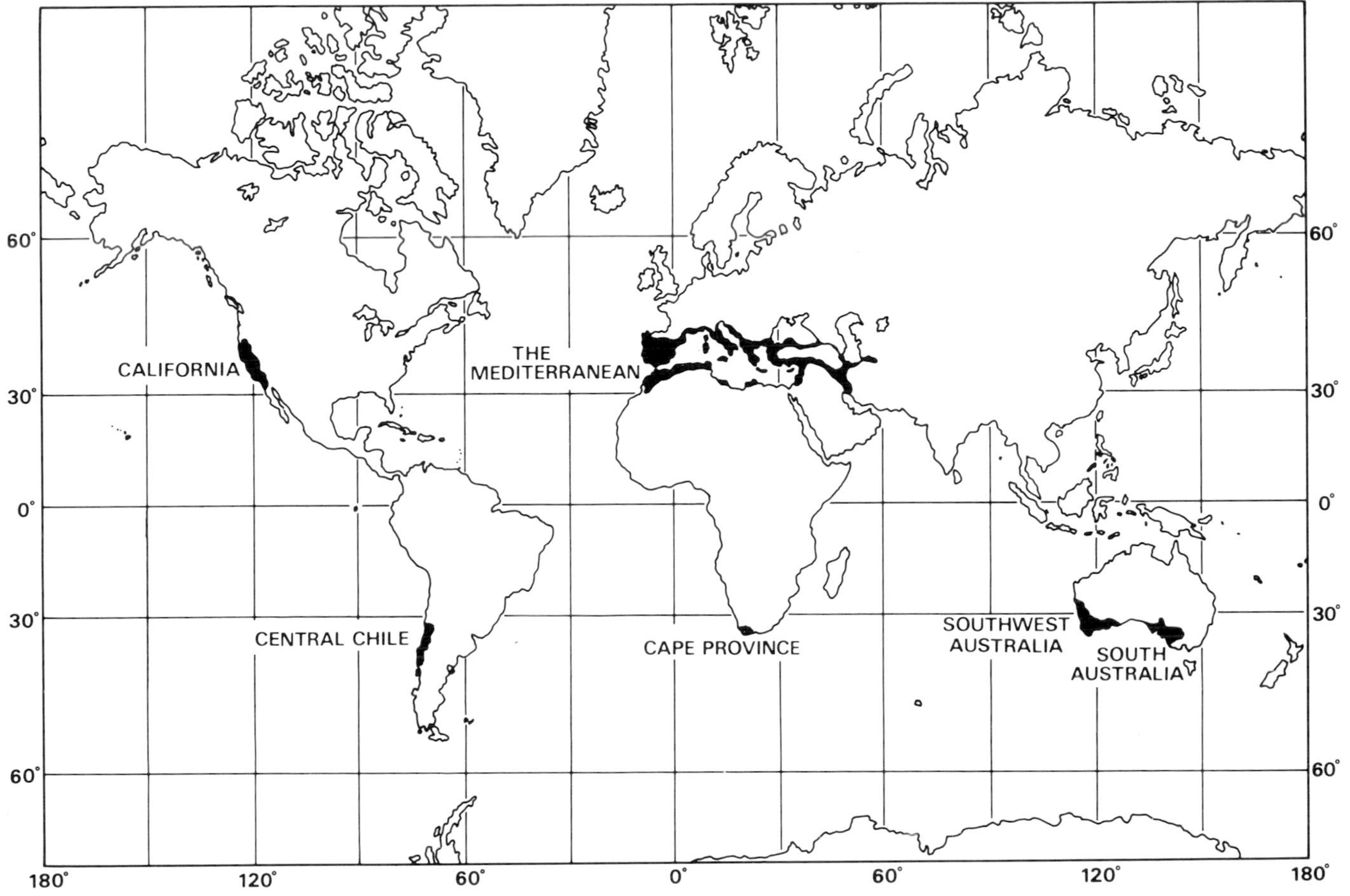

Fig. 3. World map of areas with mediterranean climate. (Thrower and Bradbury, 1977).

with an average temperature of the hottest month over 22 C (71.6 F) and *b*, cool summers with an average temperature of the hottest month less than 22 C (71.6 F). According to Russell (1926), the two subtypes of the mediterranean climate, which covers about one-half of the state, apply particularly well to California.

According to Trewartha, the *Cs* climate characteristically is located on the tropical margins of the middle latitudes (30°-40°) along the western sides of continents. In addition to the mediterranean region from which it takes its name, it is also found in California, central Chile, South Africa, western and southwestern Australia, for a total of 1.7% of the earth's surface (Fig. 3).

The *Csa* climate has its widest distribution in the borderlands of the Mediterranean Sea in southern Europe, North Africa and western Asia. Another smaller area is the more interior sections of mediterranean-Australia. In southern California and northern Baja California it occurs inland in what are often referred to as the intermediate and inland valleys.

The *Csb* climate characterized by cool summers, fronts upon cool coastal waters and it is found in mediterranean-Chile and South Africa, the Atlantic Coast portions of the European-North African *Cs*, the most southerly and coastal parts of mediterranean-Australia and certain subtropical highlands in the Middle East (Trewartha, 1954). In California and northern Baja California it is found only in what is referred to as the coastal areas. In *Cs* climates precipitation is generally less than moderate, 15-25 inches, but more characteristic is the distribution over the year, most of it occurring during the cooler months with the summers being dry. In Los Angeles, 78% of the annual precipitation falls between December and March (Trewartha, 1954). The average monthly

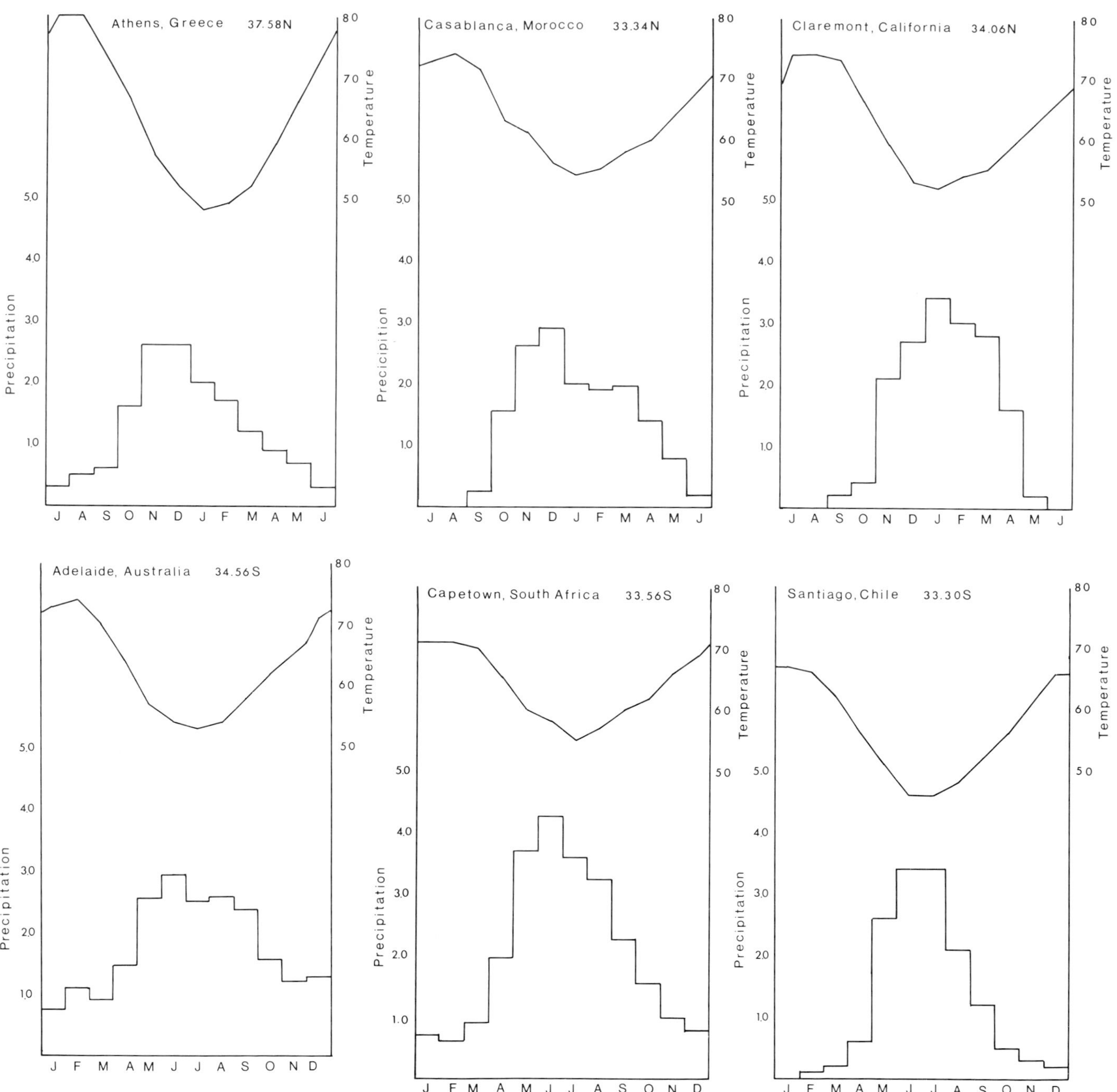

Fig. 4. Average annual temperature and precipitation for three stations in the Northern Hemisphere (top) and for three stations in the Southern Hemisphere (bottom).

temperature and precipitation at three representative stations in the Northern Hemisphere and for three stations in the Southern Hemisphere having mediterranean climate are shown in figure 4.

The daily temperature extremes for the coldest and hottest months for two pairs of paired stations, one in the Northern and one in the Southern Hemisphere, are shown in figure 5.

The most characteristic vegetation type associated with the mediterranean climate is referred to in California as chaparral and in Europe as *maqui.* The chaparral will be considered in detail in another chapter.

The dry climates (*B*), which may be either cold or hot, are characterized as areas where evaporation exceeds precipitation and they have been subdivided into six subtypes, one being *BW*, arid desert, where *W* stands for *Wüste* (German, desert). The *BW* climate is then subdivided into *BWh* (*h*, German, *heiss*; English, hot) and *BWk* (*k*, German, *kalt*; English, cold). The California deserts are classified as *BWh,* having annual temperatures over 18 C (64.6 F). Russell (1926) noting that there were obvious and consistent differences both in climate and vegetation between the Mojave and Colorado deserts proposed that the latter be coded *BWhh* (Fig. 6).

Included in the *BWh* low latitude hot deserts are some of the driest places on earth such as the Sahara, Kalahari in South Africa, Atacama-Peruvian Desert in western South America, the Thar in northwestern India and the Sonoran Desert that includes the Lower Colorado Valley, a portion of which is found in southeastern California and northeastern Baja California. Because of the great year-to-year variation in total rainfall, Tre-

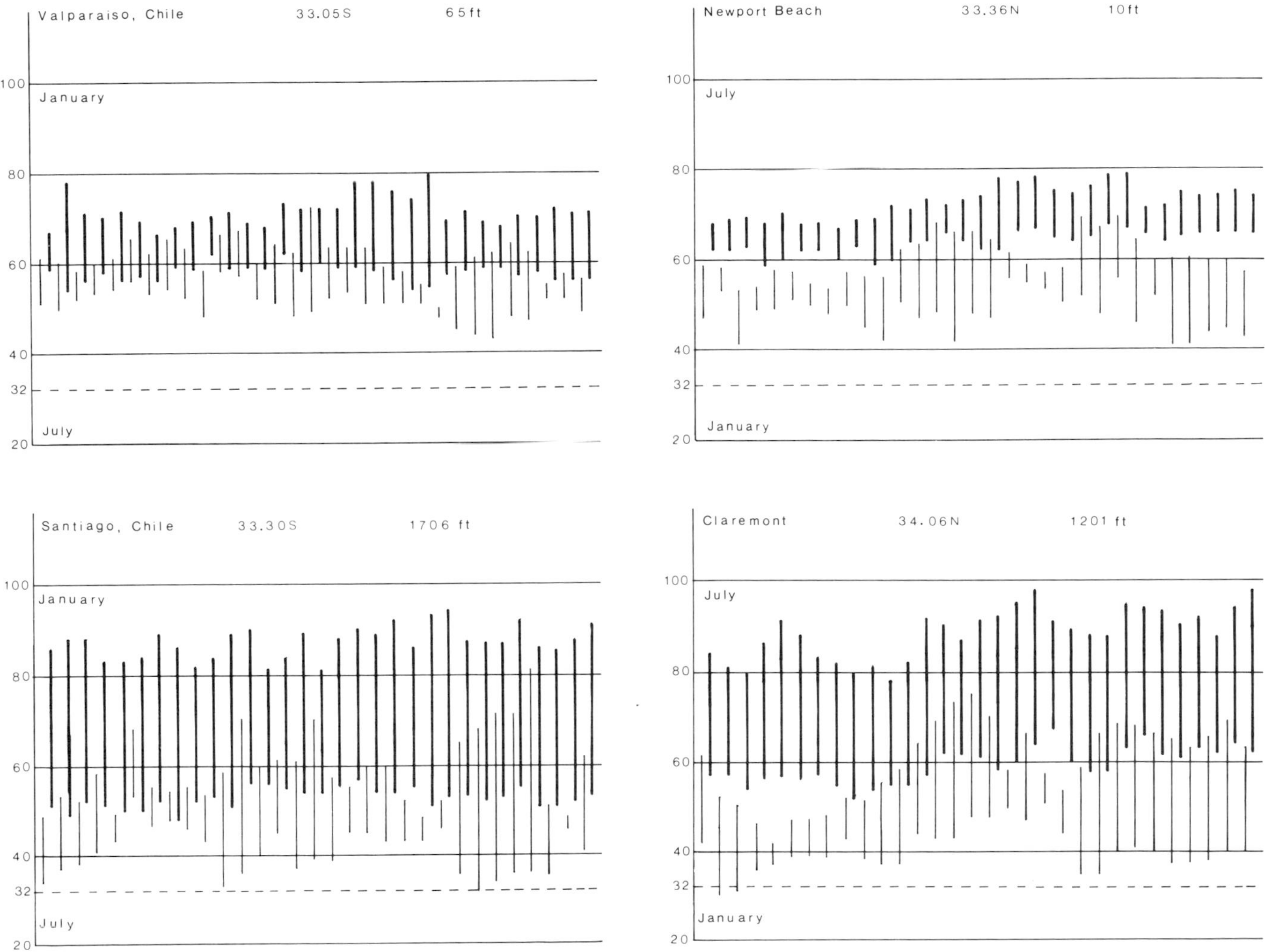

Fig. 5. Daily minimum and maximum temperatures for the hottest and coldest months for two paired coastal stations (top) and for two inland stations (bottom), one in the Northern (right) and one in the Southern Hemisphere (left). Data are for one year.

wartha says that it is almost impossible to speak of an *average* annual rainfall for desert stations. Characteristic of *BWh* climates is the abundance of sunshine. An example is Yuma, Arizona which receives for the year, 88% of the total possible sunshine. Due to high temperatures and low relative humidity, evaporation is excessive, one of the reasons for the effectiveness of the so-called evaporative coolers. At Yuma the average evaporation during the hot months is 55 inches while the average rainfall for the same period is not quite one inch (Trewartha, 1954). At the United States Date Gardens at Indio, for the year 1975, the total annual evaporation was 112.96 inches (*i.e.,* free evaporation from an open pan) and the total precipitation was 3.00 inches.

In the deserts, variation in temperature both daily and annual is great. Figure 7 shows the daily minimum and maximum temperatures for the coldest and hottest months at two stations, one in the Mojave Desert and one in the lower Colorado Valley.

The seasonal distribution of rainfall in the lower Colorado Valley and the Mojave Desert is bimodal and in this respect differs from the distribution of precipitation in the *Cs* climates of southern California. The effects on vegetation of summer rainfall are marked and can be observed almost as soon as the Colorado River is crossed. One of the most conspicuous examples being the abundance of sahuaro, *Carnegiea gigantea,* east of the Colorado River and its almost complete absence

Fig. 6. The climates of southern California. See text for explanation.

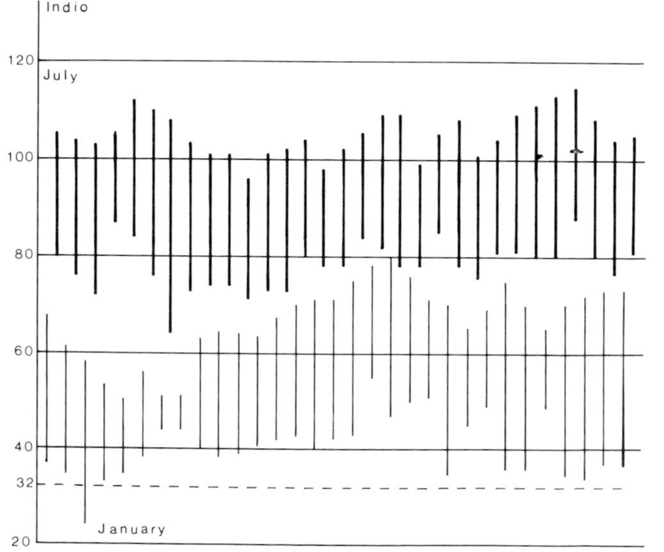

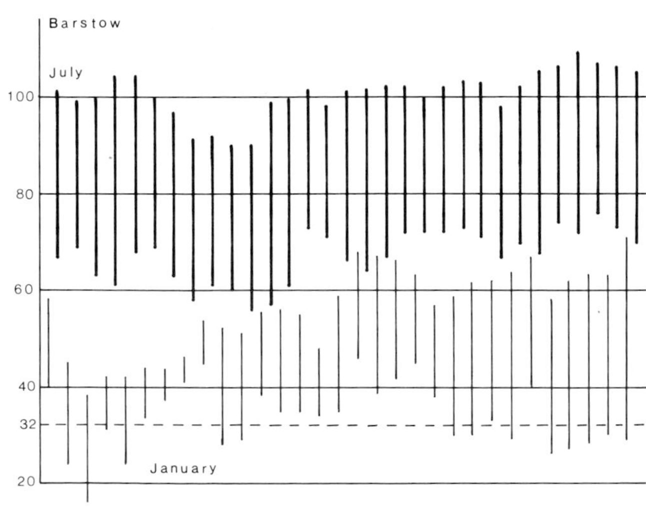

Fig. 7. Daily minimum and maximum temperatures for the coldest and hottest months for one station in the Mojave Desert (Barstow) and one station in the Lower Colorado Valley (Indio). Data are for one year.

west of the river. Figure 8 shows the amount and distribution of rainfall for five stations with *BWh* and *BWhh* climates, four of them in the Lower Colorado Valley and one in the Mojave Desert, and for one station with a *Csa* climate. It can be noted that the bimodal distribution of precipitation strongly developed at Phoenix decreases in magnitude toward the west and is entirely absent in the *Csa* climate. It can also be seen that the peak of the summer rains comes progressively later toward the west.

Figure 9.1 shows the annual precipitation for the past 100 years at three stations in southern California having either a *Csa* or a *Csb* climate.

The climate of North America from the 1930s through the 1960s has been unusually mild in certain localities and as such is not representative of conditions in the long-term past (Fritts and Gordon, 1980). According to these authors, the climate conditions to be anticipated in the future should not be evaluated exclusively from existing meteorological records but should include long-term climatic information found in proxy records. The extended record thus allows for a more realistic assessment of past long-term climatic variations.

Fritts and Gordon prepared, by use of tree ring analysis, what they called a "Reconstructed Statewide Precipitation Index For California" for the years 1600 to 1961 (Fig. 9.2). They were able to show that there have been prolonged periods of time when the state has experienced deficient rainfall. The longest period of drought extended from about 1760 to 1820 and the Fritts and Gordon data corresponds well with mission records. A drought of this length is unparalleled anywhere in the reconstruction. According to Fritts and Gordon, not only was it of great length, but the magnitude of precipitation deficit was large. Dry periods near the intensity of the 1760-1820 drought are reconstructed at other times as well; these were from about 1600 to 1625, 1665 to 1670, 1720 to 1730 and from 1865 to 1885. The authors acknowledge that additional work will be required before more solid conclusions concern-

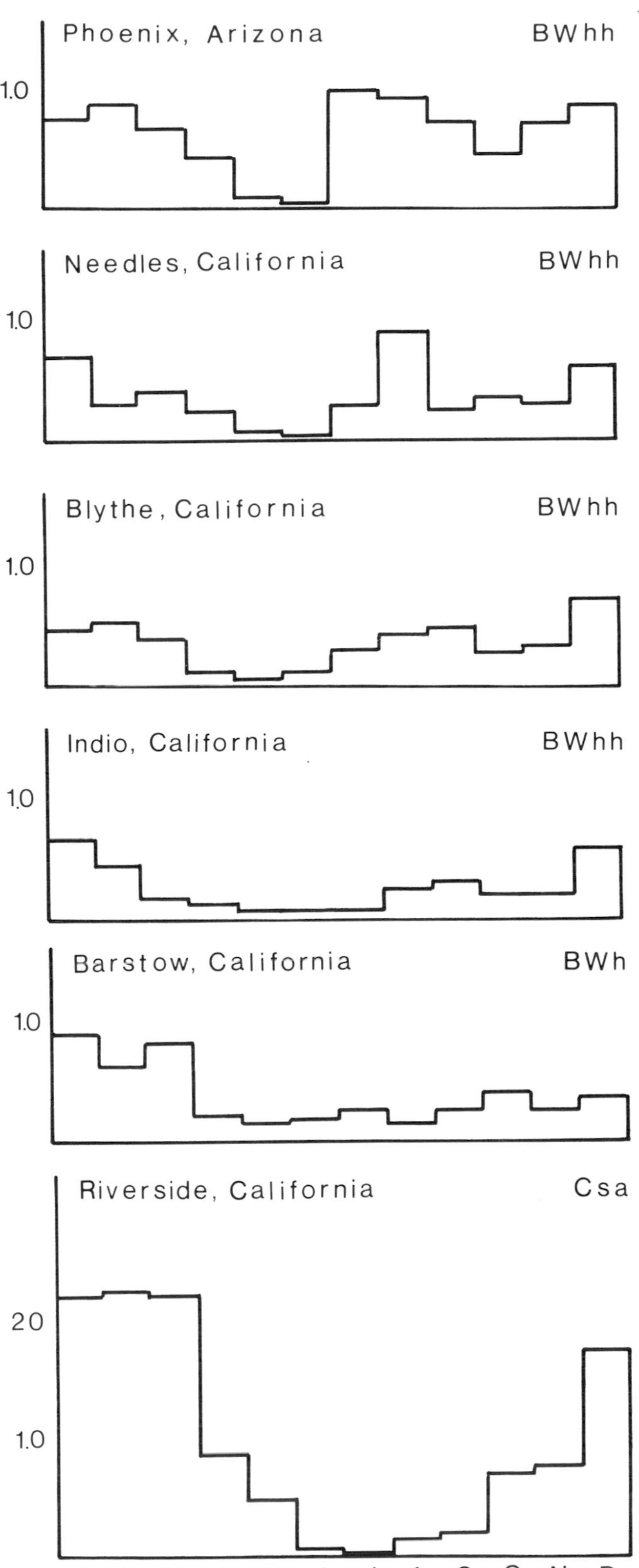

Fig. 8. Average monthly precipitation at four stations in the Lower Colorado Valley, one station in the Mojave Desert and one station in an inland valley. Precipitation in inches.

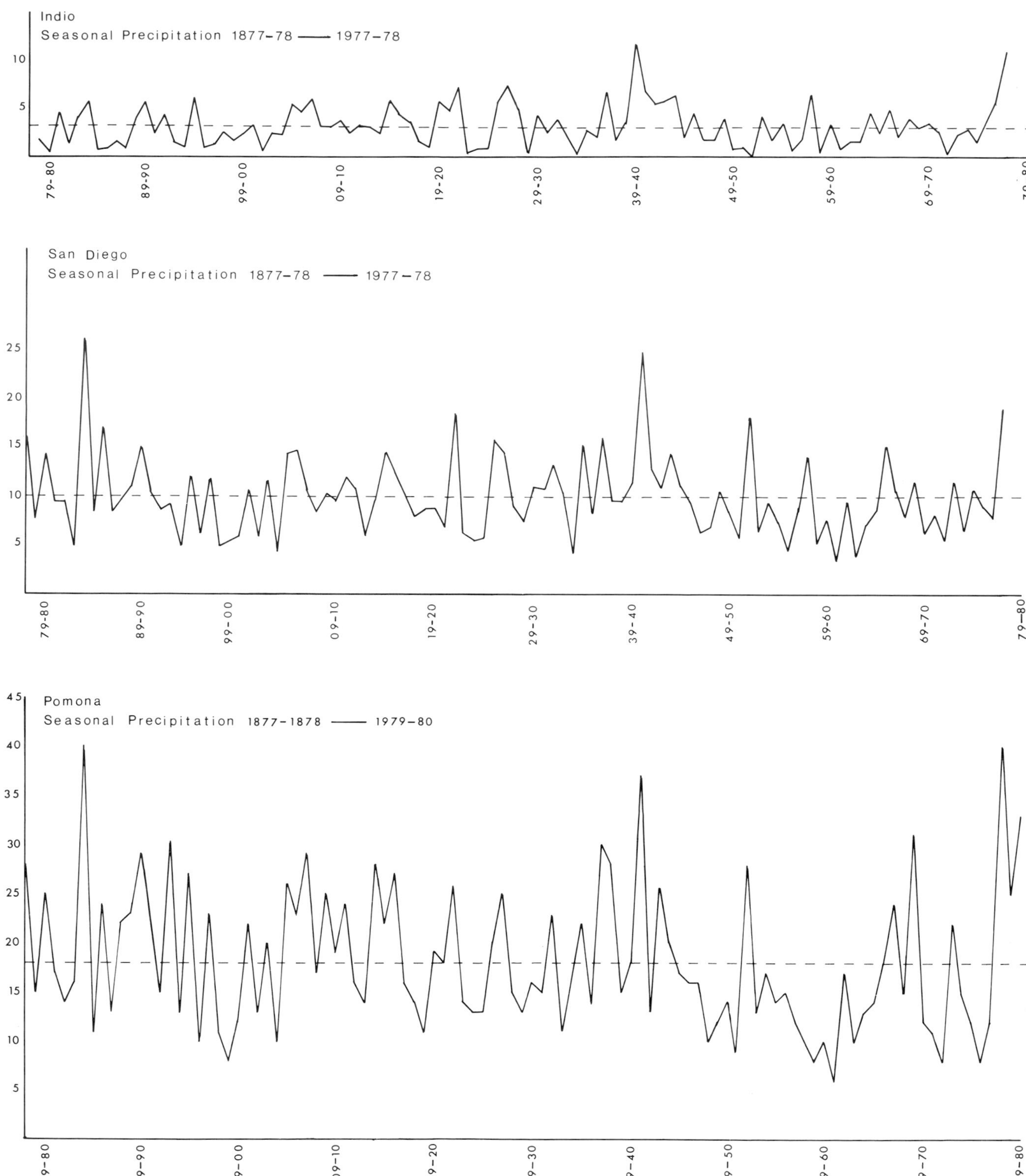

Fig. 9.1. Annual precipitation for the past 100 years at three stations in southern California. Dashed line indicates average rainfall.

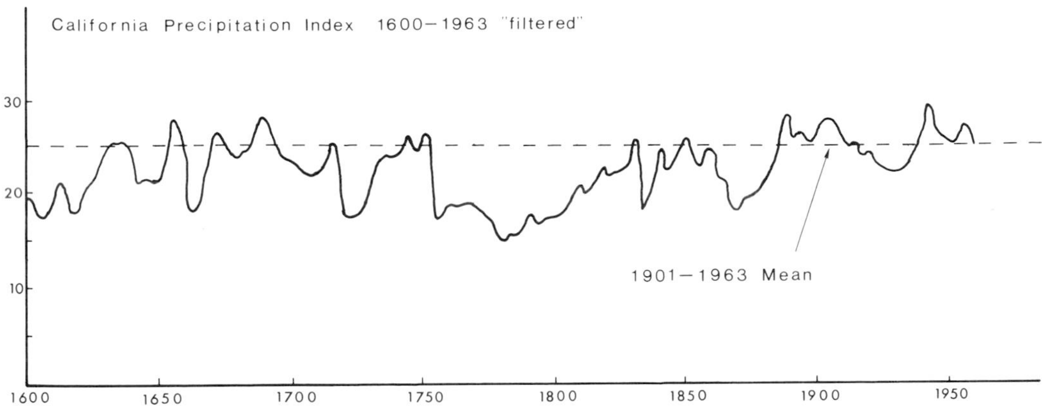

Fig. 9.2. Reconstructed precipitation index for California for the years 1600-1961. See text for explanation. (Fritts and Gordon, 1980).

ing the long-term variability of California precipitation can be drawn, but present evidence clearly demonstrates that fluctuations of precipitation sufficient in intensity and duration to have economic and social impact have probably occurred six times since 1600. In the perspective of the 360-year reconstruction of precipitation, the period from 1890 to 1960 was one of precipitation surplus.

The climate of the mountainous areas of southern California, such as Big Bear Lake, Lake Arrowhead, Idyllwild and Mt. Palomar, differs at least in degree from the other areas having a *Cs* climate. Vegetatively they are very distinct having coniferous forests rather than chaparral or coastal sage scrub characteristic of southern California's *Cs* climates. However in the Trewartha classification the mountains (at moderate altitudes of 5000-8000 ft.) would be considered *Csb* since the minimum temperature for the hottest month is below 22 C (71.6 F) (averages for five stations is 69.3 F). Russell (1926) quite rightly recognizes that the climate of these mountainous areas differs sufficiently from the *Cs* climates to be afforded recognition and he proposed another subdivision, *Cs'ab*.

A much more detailed classification of the climates of southern California is offered by Bailey (1954). By using four classes of temperature: *cold, cool, warm* and *hot*; and four classes of moisture: *humid, subhumid, semi-arid* and *arid*, he arrives at 16 possible combinations, 12 of which are recognized as occurring in southern California (Fig. 10). His temperature classes are based on annual average temperature: *hot*, > 69 F; *warm*, 58-68 F; *cool*, 44-57 F; and *cold*, < 44 F.

His moisture classes are arrived at by a complex formula which takes into consideration not only average annual temperature and annual precipitation but also the seasonal distribution of the precipitation. Bailey's classification while useful in showing the diversity of climatic types in southern California has not been used by botanists or horticulturists.

No discussion of the climates of southern California would be complete without mention of the Santa Ana winds which strike the area mainly during the period between October and March and in exceptional cases reach gale or hurricane force. These winds, which may be either cold or hot, and with relative humidities of often less than 5% and sometimes as low as 1% may cause extensive horticultural losses not only due to the mechanical damage they inflict on the plants but even more important is the stress placed on the plants through excessive transpiration caused by low humidities. This is particularly true with plants that have only recently been transplanted.

Santa Ana winds develop when there is a large high pressure system over the Great Basin and a surface low pressure trough off the California coast. As the air descends and flows seaward it is heated by compression and exits through the mountain passes, or over the mountain ranges, as strong winds that blow over the lower basins and out to sea. Figure 11 shows the surface streamlines of a strong Santa Ana wind with the air speeds indicated

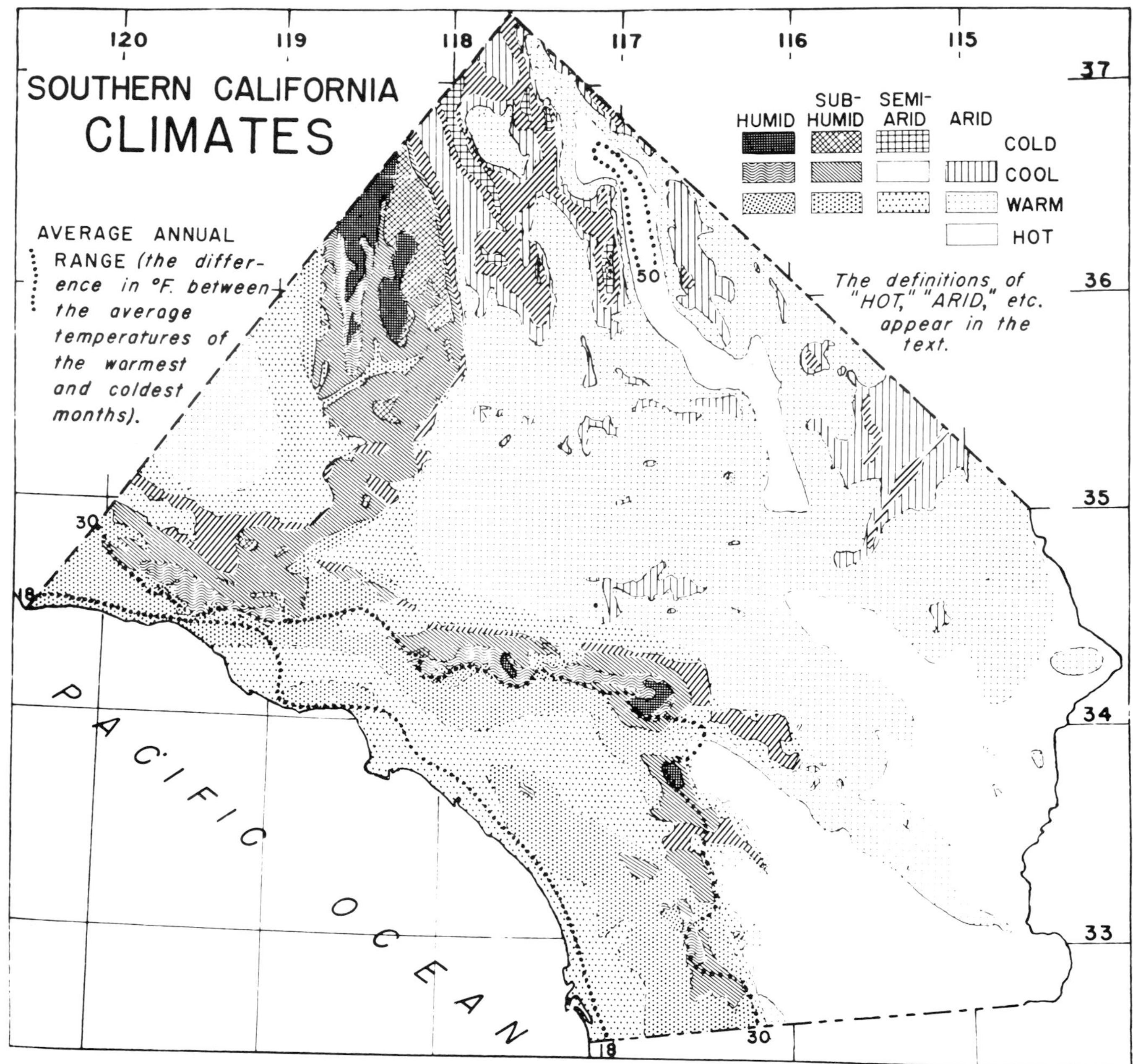

Fig. 10. Southern California climates. (Bailey, 1954).

in figures on the map. Note that the speed of the wind is greater along the coast from Santa Monica south than it is at many places inland. The strongest winds are found in the San Bernardino Valley. In weaker Santa Anas more of the wind descends through the passes and less over the mountains than is shown in figure 11.

Major Soil Types in Southern California

Probably few states can equal California in the number and diversity of soil types to be found within its borders and a large number of state soil surveys have been made and the results published. R. Earl Storie and Walter W. Weir of the University of California have prepared a

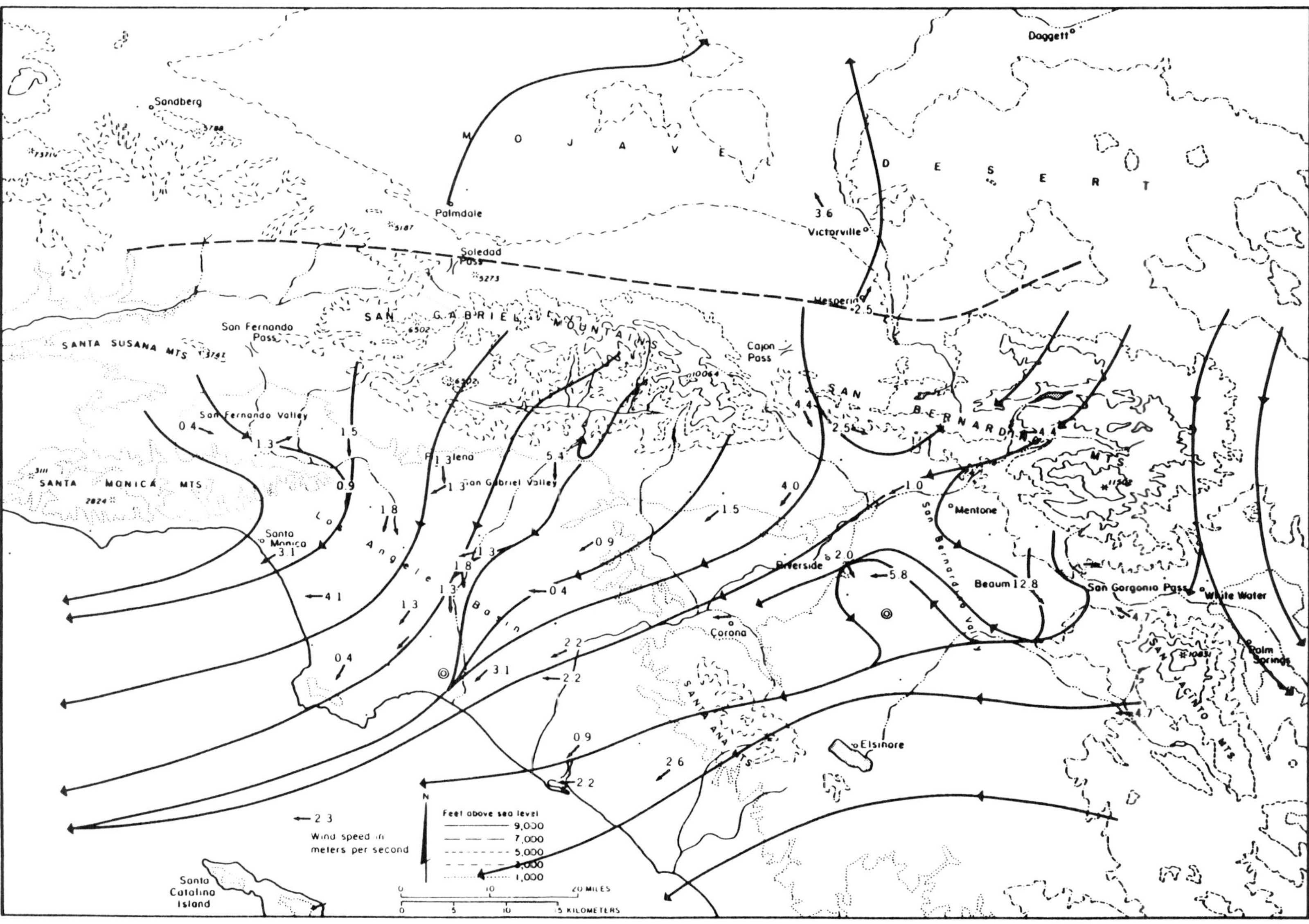

Fig. 11. Surface streamlines during a severe Santa Ana windstorm. Note that from Santa Monica south the wind is stronger near the coast than it is farther inland. During less severe storms more of the wind comes through the passes and less over the mountains. (M. A. Fosberg, et al. United States Forest Service Research Paper, PSW-30. Pacific Southwest Forest and Range Exp. Sta., Berkeley).

generalized soil map of the state (Storie and Weir, n.d. [1963]), the portion devoted to southern California is shown in map 1. In their foreword they state, "It must be kept in mind that its scale imposes certain limitations on its use. . . . Areas of less than 4,000 acres could not be shown in a map of this scale." Even with these restrictions their map and accompanying text are of considerable value in gaining an understanding of some of the edaphic factors underlying the development and distribution of the plant communities in southern California.

Storie and Weir have divided the state into four major land categories: *Valley Land, Valley Basin Land, Terrace Land* and *Upland.*

Valley Land is described as gently sloping, with smooth topography and includes the best all-purpose agricultural land in the state. The *Valley Land* is divided into six subcategories, four of which are found in southern California.

An. These are the deep alluvial fan and floodplain soils occurring in areas having intermediate amounts of rainfall but insufficient for the production of forest. They represent the most important agricultural soils in California and are highly valued for growing of many crops. In southern California this soil type is found in the Santa Clara, San Fernando, San Gabriel, Pomona and Walnut valleys and the Los Angeles, Oxnard and San Bernardino plains as well as on the Santa Rosa Plateau. Some of this area was originally southern California grassland.

Ac. These are the fan and floodplain soils of desert and semi-desert regions having low rainfall (1-7 in.). Usually light colored and low in organic matter and calcareous they are common in the Imperial and Palo Verde valleys and in parts of the Mojave Desert where they form a strip extending along the slopes of the Antelope Valley eastward to Victor, Apple and Lucerne valleys. The area is too dry to produce crops without irrigation, but under irrigation some of the area, such as portions of the Antelope Valley, is highly valued for the production of alfalfa; the Imperial Valley for the production of alfalfa, sugarbeets, cotton, wheat and sorghum and vegetable crops such as lettuce, cantaloupes, carrots, tomatoes and asparagus; and Palo Verde Valley for cotton.

At elevations of 2500-5000 ft. this soil type supports the Joshua tree woodland plant community. In the Imperial Valley where the soil is less well drained it supports the saltbush scrub community.

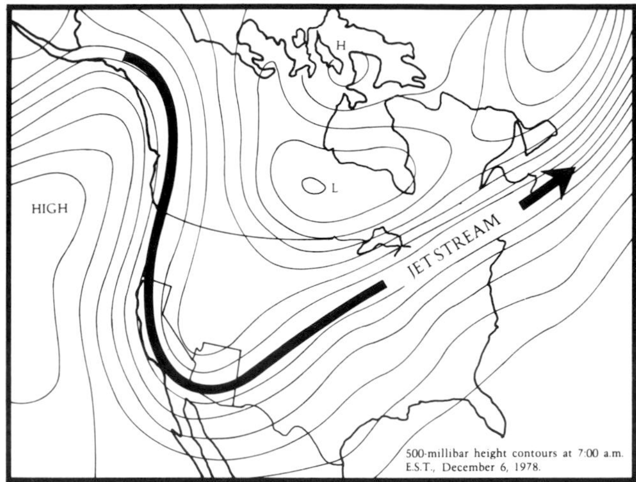

Fig. 12. The severe freeze of 1978. Position of the jet stream on the morning of 6 December 1978 bringing frigid air from Mt. McKinley south along the Pacific coast, turning eastward across Arizona and continuing northeastward to New England. Minimum temperatures for stations in southern California. Figures are degrees Fahrenheit. Arrowhead, 10; Barstow, 10; Big Bear, 03; Blythe, 25; Campo, 14; Chula Vista, 30; Claremont, 26; Corona, 27; Escondido, 25; Idyllwild, 08; Laguna Beach, 28; Los Angeles State and County Arboretum, 25; Mojave, 17; Mt. Wilson, 13; Ojai, 23; Palm Springs, 26; Palomar, 11; Pasadena, 28; Redlands, 23; Riverside, 27; San Diego, 34; Santa Paula, 28; Tustin, 32; Twentynine Palms, 20; Victorville, 12; Yorba Linda, 26. Due to local conditions temperatures may vary widely even within a single geographical area; the temperature recorded for Claremont was 26, at the botanic garden in low-lying areas the temperature was 18. (Map from *Desert Plants* 1: 38, 1979).

Acw. These are the aeolian or wind modified sandy soils found in areas of low rainfall (0-3 in.). They are calcareous, low in organic matter and in nitrogen and have low water-holding capacity. They occupy extensive areas in Imperial Co., in the Coachella Valley and on the Palo Verde Mesa in eastern Riverside Co. They are low in plant nutrients and must be irrigated in order to produce crops such as dates, grapes and grapefruit in the Coachella Valley and citrus on the Palo Verde Mesa. This soil type is also found in an extensive area on East Mesa, east of the Highline Canal between Holtville and Yuma as well as on the West Mesa west of Brawley.

Ang. These are deep alluvial fan soils of very sandy or very gravelly texture. These raw alluvial soils are derived from granitic material and are of very low value for agricultural purposes. Found principally in Los Angeles, Riverside and San Bernardino counties.

Asnw. Sandy, wind-modified soils in areas with intermediate amounts of rainfall. In southern California they are mostly restricted to coastal areas in Los Angeles and Orange counties. These soils are light brown and about neutral. They are subject to wind erosion, have low water-holding capacity, and are somewhat deficient in plant nutrients.

Valley Basin Land is described by Storie and Weir as occupying the lowest parts of the valleys and are usually imperfectly or poorly drained. Of the three subcategories, two are found in southern California.

Bnc. Imperfectly drained basin soils. These are natural grassland soils, generally dark-colored clays with a high water table and subject to overflow. In the Sacramento Valley these soils are used extensively for the production of rice and for pasture. In southern California they are restricted to small areas in coastal Santa Barbara and Ventura counties as well as in a few places in Los Angeles Co.

Bck. Saline and alkali soils. They are characterized by a moderate to high content of soluble salts and have poor drainage. This soil type is widely distributed in the playas in the Mojave Desert, at both the north and south end of the Salton Sea as well as many other small areas in Imperial and Riverside counties.

Terrace Land is generally gently sloping to undulating and is usually found along the edges of valleys at eleva-

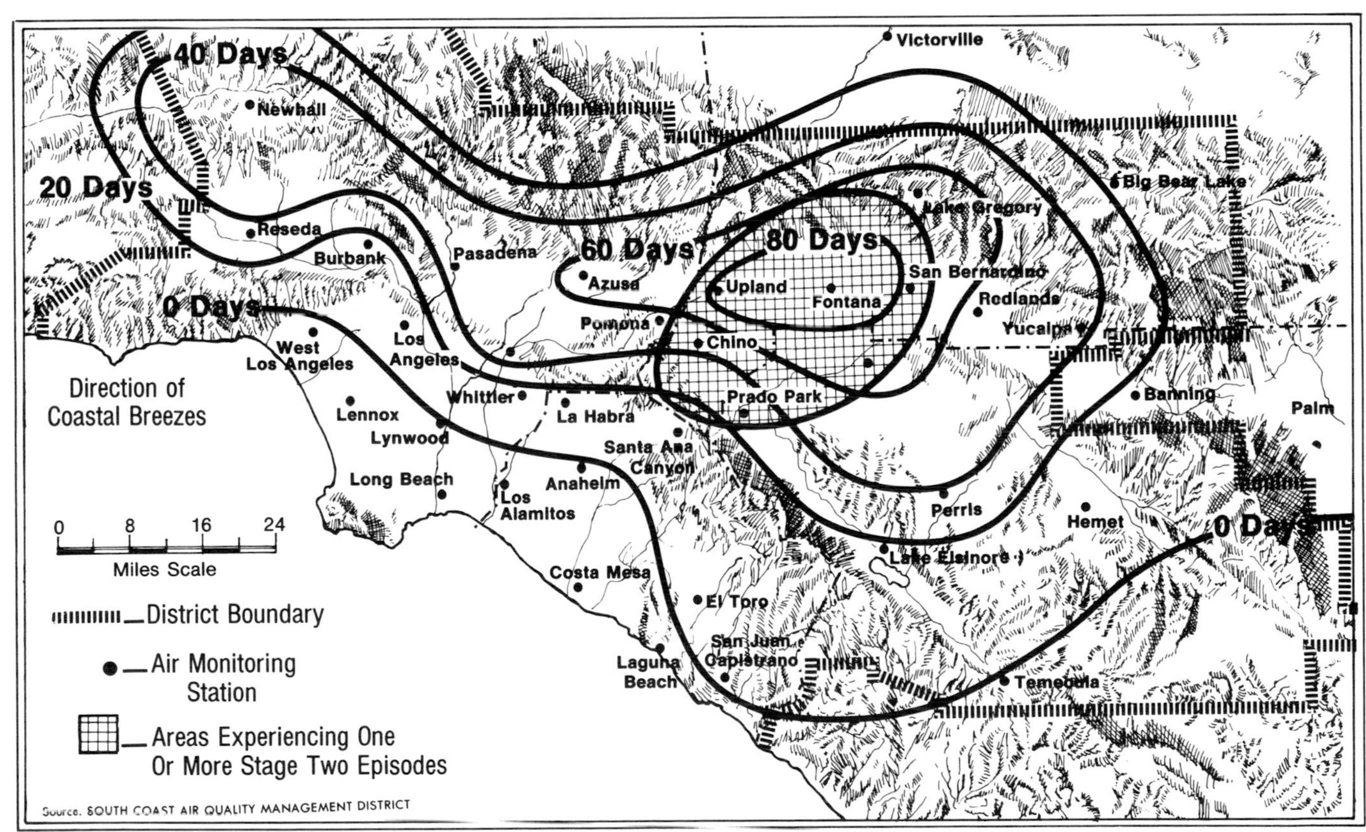

Fig. 13. In southern California air pollution is a serious threat to the successful cultivation of many plants, native and exotic. The map shows the pattern of smog episodes for 1977. The lines indicate areas experiencing 0, 20, 40, 60, and 80, Stage 1 alerts during the year. The crosshatched portion indicates the areas experiencing one or more Stage 2 alerts during 1977. (*Los Angeles Times*, used by permission).

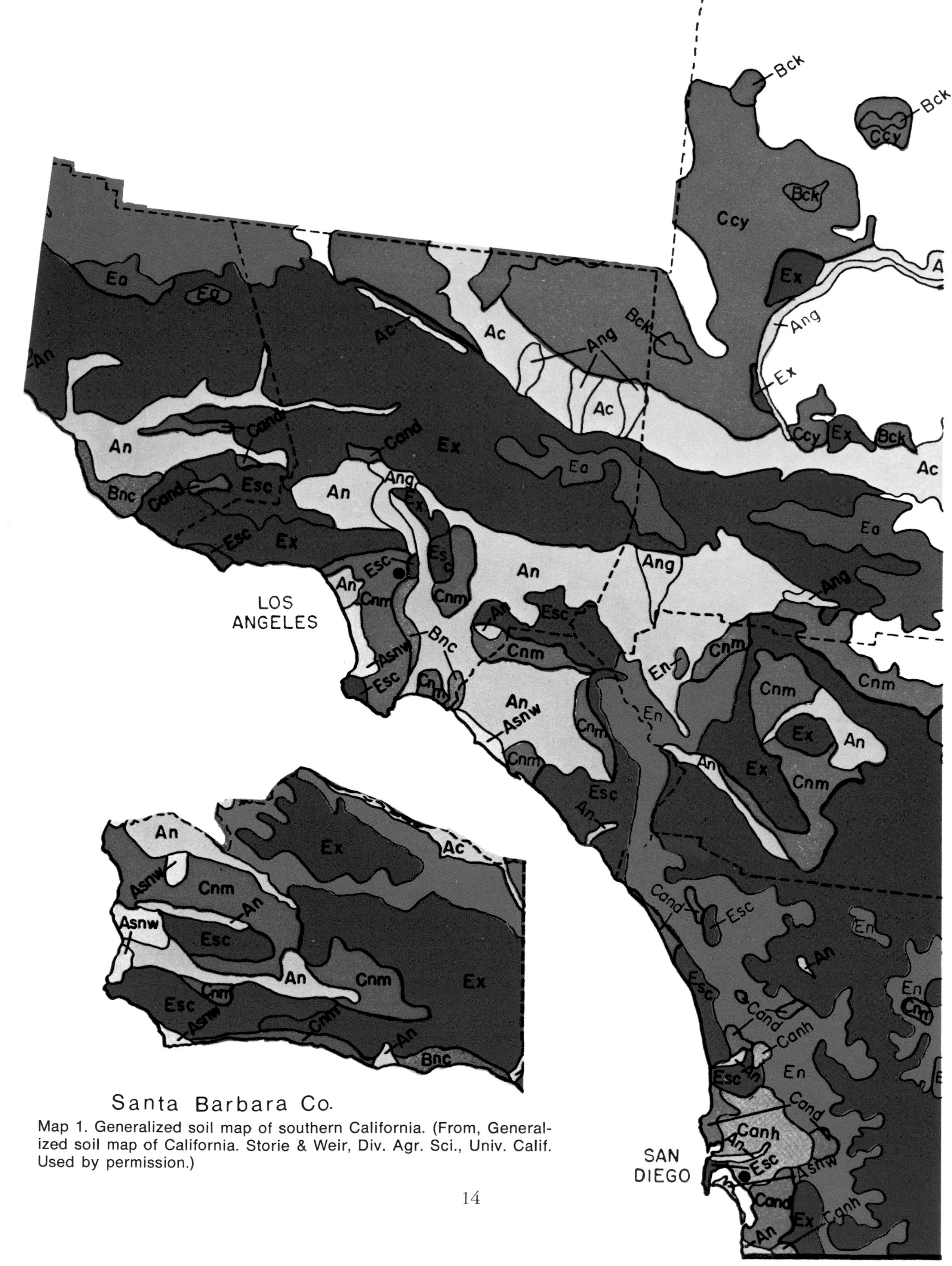

Map 1. Generalized soil map of southern California. (From, Generalized soil map of California. Storie & Weir, Div. Agr. Sci., Univ. Calif. Used by permission.)

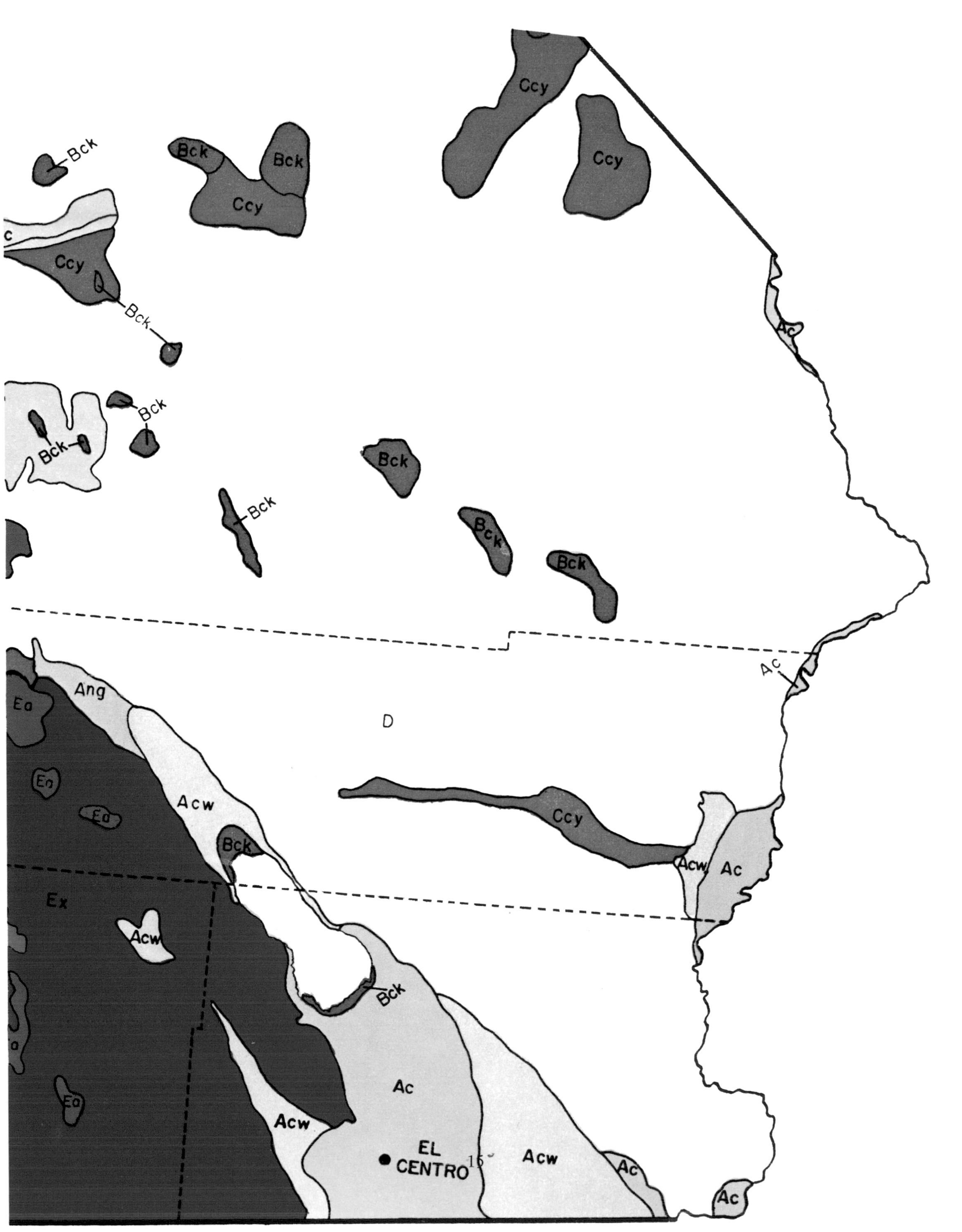

Bck
Bck
Bck
Ccy
Ccy
Ccy
c
Ccy
Bck
Ac
Bck
Bck
Bck
Bck
Bck
Bck
Ac
Ang
Ea
D
Ea
Ea
Acw
Ccy
Bck
Acw
Ac
Ex
Acw
Bck
Ac
Acw
EL
CENTRO
Acw
Ac
Ac

tions of from 5-100 ft. above the valley floors. The soils consist of older secondary deposits or old valley filling material. Associated with these soils are moderately dense to dense subsoils or those with lime or iron hardpan subsoils. Storie and Weir recognize five subcategories, four of which occur in southern California.

Cnm. Terrace lands having moderately dense subsoils, usually with brownish soils of neutral reaction. They occupy low terrace positions in areas with 10-20 inches of precipitation and the native vegetation is grassland or woodland and grass and coastal sage scrub. In southern California this soil type is more or less restricted to southern Los Angeles, Orange and western Riverside counties with a few locations in San Bernardino and San Diego counties. This is the soil type that is found on Indian Hill Mesa at the Rancho Santa Ana Botanic Garden in Claremont.

Ccy. Desert soils. This subcategory includes the desert terrace soils in low rainfall areas (1-6 in.). They are light-colored and are low in organic matter and usually high in lime. They are found mainly in San Bernardino Co. with smaller areas in eastern Riverside Co.

Cand. Terrace land having soils with dense clay subsoils. These soils have medium textured surface soils underlaid by a very dense clay subsoil with the change from surface to subsoil generally being very abrupt. Suitable for grasses and shallow-rooted plants, these soils are widely scattered in San Diego Co. with a few locations in Ventura and Los Angeles counties.

Canh. Terrace soils with red-iron hardpan subsoils. These soils are often referred to as red hogwallow lands and are very common along the eastern side of the San Joaquin and Sacramento valleys. In southern California this soil type is restricted to mesas in San Diego Co., particularly the Kearny Mesa which is well known for its vernal pools which are rapidly disappearing due to urban development. The soils are characterized by reddish-colored surface soils with dense clay subsoils that rest on a silica-iron cemented hardpan that is generally more than a foot in thickness and is impermeable to roots and water.

Upland Soils are found in upland areas that have rolling, hilly to mountainous topography. Most of the soils are residual having been formed in place through decomposition and disintegration of underlying parent rock. Nearly all the timberlands and most of the foothill and mountain grazing lands fall into this category which Storie and Weir subdivide into four subcategories all of which occur in southern California.

Ea. Rolling, hilly to steep upland with acid residual soils of good depth and with high rainfall. They are characterized by a moderate to strongly acid reaction, especially in the subsoils and with depths of three to six feet to bedrock. In southern California this soil type is found at higher elevations, principally in the San Gabriel, San Bernardino, San Jacinto, Palomar and Laguna mts. where it supports the coniferous forests. It is not found in the Santa Ynez, Santa Monica or Santa Ana mts.

Esc. Rolling hilly to steep upland with residual soils of moderate depth and with medium to moderately high rainfall. These are the natural grassland soils, moderately dark in color, fairly high in organic matter and usually of a medium to fine texture. Soils of this type have in the past constituted some of the best grazing land in the state. In southern California *Esc* soils are found on the rolling hills of southeastern Ventura Co. and adjacent Los Angeles Co. as well as southeastern Los Angeles Co. and a portion of adjacent Orange and San Bernardino cos. It also occurs in areas near the coast in San Diego Co.

En. Rolling, hilly to steep upland topography with residual soils of medium to shallow depth and with intermediate to low amounts of rainfall. These are the soils which support much of the southern oak woodland community found in eastern Orange Co. and adjacent Riverside Co. but most abundant in western San Diego Co.

Ex. Residual soils of very shallow depth. Common on steep slopes the soil is usually rocky or gravelly, usually well drained but at times quite heavy. Very widespread in southern California this is the soil type that supports the coastal sage scrub and chaparral plant communities.

Chapter 2

PLANT CLIMATES OF SOUTHERN CALIFORNIA

HORTICULTURISTS have long attempted to devise a system which would make it possible to predict with some degree of accuracy whether a particular plant could be expected to survive in a particular region. Many systems have been proposed but at present no one is universally acceptable. Some fail because, geographically, they cover such a large area that it is impossible to treat in adequate detail all the climates and microclimates that may occur within an area of a few miles. This is particularly true in California which is known for its great number of microclimates.

Marston H. Kimball, Extension Horticulturist of the University of California, Los Angeles, developed a series of tentative plant climate maps for southern and central California. His first tentative statewide map was published in 1959. Within California, Kimball recognized seven major zones, some of which were then subdivided. Playing an important role in the Kimball system was what he termed effective day-night temperatures which, he said, fits in with the four-dimensional relationship of: 1, photoperiod; 2, effective day temperature — temperature during photosynthesis; 3, effective night temperature — temperature during growth and sugar translocation; and 4, state of growth. Kimball's work has been influential in the development of other plant climate maps published in recent years.

Henry T. Skinner, then director of the National Arboretum, Washington, D.C. prepared a plant hardiness map for the United States and southern Canada in which he divided the area into 13 Zones of Plant Hardiness based upon minimum temperatures (Skinner, 1960). Though widely used by horticulturists, it is not suitable for southern California since it recognizes only three Zones of Hardiness and places the two most extreme plant climates, the cool coastal area and the very hot and dry Lower Colorado Valley, together in Zone 13.

In both the first (1954) and second (1961) editions of Sunset's *Western Garden Book,* 13 western garden climates were described as occurring in the three Pacific Coast states and a portion of Arizona. In the third edition (1967) the number of plant climates was increased to 24, of which 14 are found in southern California, the same number listed in the fourth edition (1979). Listed as important factors in making up western plant climates are (1) distance from the equator, (2) elevation, (3) influence of the Pacific Ocean, (4) influence of the continental air mass, (5) mountains and hills and (6) local terrain. The editors carefully point out that the lines on their maps that delimit climates cannot be considered as rigid and that conditions in a single garden or neighborhood can create microclimates. Also, there are small thermal belts, canyons and fog-belt fingers that are too small to register on the maps used.

Bailey (1966), in *Weather In Southern California,* recognizes five climatic regions, chosen according to the author, to portray regional differences important to both natural and man-made landscapes, these he then relates to geographical features.

In the *World Of Trees* (Western edition), (Walheim, 1977), 16 tree zones are recognized for the area covering the three western states and a portion of British Columbia. Of the 16 zones, seven are found in southern California.

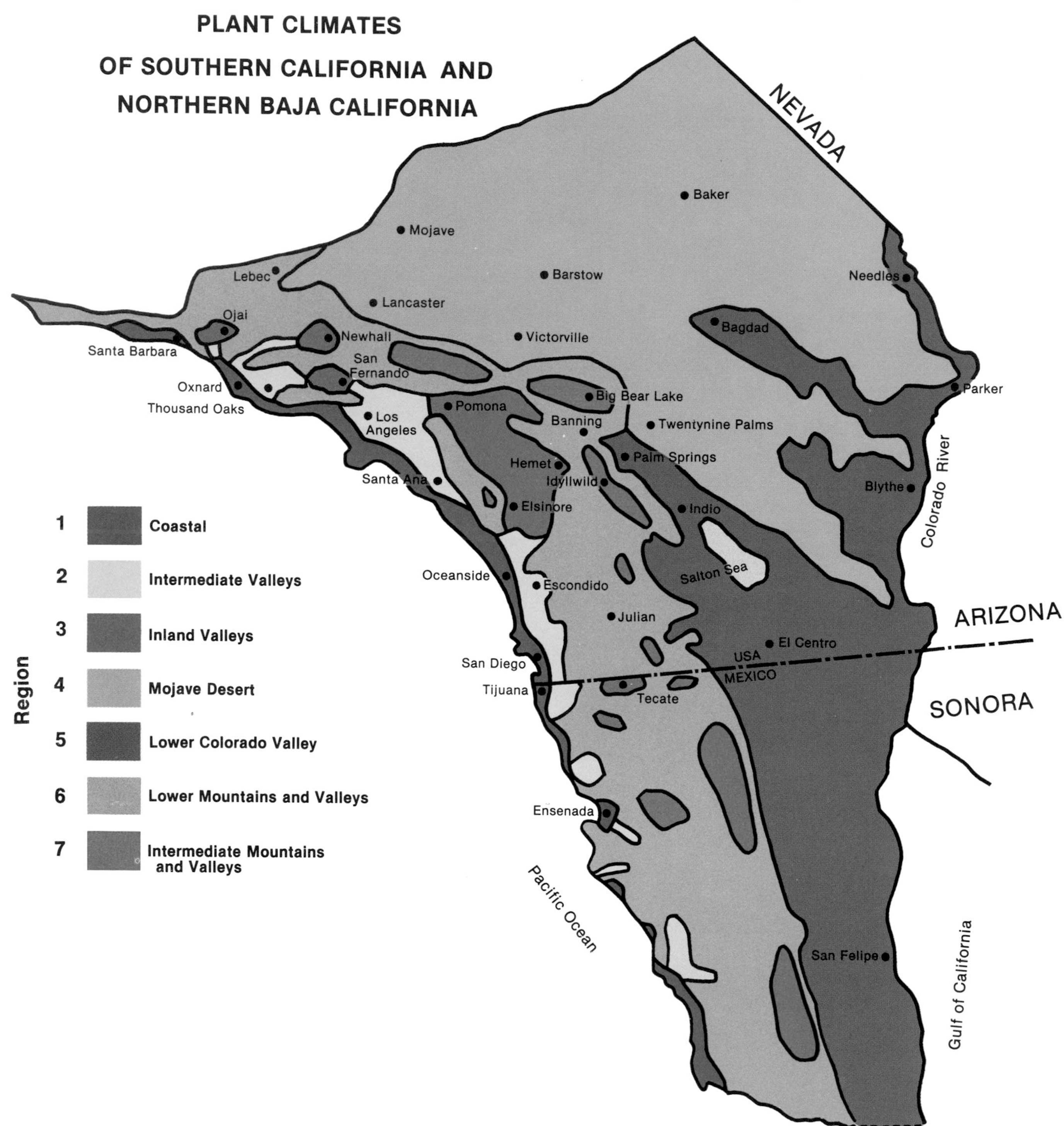

Map 2. Plant climates of southern California and adjacent Baja California. See text for explanation.

The California portion of the plant climate map presented here (Map 2) is adapted from Kimball's map of 1959.

The term *Zone* has been used by various authors to mean different things and the word has therefore lost its usefulness. In our plant climate system we will use the term *Region.* A correlation of the five plant climate systems is shown in figure 14.

Region 1. *Coastal.* A narrow strip of land along the Pacific coast from near Santa Barbara south, rarely does it extend inland more than a few miles. The area is dominated by the ocean and has a *Csb* climate. Winters are mild with an average January temperature of 54.6 F (11 stations). The summers are cool with an average July temperature of 67.7 F and there is usually considerable cloudiness. The differences between the average temperature for the coldest month and the hottest month is less than 18 degrees (Fig. 15). The daily maximum and minimum temperatures for the coldest and hottest months for two stations is shown in figure 16. The annual precipitation (18 stations) is 11.45 inches, somewhat less than that received farther inland.

Region 2. *Intermediate Valleys.* In California, Region 2 consists of the valleys and adjacent lower hills bounded on the west by Region 1 and in the east by Regions 3 and 6. Included is much of the Los Angeles basin and the adjacent valleys and hills east to, and including, the San Gabriel Valley as well as the area west of the Santa Ana Mts. south to about Mission Viejo. To the north, Region 2 includes the areas around Thousand Oaks, Camarillo, Moorpark and the Santa Clara Valley, to the south the valleys and hills around Rainbow, Fallbrook,

<table>
<tr><th>Lenz & Dourley</th><th>Western Garden Book</th><th>World of Trees</th><th>Climates of S. Cal.</th><th>USDA Hardiness Map</th></tr>
<tr><td rowspan="2">Region 1.
Coastal</td><td>Zone 24</td><td>Zone 12</td><td rowspan="2">Maritime Fringe</td><td rowspan="2">Zone 13</td></tr>
<tr><td>Zone 22</td><td>Zone 11</td></tr>
<tr><td rowspan="3">Region 2.
Intermediate Valleys & Adjacent Hills</td><td>Zone 23</td><td rowspan="3">Zone 10</td><td rowspan="3">Intermediate Valleys</td><td rowspan="6">Zone 9</td></tr>
<tr><td>Zone 21</td></tr>
<tr><td>Zone 20</td></tr>
<tr><td rowspan="2">Region 3.
Inland Valleys & Adjacent Hills</td><td>Zone 19</td><td rowspan="2">Zone 9</td><td rowspan="2">?
Transition & Mountain</td></tr>
<tr><td>Zone 18 (in part)</td></tr>
<tr><td>Region 4.
Mojave Desert</td><td>Zone 11</td><td>Zone 13</td><td>High Desert</td></tr>
<tr><td>Region 5.
Lower Colorado Valley</td><td>Zone 13</td><td>Zone 14</td><td>Low Desert</td><td>Zone 13</td></tr>
<tr><td>Region 6.
Lower Mountains</td><td>Zone 3
Zone 18 (in part)</td><td rowspan="2">Zone 1</td><td rowspan="2">Mountains</td><td rowspan="2">Zone 8</td></tr>
<tr><td>Region 7.
Intermediate Mountains</td><td>Zone 2</td></tr>
</table>

Fig. 14. Correlation of five plant climate systems.

Bonsall, Vista, San Marcos, Escondido, Rancho Santa Fe, Rancho Bernardo, Ramona, Lakeside and La Mesa. Portions of the latter area are considered among the most favorable places in North America for the growing of subtropical plants and they are often referred to as the Avocado Belt.

Although inland and with a *Csa* climate much of Region 2 is under the occasional influence of the Pacific Ocean, in places as much as 85% of the time (*Western Garden Book,* 1967). To the north, around Moorpark, Santa Paula and Fillmore the marine influence is reduced and citrus has been one of the major crops.

The average temperature for January (8 stations) is 53.9 F, the average July temperature is 72.6 F and the average annual temperature is 63.1 F. The difference between the average temperature for the coldest and the hottest months is 18.5 degrees. The average annual precipitation (7 stations) is 15.48 inches. Daily minimum and maximum temperatures for the coldest and hottest months for two stations is shown in figure 17.

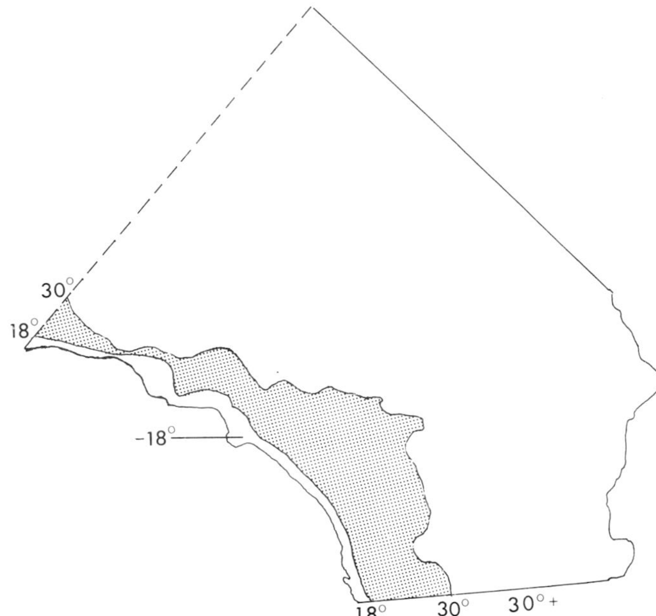

Fig. 15. Average temperature differences between coldest and hottest months. (Adapted from Bailey, 1954).

Region 3. *Inland Valleys.* Except for three relatively small isolated areas, Region 3 in southern California is made up of a network of valleys and adjacent low-lying hills that occupy the area east of the Santa Ana Mts. and west of Region 6. To the north it adjoins Region 2 west of Pomona and its southern border is near Rancho California. Region 3 also includes the San Fernando Valley and areas around Newhall and Ojai. In Mexico it includes Valle de la Palmas, Valle de Santo Tomás and Valle de Trinidad.

The *Csa* climate is influenced by continental air masses and the marine influence is felt but rarely. Historically much of this area in southern California has been prime citrus land, now much reduced by urbanization. Some of the colder areas have been utilized for growing apricots, peaches, apples and walnuts, crops that require some

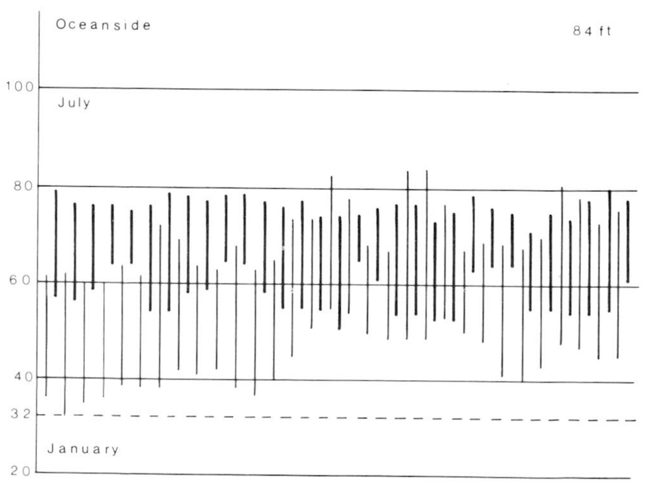

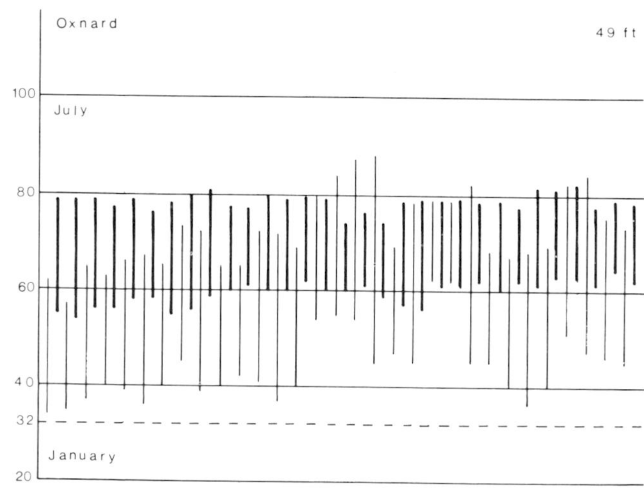

Fig. 16. Daily minimum and maximum temperatures for coldest and hottest months at two stations in the Coastal Region. Data are for one year.

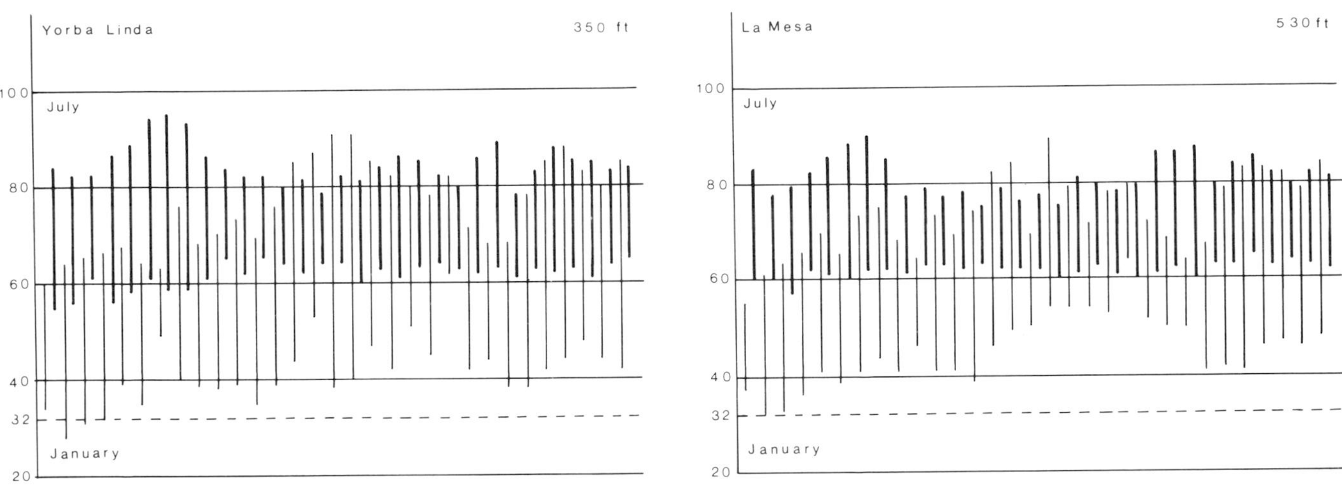

Fig. 17. Daily minimum and maximum temperatures for coldest and hottest months at two stations in the Intermediate Valleys. Data are for one year.

winter chilling. The average January temperature (11 stations) is 50.8 F, that for July 75.4, and the annual average temperature is 62 F. The difference in average temperature for the coldest month and the hottest month is 27.2 degrees. The average annual precipitation (11 stations) is 14.82 inches. Daily minimum and maximum temperatures for the hottest and coldest months for two stations is shown in figure 18.

Region 4. *Mojave Desert.* Region 4 which is often called the high desert occupies a vast area of the eastern portion of southern California, being bordered on the northwest by Region 6 and in the south by Region 5. Within the Mojave Desert are numerous desert ranges, the New York and Providence mts. being the highest with elevations of over 6000 ft. Because of the small extent of the area lying above 5000 ft. and its lack of significance from the standpoint of horticulture, these areas are not shown on the map (Map 2). The *BWh*

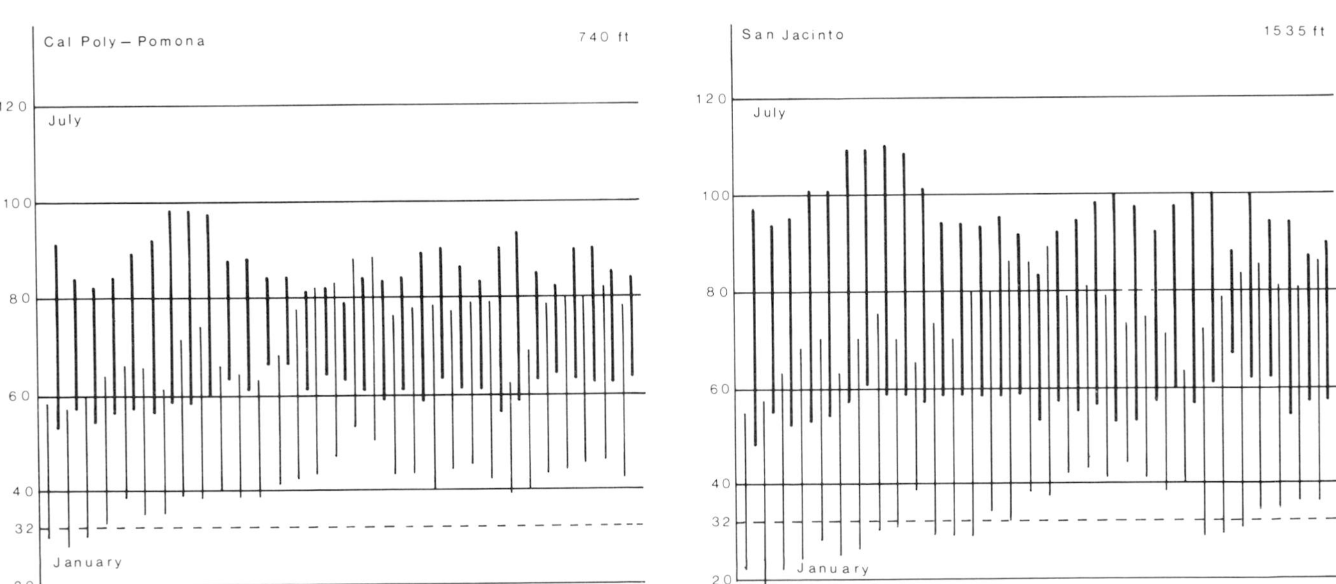

Fig. 18. Daily minimum and maximum temperatures for the coldest and hottest months at two stations in the Inland Valleys. Data are for one year.

climate of the Mojave Desert is characterized by wide ranges in daily temperatures as well as great differences between winter and summer temperatures (Fig. 20). Elevations of the main centers of population are between 2000-3000 ft. (Barstow, 2160; Lancaster, 2340; Victorville, 2853; Mojave, 2735; Apple Valley, 2935; Boron, 2460). The average January temperature (8 stations) is 45.4 F, the average July temperature is 84 F with the average annual temperature being 63.5 F. The difference between the average January and July temperatures is 38.6 F. The average rainfall (7 stations) is 4.91 inches.

A comparison of the daily minimum and maximum temperatures for the coldest and hottest months of the year for Barstow in the creosote bush scrub community and for Redding in the northern foothill woodland community is shown in figure 21.

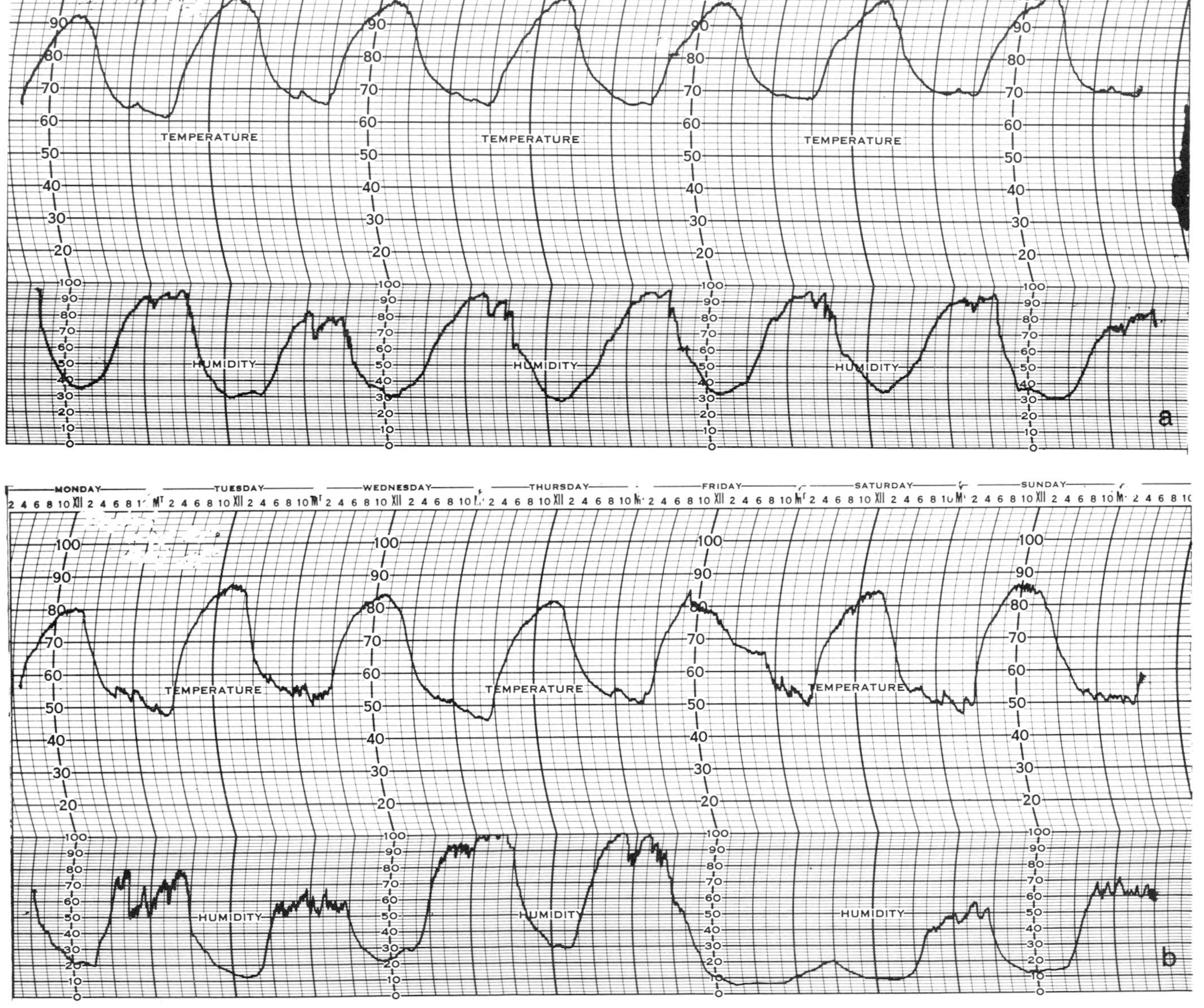

Fig. 19. Hygro-thermograph charts for Claremont: a, 21-27 July 1980; b, 7-13 April 1980. Note low humidity during Santa Ana winds toward the end of the week. Region 3.

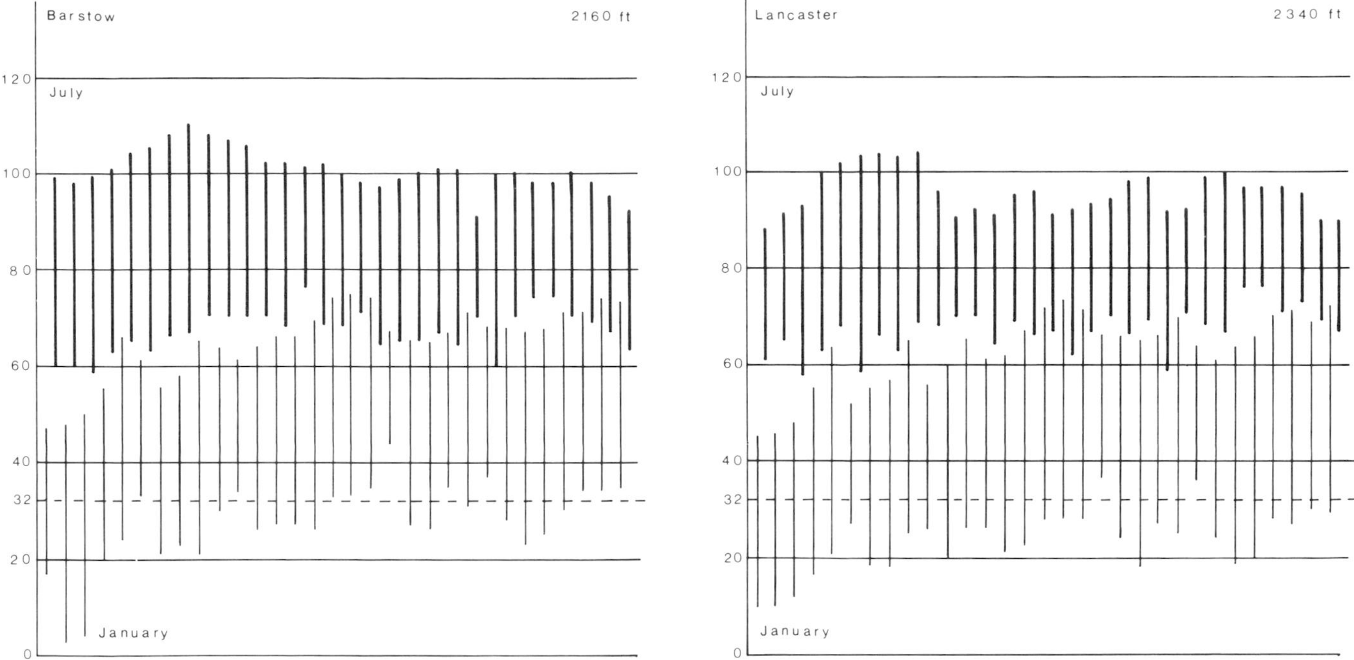

Fig. 20. Daily minimum and maximum temperatures for the coldest and hottest months at two stations in the Mojave Desert. Data are for one year.

The two have a very similar temperature profile and the average annual temperature for the two differs by only 1.2 degrees. In addition to soil differences, the major factor is that of moisture with Redding receiving 37.81 inches of precipitation a year compared with 4.27 inches at Barstow.

Region 5. *Lower Colorado Valley.* This area is often referred to as the low desert or the Colorado Desert. Munz and Keck (1949) combined the Colorado and Mojave deserts into what they called the Southern Deserts on the basis that the creosote bush scrub community occupies the largest single area in both. The Sonoran

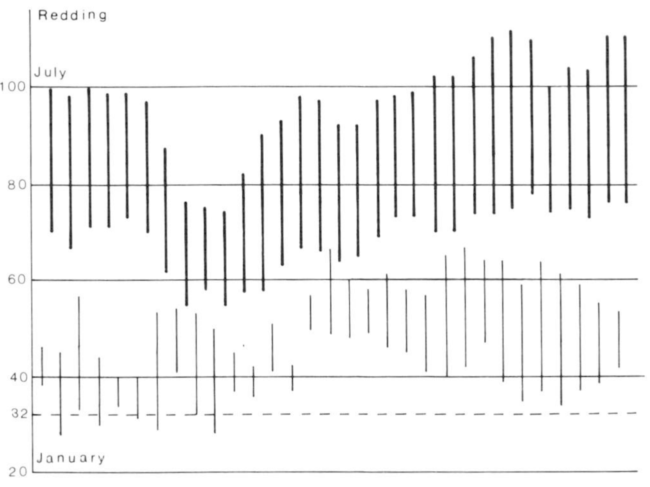

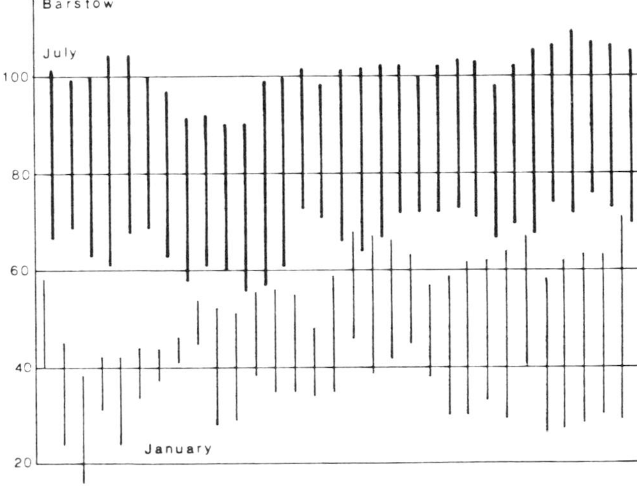

Fig. 21. Daily minimum and maximum temperatures for the coldest and hottest months at one station in the creosote bush scrub community (Barstow) and for one station in the northern oak woodland community (Redding). Data are for one year. See text for explanation.

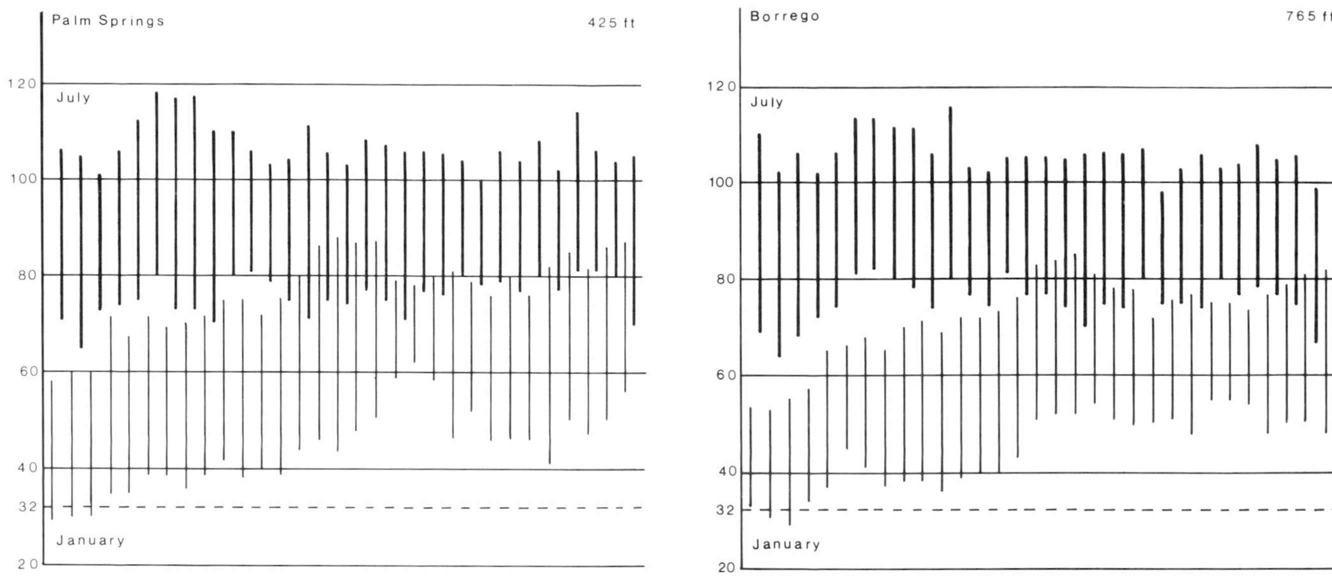

Fig. 22. Daily minimum and maximum temperatures for the coldest and hottest months at two stations in the Lower Colorado Valley. Data are for one year.

Desert, according to Shreve and Wiggins (1964) is composed of seven distinct units, one of which they call the Lower Colorado Valley which occupies both sides of the Colorado River, east as far as Phoenix and south into Sonora, Mexico; to the west it occupies a portion of southeastern California and adjacent northeastern Baja California. In following Shreve and Wiggins we shall refer to the area as the Lower Colorado Valley. The climate is designated as *BWhh.* This area, some of it below sea-level, is known for its excessively hot summer temperatures, which in California are only exceeded by those in Death Valley, and for its low annual rainfall. Average

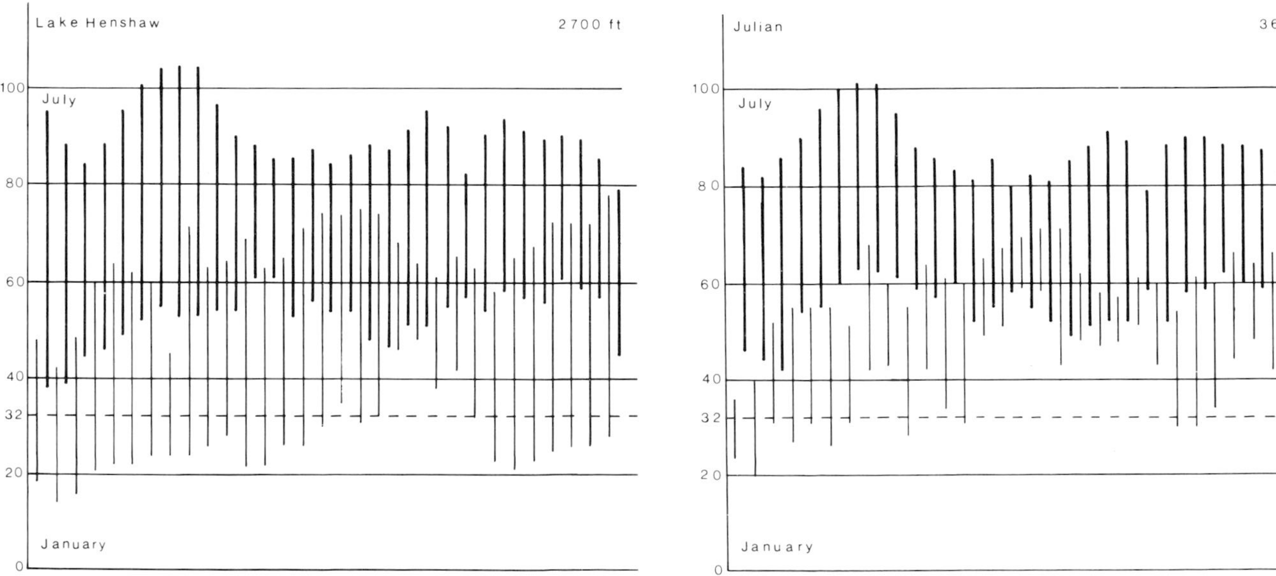

Fig. 23. Daily minimum and maximum temperatures for the coldest and hottest months at two stations in the Lower Mountains and Mountain Valleys. Data are for one year.

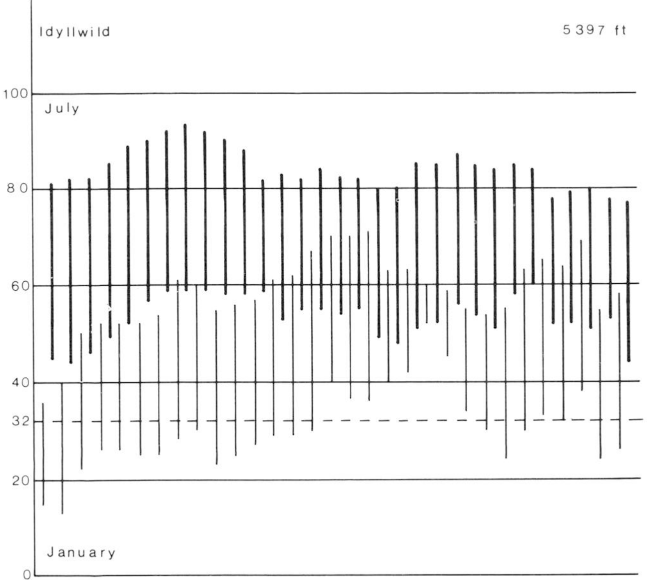

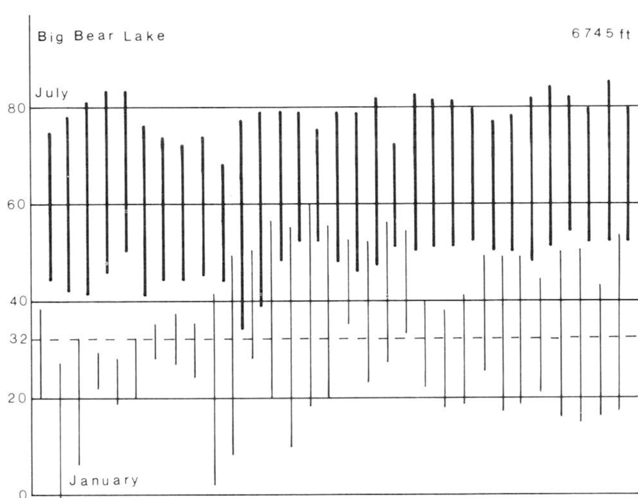

Fig. 24. Daily minimum and maximum temperatures for the coldest and hottest months at two stations in the Intermediate Mountains and Mountain Valleys. Data are for one year.

January temperature (10 stations) is 53.6 F, average July temperature is 92 F, and the average annual temperature is 72.6 F. The difference between average January and July temperatures is 38.4 degrees. Daily minimum and maximum temperatures for the coldest and hottest months for two stations is shown in figure 22. Average annual rainfall is 3.12 inches.

Region 6. *Lower Mountains and Mountain Valleys.* This region might be called the chaparral region since that is the dominant plant community throughout the area. The region occupies a vast area in southern California and northern Baja California and varies in elevation from as low as 1000 ft. to around 5500 ft. Over much of the area the limits of elevation might be set at approximately 2500-5000 ft. Because chaparral is a fire prone community relatively few people live in Region 6, and climatic data is not as available as it is in most of the other regions. In general, rainfall is between 14-30 inches, summers are dry and hot and winters cool but not cold (Munz and Keck, 1949). From the standpoint of the horticulturist this region is one of the most productive of useful or potentially useful plant materials. Daily minimum and maximum temperatures for the hottest and coldest months for two stations is shown in figure 23.

Region 7. *Intermediate Mountains and Mountain Valleys.* This region occupies the area above the chaparral of Region 6 and consists to a large extent of yellow pine forest and associated communities. Included are the population centers of Lake Arrowhead (5205 ft.), Big Bear Lake (6815 ft.), Idyllwild (5397 ft.), Mt. Palomar (5500 ft.), and Mt. Wilson (5709 ft.). The upper limit of Region 7 is considered here to be the beginning of the red fir forest at about 8000 ft. The *Cs'ab* climate has an average of 32.65 inches of precipitation (4 stations), much of it coming as snow. There may also be occasional summer thunderstorms. The average January temperature is 40.4 F and that for July 70.6 F (4 stations). The average annual temperature is 53.5 F. Because of the rapid development of residential communities in Region 7 this area is becoming increasingly important to the horticulturist.

Daily minimum and maximum temperatures for the hottest and coldest months at two stations in Region 7 is shown in figure 24.

Chapter 3

MAJOR PLANT COMMUNITIES OF SOUTHERN CALIFORNIA

ONE OF THE EARLIEST attempts to correlate climatic conditions to the distribution of plants and animals was that of C. Hart Merriam, Chief of the United States Biological Survey, who concluded that apart from obvious mechanical barriers, temperature was the most important single factor in fixing the limits beyond which particular species of animals and plants cannot go. Merriam's Life Zone System found great favor with biologists during the early years of this century and was widely used but its limitations have long been known.

In contemplating a new flora for California, Philip A. Munz and David D. Keck, feeling a need for a more precise classification to describe the distribution of plants than that provided by Merriam, devised a system which they published in 1949 with a supplement in 1950. In their "California Plant Communities" Munz and Keck divided the state into five biotic provinces further divided into 14 vegetation types. Within the 14 vegetation types they recognized 24 plant communities, adding four more in their 1950 supplement.

Oosting (1948) had defined a community as an aggregation of living organisms having mutual relationships among themselves and to their environment. Munz and Keck used the term plant community for each regional element of the vegetation that is characterized by the presence of certain dominant species. A vegetation type, according to these authors, may then consist of one to several communities (Table 1). The Munz and Keck system was essentially a practical one that would allow them to describe "with some exactness the type of ecological niche in which a species belongs, where it is found, and with what other species it is associated."

Since Munz and Keck first published their plant communities, other botanists have shown interest in investigating and describing natural plant associations as they occur in California, and at the present time the literature on the subject is extensive. Particularly valuable is the work on terrestrial vegetation of California edited by Barbour and Major (1977), and that of a symposium on the plant communities of southern California edited by Latting (1976). As might be expected, authors do not always agree, and Johnson (1976) noted that "the problem of 'lumpers' and 'splitters,' so renowned in taxonomy, is in evidence in community nomenclature as well." In this work we will take a pragmatic approach to plant communities and will basically follow Munz and Keck whose system has become familiar through its use in the most recent state floras (Munz and Keck, 1959; Munz, 1974). We will however not consider high elevation communities or those with few or no species of horticultural interest. We also recognize two communities not recognized by Munz and Keck.

Valley Grassland (Plate 1a). This easily recognized plant community occurs south of the Transverse Ranges along the coast and inland in the hot valleys and on some plateaux. In southern California the treeless expanses are now mostly under cultivation or covered with weedy introduced grasses and other herbs, but originally they supported a rich flora of native bunch grasses and showy

Table 1

Major Vegetation Types and Plant Communities of California

Vegetation Type	Plant Community
I. Strand	1. Coastal Strand
II. Salt Marsh	2. Coastal Salt Marsh
III. Freshwater Marsh	3. Freshwater Marsh
IV. Scrub	4. Northern Coastal Scrub
	5. Coastal Sage Scrub*
	6. Sagebrush Scrub
	7. Shadscale Scrub
	8. Creosote Bush Scrub*
	9. Alkali Sink
V. Coniferous Forest	10. Northern Coastal Coniferous Forest
	11. Closed-cone Pine Forest
	12. Redwood Forest
	13. Douglas-fir Forest
	14. Yellow Pine Forest*
	15. Red Fir Forest
	16. Lodgepole Forest
	17. Subalpine Forest
VI. Mixed Evergreen Forest	18. Mixed Evergreen Forest
VII. Woodland-Savanna	19. Northern Oak Woodland
	20. Southern Oak Woodland*
	21. Foothill Woodland
VIII. Chaparral	22. Chaparral*
IX. Grassland	23. Coastal Prairie
	24. Valley Grassland*
X. Alpine Fell-fields	25. Alpine Fell-fields
XI. Desert Woodland	26. Northern Juniper Woodland
	27. Pinyon-Juniper Woodland*
	28. Joshua Tree Woodland*

The Munz and Keck plant community system as revised in 1950. Those marked with an asterisk are treated in detail here.

annuals representing a wide variety of genera. deAnza, in describing the San Jacinto Valley in March 1774, wrote, "All the plain is full of flowers, fertile pastures and other plants, useful for the breeding of cattle." Today the colorful spring annuals are often only to be seen in undisturbed areas along roadsides.

At higher elevations the grassland merges with the oak woodland or chaparral. Rainfall varies from 6-20 inches per year (Munz and Keck, 1949). The soils are usually Valley Land, *An*; Upland, *Esc*; or Terrace Land, *Cnm.*; with small poorly drained areas in Santa Barbara, Ventura and Los Angeles cos. of Valley Basin Land, *Bnc.*

The average monthly precipitation and temperature for two stations in the valley grassland community are shown in figure 25.

Coastal Sage Scrub (Plate 1b). Although this community has been referred to as southern coastal sage scrub (Mooney, 1977) and the southern coastal sage scrub of the coastal scrub (Thorne, 1976), all authors agree that it is a community occupying coastal sites as well as the

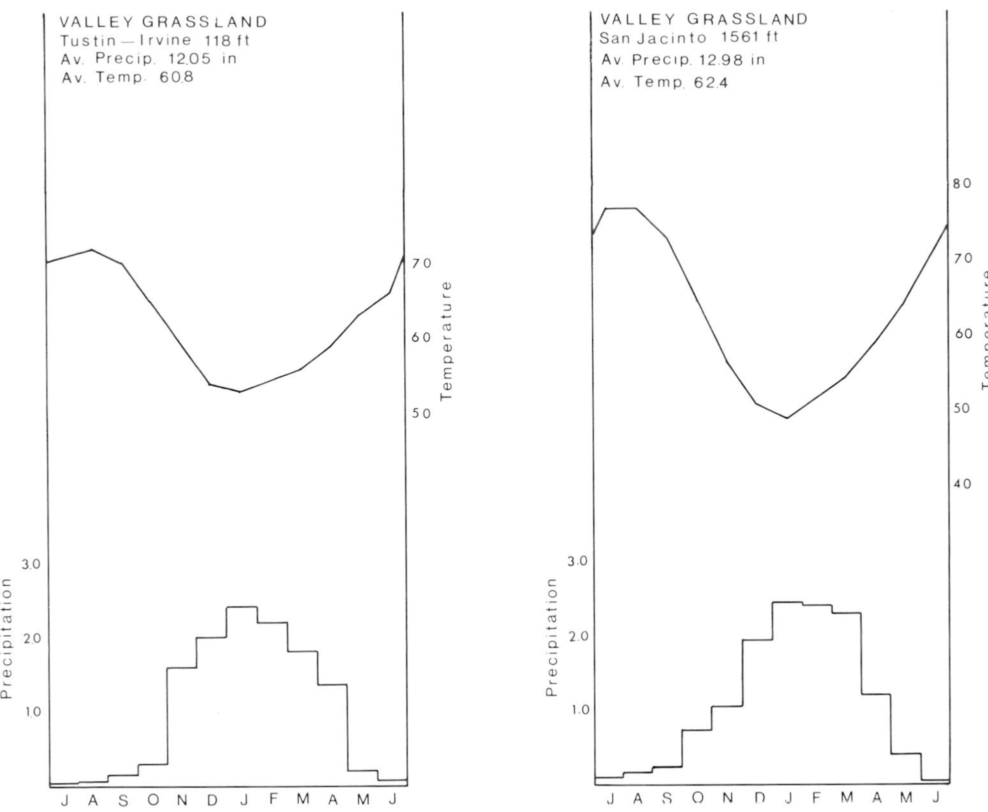

Fig. 25. Average monthly precipitation and temperature for two stations in the valley grassland community.

cismontane slopes of the mountains mostly below 3000 ft. and below the chaparral. Often but not always occupying rocky or gravelly slopes it is a low, open scrubby vegetation less dense, evergreen and thick-leaved than the chaparral (Thorne, 1976). Precipitation is relatively light varying from 10-20 inches a year and the temperature extremes are not great (Fig. 26).

Plants characteristic of this community include coastal sagebrush, *Artemisia californica*; white sage, *Salvia apiana*; black sage, *Salvia mellifera*; purple sage, *Salvia leucophylla*; California buckwheat, *Eriogonum fasciculatum*; lemonade berry, *Rhus integrifolia*; California encelia, *Encelia californica*; *Horkelia cuneata*; sawtooth goldenbush, *Haplopappus squarrosus*; coast goldenbush, *H. venetus*; golden yarrow, *Eriophyllum confertiflorum.*

Although occurring in a very limited area in southern California and more extensively in northwestern Baja California, a word should be said about what Mooney (1977) calls the coastal sage succulent scrub and Thorne (1976) calls the maritime desert scrub. According to Thorne, it is an arid, desert-like scrub dominated by spinescent and succulent plants. In addition to many of the species listed as occurring in the coastal sage scrub it includes some interesting endemics such as the Shaw agave, *Agave shawii*; Parry buckeye, *Aesculus parryi*; *Rosa minutifolia*; and palo blanco, *Ornithostaphylos oppositifolia.* Other species include *Salvia munzii*; *Adolphia californica*; and San Diego sunflower, *Viguiera laciniata*, all of which while known from San Diego Co. have their main distribution in northern Baja California. Other species such as the San Diego bursage, *Franseria chenopodiifolia*, and jojoba, *Simmondsia chinensis,* are more commonly encountered in desert communities.

One vegetation type recognized by Munz and Keck (1949) is the woodland savanna which they divided into three plant communities: foothill woodland, northern

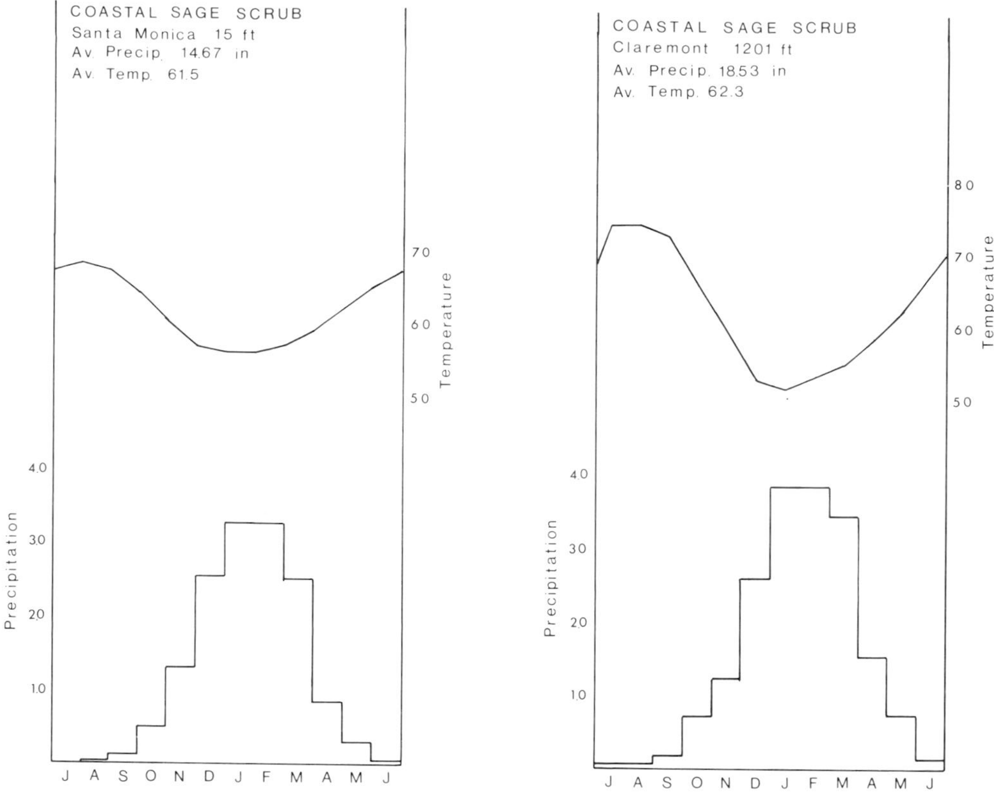

Fig. 26. Average monthly precipitation and temperature for two stations in the coastal sage scrub community.

oak woodland and southern oak woodland. Only one of these, the southern oak woodland, is well developed in southern California.

Southern Oak Woodland (Plate 1c). According to Griffin (1977), Munz and Keck's southern oak woodland is a convenient name for the variable woodlands in coastal southern California within which he recognized two phases: the savanna portion (isolated trees) with Engelmann oak, *Quercus engelmannii,* prominent, which he calls the Engelmann oak phase; and the denser, more widespread woodlands (> 30% cover) with the coast live oak, *Q. agrifolia,* always important, and the California walnut, *Juglans californica,* sometimes dominant, which he calls the coast live oak phase. We will follow Munz and Keck (1949) and Thorne (1976) and consider the two phases together as representing the live oak woodlands of southern California which extend from the Santa Ynez Mts. to southern San Diego Co., reaching elevations of almost 5000 ft. in the San Jacinto Mts. (Thorne, 1976). In the savanna or park-like woodland the trees are surrounded by grassland and in the more wooded areas by shrubs such as California lilac, *Ceanothus* spp.; sumac, *Rhus* spp.; currants or gooseberries, *Ribes* spp., which intrude from the surrounding chaparral. According to Thorne, Engelmann oak is usually found on drier soils and coast live oak on moister slopes. Between Santa Barbara and Orange cos. the California walnut, *Juglans californica,* may be locally dominant usually on north-facing slopes, and toyon, *Heteromeles arbutifolia,* a conspicuous subdominant. According to Griffin, the best remaining stands of California walnut are to be found in the San Jose Hills south and east of Covina.

Precipitation is from 15-25 inches a year often with considerable runoff. The soils are usually rather deep Terrace Land *Cnm* or Upland *Esc* and *En,* the latter very common in portions of Orange, Riverside and San Diego cos.

Although not an important plant community in southern California, mention should be made of the foothill woodland community which extends as far south as northwestern Los Angeles Co. Of the three dominant trees endemic to the community, blue oak, *Quercus douglasii*; valley oak, *Q. lobata*; and digger pine, *Pinus sabiniana*, only the valley oak extends any distance into southern California where it reaches its southern limit at San Marino. The foothill woodland community occupies a vast area of central and northern California forming a ring surrounding the great central valley and contains in addition to the species just mentioned, a variety of plants of particular interest to the horticulturist.

Riparian Woodland. According to Thorne (1976), who accorded formal recognition to this community, "Most permanent streams and springy areas in the state support along their margins a riparian community of semiaquatic trees and herbs." As such it can be found within the geographical limits of many plant communities but, according to Thorne, it is most conspicuous in the valley grassland, oak woodland and chaparral. The component species vary from place to place reflecting at least in part climatic differences. No single soil type is indicated for this community but it might be expected, that since canyon bottoms receive an accumulation of materials through erosion from higher land, that the composition of these materials would reflect the character of the surrounding slopes. These materials are then deposited upon, or incorporated into the sands and gravels that line the waterway. At lower elevations the woody species include bigleaf maple, *Acer macrophyllum*; California box elder, *A. negundo* ssp. *californicum*; Arizona ash, *Fraxinus velutina* var. *velutina*; sycamore, *Platanus racemosa*; Fremont cottonwood, *Populus fremontii*;

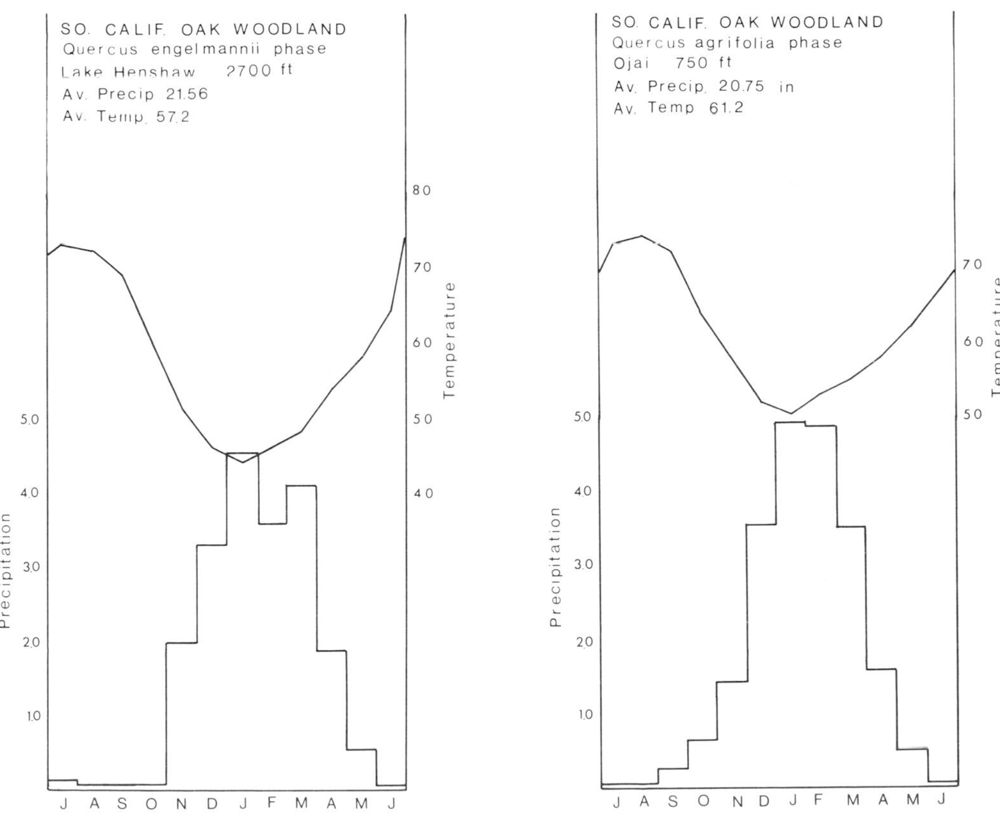

Fig. 27. Average monthly precipitation and temperature for two stations in the southern California oak woodland community.

black cottonwood, *P. trichocarpa* var. *trichocarpa* and var. *ingrata*; coast live oak, *Quercus agrifolia*; valley oak, *Q. lobata*; *Baccharis* spp.; dogwood, *Cornus* spp.; willows, *Salix* spp.; and grape, *Vitis girdiana.* Along the Colorado and Mojave rivers, as well as along many irrigation canals in Imperial Co. are cottonwoods, *Populus* spp.; willows, *Salix* spp.; tamarisk, *Tamarix* spp.; arrow-weed, *Pluchea sericea*; common reed, *Phragmites australis*; and giant reed, *Arundo donax* (Thorne, 1976).

Chaparral. The most extensive vegetation type in California, the chaparral is composed mainly of evergreen woody shrubs forming extensive shrublands that occupy most of the hills and lower mountain slopes of the state. It is adapted to drought and fire, passing endlessly through cycles of burning and regrowth (Hanes, 1977).

The word chaparral evolved from *chabarra,* the Basque word for the scrub oak of the Pyrenees. The Spanish adapted it as "a dwarf evergreen oak" and spelled it *chaparro.* When the Spanish arrived in the New World they were faced with a tremendous job of inventing place names. The saints furnished an abundance of names for important places and descriptive terms were used for places of secondary importance. The suffix *-al,* meaning "place of," was often used, and for the vegetation consisting of dense evergreen scrub oaks the word *chaparral* was invented. Soon the term came to be applied to other similar types of vegetation (Cronemiller, 1942). In other areas with a mediterranean climate, similar types of vegetation are referred to as *maqui, macchia, garigue* and *matorral.* In order to designate various kinds of chaparral the words soft, hard and montane have been used. Soft chaparral is now correctly known as coastal sage scrub and the term montane chaparral is used for one distinct type of chaparral. The leaf is the most distinctive feature of chaparral shrubs, being small, thick, stiff and evergreen. Schimper (1903) used the term *sclerophyll* to apply to the hard-leaved evergreen shrub communities adapted to mediterranean climates of the world, and some authors now refer to the chaparral as broad-leaved sclerophyllous scrub.

	Community	Region(s) Where Found
1.	Valley Grassland	1, 2, 3
2.	Coastal Sage Scrub	1, 2, 3
3.	Southern Oak Woodland	2, 3
4.	Riparian Woodland*	1, 2, 3, 4, 5, 6
5.	Chaparral	2, 3, 4, 6, 7
6.	Creosote Bush Scrub	4, 5
7.	Joshua tree Woodland	4
8.	Pinyon-Juniper Woodland	4
9.	Yellow Pine Forest	7
10.	Desert Oasis*	5

*Not recognized by Munz and Keck.

Areas with chaparral are usually regions with mild winter temperatures and with limited winter rainfall (12-35 inches) often occurring in a few intense storms, frequently accompanied by considerable runoff, followed by hot summers with prolonged drought. The community is subject to frequent fires and many of the shrubs are capable of stump-sprouting and many of the annuals are fire-annuals. The soils are rocky, gravelly or sometimes heavy — Upland Soil, *Ex.*

This community contains more species of importance, or potential importance, to the horticulturist than any other community in the state. Munz and Keck (1949) treated the community as a single unit, but more recent authors (Thorne, 1976; Hanes, 1977), recognizing the great diversity of vegetation types to be found within the chaparral, have separated it into subunits or chaparral types. Because of its widespread occurrence and its interest to the horticulturist we will recognize seven types of chaparral in southern California. No critical studies of the chaparral of northern Baja California have yet been published.

Chamise Chaparral (Plate 1e). According to Hanes (1977), this is the most common type of chaparral in the state and it is dominated by chamise, sometimes also called greasewood, *Adenostoma fasciculatum.* It is a dense interwoven vegetation found on hot, xeric sites, and according to Hanes, it is the predominant chaparral type throughout the mountains of Ventura, Los Angeles, Orange, San Bernardino, Riverside and San Diego cos. Associated species of low frequency include manzanitas, *Arctostaphylos* spp.; California lilacs, *Ceanothus* spp.; California buckwheat, *Eriogonum fasciculatum*; scrub oak, *Quercus dumosa*; sugarbush, *Rhus ovata*; laurel sumac, *Malosma laurina*; white sage, *Salvia apiana*; black sage, *S. mellifera*; and *Yucca whipplei.* According to Hanes, the overall appearance of chamise chaparral is a

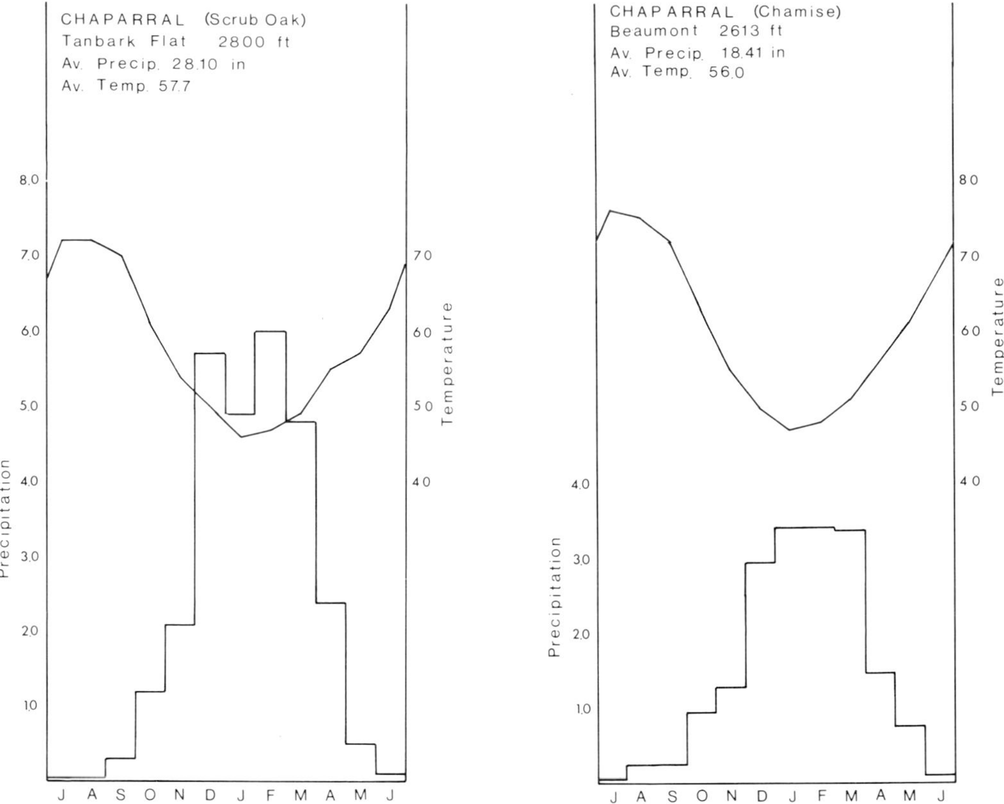

Fig. 28. Average monthly precipitation and temperature for two stations in the chaparral community.

uniform sea of chamise with an occasional individual of an associated species.

Ceanothus Chaparral. In southern California, ceanothus chaparral is rarely found over 4000 ft. and it occurs in more mesic sites than does chamise chaparral (Horton, 1960). The dominant species are often hoaryleaf ceanothus, *Ceanothus crassifolius*; and chaparral whitethorn, *C. leucodermis*; with occasional individuals of scrub oak, *Quercus dumosa*; toyon, *Heteromeles arbutifolia*; sugarbush, *Rhus ovata*; as well as *Ceanothus spinosus, C. megacarpus* and *C. tomentosus* ssp. *olivaceus.*

Scrub Oak Chaparral (Plate 1g). Termed scrub oak chaparral by Hanes (1977), mixed chaparral by Thorne (1976) and woodland chaparral by Horton (1960) it is a mesic type of chaparral composed of a great many species of large shrubs. In southern California it occupies north-facing slopes below 3000 ft. and all slopes above 3000 ft. (Hanes, 1977). In southern California the dominant species is scrub oak, *Quercus dumosa*; but also present are chaparral whitethorn, *Ceanothus leucodermis*; western mountain mahogany, *Cercocarpus betuloides*; flowering ash, *Fraxinus dipetala*; silktassel bush, *Garrya* spp.; toyon, *Heteromeles arbutifolia*; honeysuckle, *Lonicera* spp.; hollyleaf cherry, *Prunus ilicifolia*; hollyleaf coffeeberry, *Rhamnus ilicifolia*; coffeeberry, *R. californica*; sugarbush, *Rhus ovata*; poison oak, *Toxicodendron diversilobum*; currants and gooseberries, *Ribes* spp.; and elderberry, *Sambucus* spp. The average temperature and precipitation for one site with scrub oak chaparral is shown in figure 28.

Manzanita Chaparral. Rather restricted in southern California, this chaparral type is dominated by species of manzanita, *Arctostaphylos.* It occurs at higher elevations than the chamise chaparral and on deeper soils. Manzanita chaparral site requirements are intermediate between those for chamise chaparral and coniferous forest (Cooper, 1922; Wilson and Vogl, 1965). The latter authors consider manzanita chaparral "cold chaparral" since the major forms of precipitation it receives are fog, drip, freezing moisture and snow. According to Hanes (1977), the appearance from a distance is that of a light-green, velvety mantle.

Montane Chaparral (Plate 1h). The montane chaparral, also referred to as yellow pine chaparral, or timberland chaparral, occurs at the same elevations and under the same general climatic and soil conditions as the coniferous forest; cool to cold temperatures and plentiful precipitation (22-45 inches), mainly as snow (Hanes, 1977). In southern California it occurs at 5000-11,000 ft. in the Transverse and Peninsular ranges of Los Angeles, San Bernardino, Riverside and San Diego cos. The most common species include Eastwood manzanita, *Arctostaphylos glandulosa*; Otay manzanita, *A. otayensis*; pinkbract manzanita, *A. pringlei* var. *drupacea*; Parry manzanita, *A. parryanna*; deerbrush, *Ceanothus integerrimus*; chaparral whitethorn, *C. leucodermis*; mountain whitethorn, *C. cordulatus*; wavyleaf ceanothus, *C. foliosus*; thickleaf yerba santa, *Eriodictyon crassifolium*; hairy yerba santa, *E. trichocalyx*; Fremont tasselbush, *Garrya fremontii*; Veatch silktassel, *G. veatchii*; toyon, *Heteromeles arbutifolia*; hollyleaf cherry, *Prunus ilicifolia*; dwarf interior live oak, *Quercus wislizenii* var. *frutescens*; and California coffeeberry, *Rhamnus californica.* As we shall later see, many of these species are of prime importance to the ornamental horticulturist.

Redshanks Chaparral (Plate 1f). This chaparral type is dominated by redshanks, *Adenostoma sparsifolium,* which often forms pure stands. It is also in places associated with chamise, *A. fasciculatum*; California lilac, *Ceanothus* spp.; and sugarbush, *Rhus ovata.* Redshanks chaparral is centered in the San Jacinto and Santa Rosa mts. and in the interior valleys of Riverside and San Diego cos. but it is also found in the western Santa Monica Mts. It occurs on granitic soils at elevations of 2000-6000 ft. on both coastal and desert slopes. According to Hanes, many consider this the most attractive form of chaparral. Redshanks is an open shrub or small tree with multiple branches, the foliage is feather-like and chartreuse in color and the plants are open enough to allow major branches to be seen (Hanes, 1977).

Desert Chaparral (Plate 1d). In southern California this chaparral type is found on the lower, north-facing slopes of the San Gabriel Mts. above the Mojave Desert and on the lower slopes of the San Jacinto, Santa Rosa and Laguna mts. facing the Lower Colorado Valley. It is noticeably more open than the other types of chaparral and does not burn as often due in part to the greater spacing of the plants. Precipitation averages 12-25 inches. Species of common occurrence include chamise, *Adenostoma fasciculatum*; bigberry manzanita, *Arctostaphylos glauca*; California lilacs, *Ceanothus* spp.; western mountain mahogany, *Cercocarpus betuloides*; bush poppy, *Dendromecon rigida*; mormon tea, *Ephedra* spp.; hairy yerba santa, *Eriodictyon trichocalyx*; California buckwheat, *Eriogonum fasciculatum*; fremontia, *Fremontodendron californicum*; squaw bush, *Rhus trilobata*; pale tasselbush, *Garrya flavescens* var. *pallida*; California juniper, *Juniperus californica*; desert apricot, *Prunus fremontii*; and *Yucca whipplei.* Spaces between the shrubs in this community are occupied by various annual grasses and herbs.

Island Chaparral. Although not otherwise considered by us, the southern California offshore islands form a part of southern California and possess a distinct type of chaparral referred to by Philbrick and Haller (1977) and Thorne (1976) as island chaparral. The climate of the islands is maritime, characterized by mild temperatures, mostly above freezing, with little fluctuation during the year. Average annual precipitation varies from island to island being about 20 inches on Santa Cruz Island and 11 inches on San Clemente Island. The island chaparral contains some very interesting endemics such as the Trask mountain mahogany, *Cercocarpus traskiae,* now nearly extinct in its native habitat; *Crossosoma californicum,* also known from the Palo Verde peninsula on the mainland; Trask yerba santa, *Eriodictyon traskiae* ssp. *traskiae*; Catalina cherry, *Prunus lyonii*; and island coffeeberry, *Rhamnus pirifolia.* Non-endemic species include chamise, *Adenostoma fasciculatum*; scrub oak, *Quercus dumosa*; toyon, *Heteromeles arbutifolia*; laurel

→

Plate 1. Plant communities of southern California: a, Valley grassland; b, Coastal sage scrub; c, Southern oak woodland; d, Desert chaparral; e, Chamise chaparral; f, Redshanks chaparral; g, Scrub oak chaparral; h, Montane chaparral; i, Creosote bush scrub; j, Joshua tree woodland; k, Yellow pine forest; l, Pinyon-juniper woodland.

Plate 1

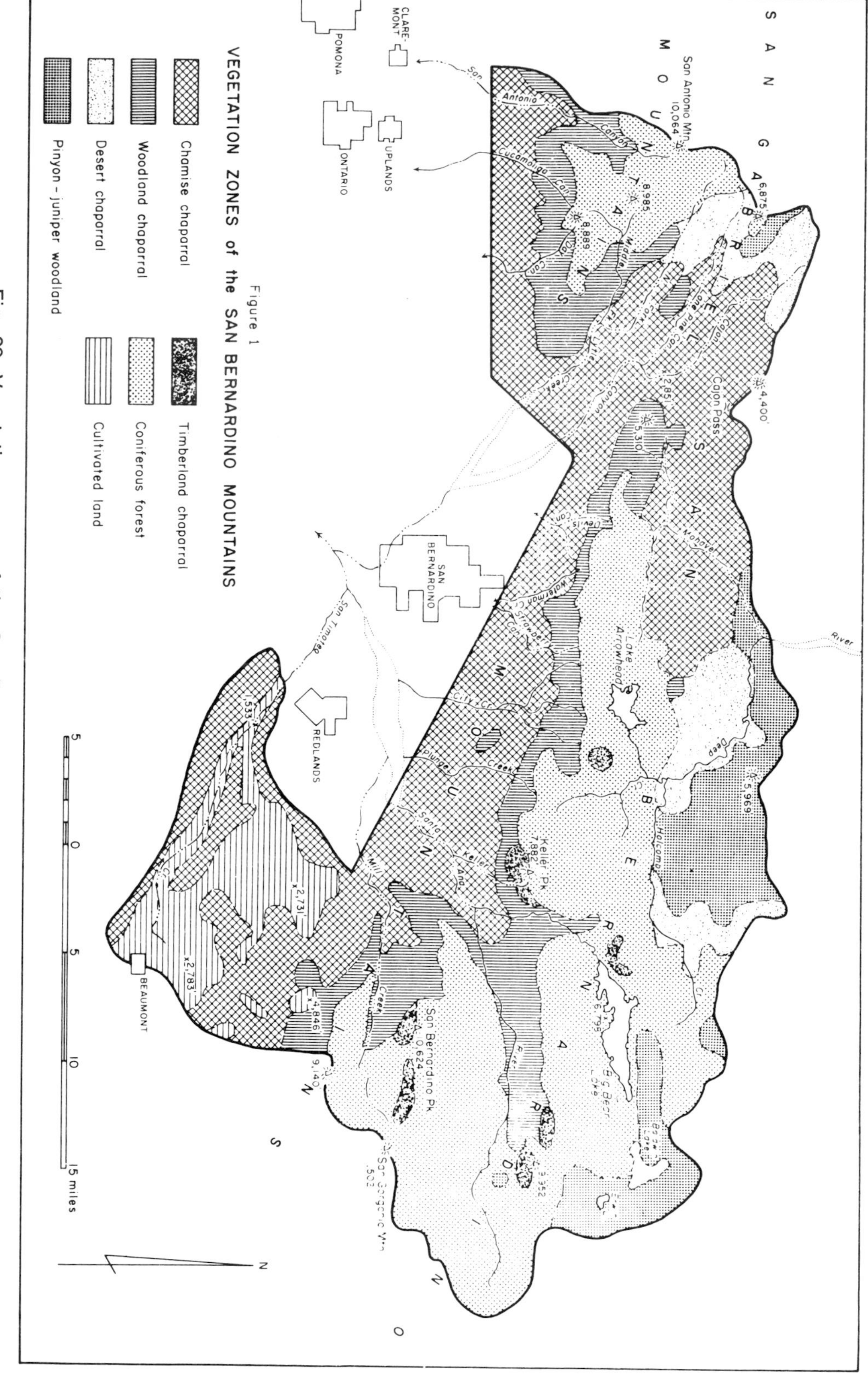

Fig. 29. Vegetation zones of the San Bernardino Mountains. (Horton, 1960).

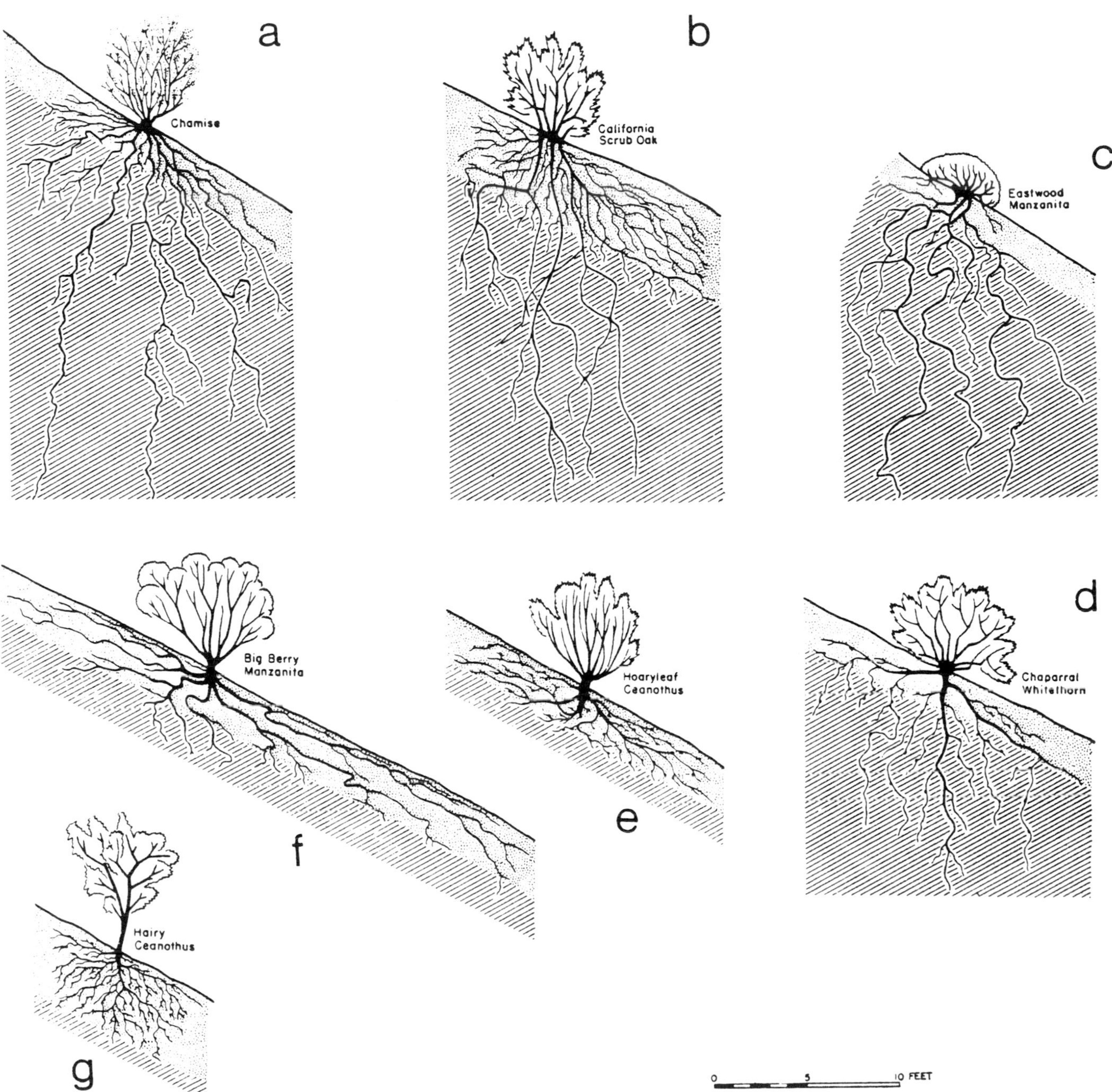

Fig. 30. Diagrammatic sketch of typical growth forms and root systems of seven chaparral shrubs: a-d, woody shrubs with deeply penetrating root systems; e-g, woody shrubs with shallow root systems; a, chamise, *Adenostoma fasciculatum*; b, scrub oak, *Quercus dumosa*; c, chaparral whitethorn, *Ceanothus leucodermis*; d, Eastwood manzanita, *Arctostaphylos glandulosa*; e, hoaryleaf ceanothus, *Ceanothus crassifolius*; f, bigberry manzanita, *Arctostaphylos glauca*; g, hairy ceanothus, *Ceanothus oliganthus*. Shrubs a-d are optimum for soil stabilization because the root systems bind the soil and anchor it to the underlying substrate; shrubs e-g bind the soil but do not anchor it to the underlying substrate. Chaparral shrubs with deep penetrating roots are often stump-sprouters, those with shallow root systems often non-stump-sprouters. (Adapted from Hellmers, *et al.*, 1955).

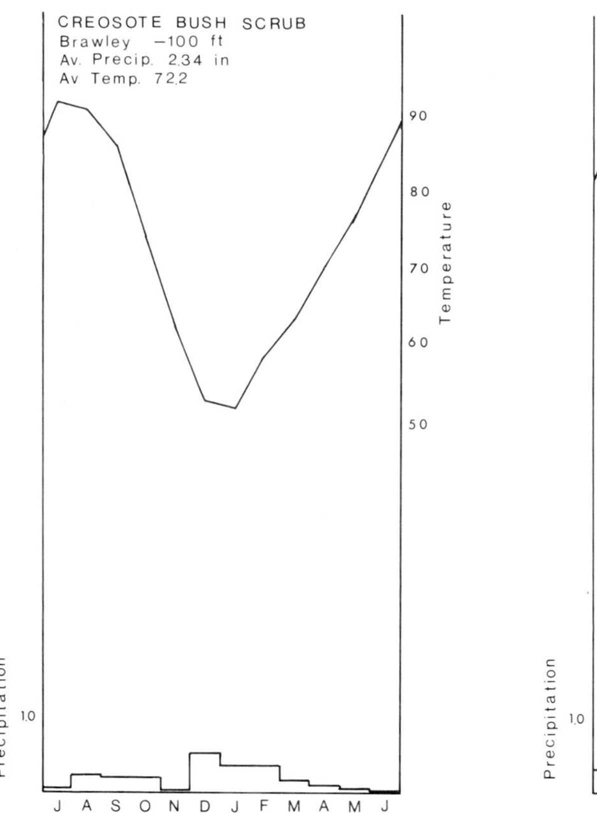

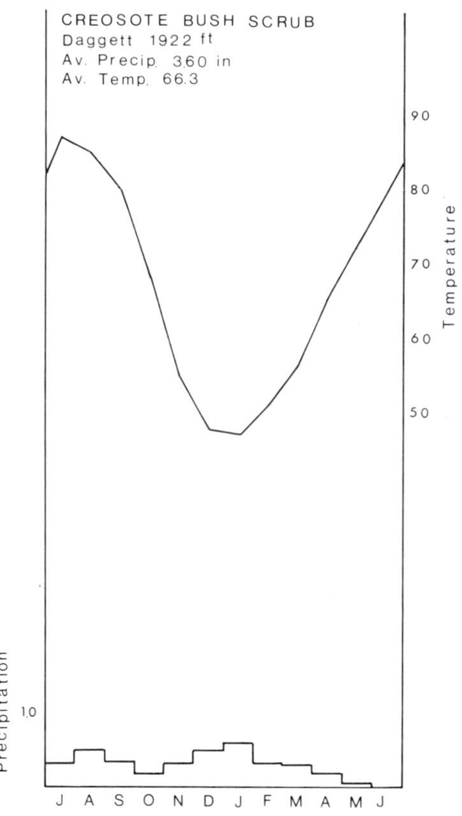

Fig. 31. Average monthly precipitation and temperature for two stations in the creosote bush scrub community.

sumac, *Malosma laurina*; lemonade berry, *Rhus integrifolia*; sugarbush, *R. ovata*; and mission manzanita, *Xylococcus bicolor.*

Creosote Bush Scrub (Plate 1i). The desert vegetation has suffered more from lack of uniformity in its treatment by botanists than has any other major vegetation type in southern California, but the creosote bush scrub is one community on which there is general agreement. It is a community that covers vast expanses of land in both the Mojave and Colorado deserts. It occurs on well-drained sandy flats, bajadas and upland slopes to about 5000 ft. with the plants widely and often even spaced. In addition to the creosote bush, *Larrea tridentata,* the community includes burroweed, *Ambrosia dumosa*; ocotillo, *Fouquieria splendens*; *Dalea* species including the smoke tree, *Dalea spinosa*; box thorn, *Lycium* spp.; burrobrush, *Hymenoclea salsola* var. *salsola*; incienso, *Encelia farinosa*; bush encelia, *E. frutescens*; desert or apricot mallow, *Sphaeralcea ambigua*; squaw waterweed, *Baccharis sergiloides*; mesquite, *Prosopis glandulosa* var. *torreyana*; desert ironwood, *Olynea tesota*; arrowweed, *Pluchea sericea*; and along water courses, desert willow, *Chilopsis linearis.*

In the creosote bush scrub precipitation is light varying from 2-8 inches with some of it coming as summer showers. Temperatures vary widely both seasonally and diurnally.

Joshua Tree Woodland (Plate 1j). A very conspicuous community in the Mojave Desert well known to botanists and non-botanists alike. This community is located be-

→

Plate 2. a, *Acacia greggii*; b, c, *Adenostoma fasciculatum*; d, e, *Adenostoma sparsifolium*; f, *Adolphia californica*; g, *Aesculus californica*; h, *Agave deserti* ssp. *deserti, Cercidium floridum* in background; i, *Arbutus menziesii*; j, *Amorpha californica*; k, *Amorpha californica* colonizing raw roadcut in the San Bernardino Mts.; l, *Atriplex canescens.*

Plate 2

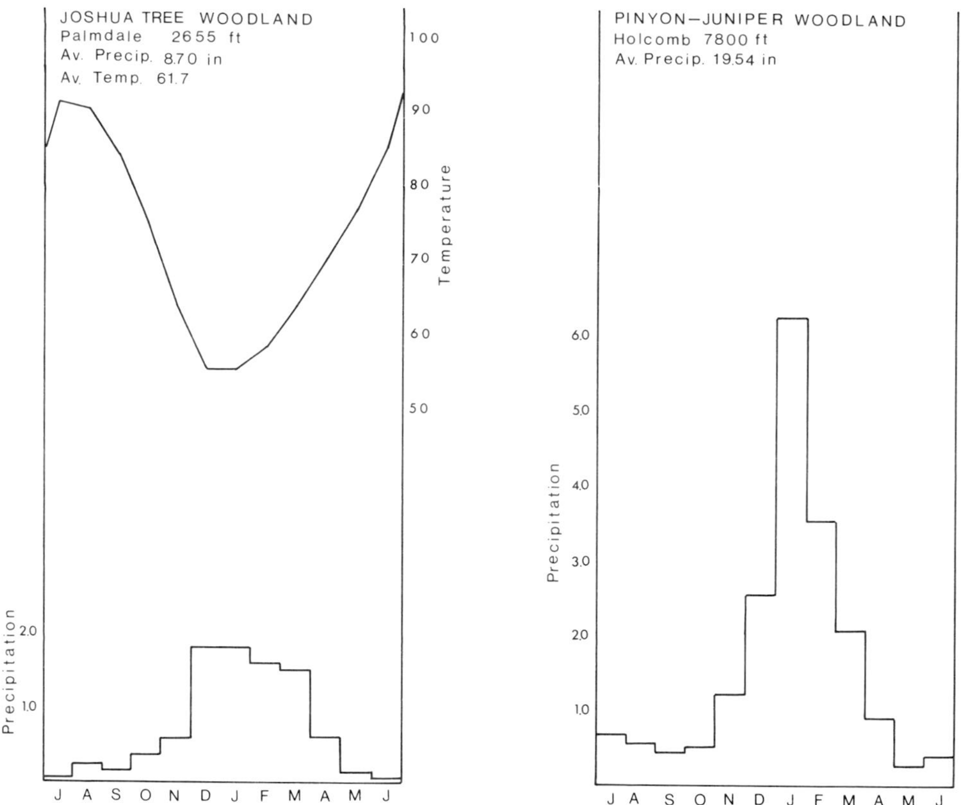

Fig. 32. Average monthly precipitation and temperature for one station in the Joshua tree woodland community and average monthly precipitation for one station in the pinyon-juniper woodland community.

tween the creosote bush scrub and the pinyon-juniper woodland where it is found on sandy, loamy or fine gravelly soils, usually on fairly gentle slopes at elevations of 2500-4000 ft. The average rainfall is about 6-15 inches with summer showers. Besides Joshua tree, *Yucca brevifolia*, and its variety *jaegeriana*, other species include bladdersage, *Salazaria mexicana*; Anderson desert thorn, *Lycium andersonii*; Cooper desert thorn, *L. cooperi*; California buckwheat, *Eriogonum fasciculatum* var. *polifolium*; cotton thorn, *Tetradymia spinosa* var. *longispina*; and Spanish bayonet, *Yucca baccata* var. *baccata.* At low elevations creosote bush, *Larrea tridentata,* may be present and toward the upper limits of the community it occurs with California juniper, *Juniperus californica*; and Utah juniper, *J. osteosperma.*

Pinyon-Juniper Woodland (Plate 11). According to Vasek and Thorne (1977), the California pinyon-juniper woodland represents the westernmost extension of widespread vegetation types occurring in the Great Basin and Colorado Plateau regions. It is common on the higher mountains of the Mojave Desert and the desert slopes of the San Gabriel, San Bernardino and Little San Bernardino mts. where a familiar pattern prevails of juniper on gentle slopes and oneleaf pinyon pine, *Pinus monophylla,* on higher, steeper slopes. This community is also common on mountain slopes in northern Baja California. California juniper, *Juniperus californica,* is most common below 5000 ft. where precipitation is very low, and Utah juniper, *J. osteosperma,* at elevations of 4800-8500 ft. where precipitation is more abundant, as around Baldwin Lake. This community will often be found above the creosote bush scrub and below the yellow pine forest. At the Boyd Deep Canyon Research Center in Riverside Co., in addition to California juniper, *J. californica*; and oneleaf pinyon pine, *Pinus monophylla*; associated species include gray oak, *Quercus turbinella*; bigberry manzanita, *Arctostaphylos glauca*; Parry nolina, *Nolina parryi*; desert agave, *Agave deserti* ssp. *deserti*; Mojave yucca, *Yucca schidigera*; and California buckwheat, *Eriogonum fasciculatum*. On the desert-facing slopes of the San Bernardino Mts., in addition to oneleaf pinyon, *Pinus monophylla*; and Utah juniper, *Juniperus osteosperma*; are gray oak, *Quercus turbinella*; narrowleaf goldenbush, *Haplopappus linearifolius*; waxy bitterbush, *Purshia glandulosa*; and California buckwheat, *Eriogonum fasciculatum*. Other species which may be encountered in this community include great basin sagebrush, *Artemisia tridentata*; rabbitbrush, *Chrysothamnus nauseosus*; crossosoma, *Crossosoma bigelovii*; and desert tea, *Ephedra nevadensis*.

Precipitation, which includes both winter snow and summer showers, varies from 10-30 inches.

Yellow Pine Forest (Plate 1k). The lower coniferous forests of southern California have been accorded different treatment by different authors. Thorne's (1977) lower montane coniferous forests include what he considers the Coulter pine and yellow pine forests, the latter he then divides into the ponderosa and Jeffrey pine forests. We follow Munz and Keck and recognize a single community, the yellow pine forest which is common throughout the San Gabriel, San Bernardino, San Jacinto, Palomar, Laguna, Sierra Juárez and Sierra San Pedro Mártir mts. at elevations of from 5000-8000 ft. Precipitation, much of it coming in the form of snow, varies from 25-80 inches. The soils are mostly residual upland soils of type *Ea* and are moderately to strongly acid in reaction, especially in subsoils, and have depths of from three to six feet to bedrock (Storie and Weir, 1963).

Plants characteristic of the community include yellow pine, *Pinus ponderosa*; Jeffrey pine, *P. jeffreyi*; sugar pine, *P. lambertiana*; incense cedar, *Calocedrus decurrens*; white fir, *Abies concolor*; bigcone spruce, *Pseudotsuga macrocarpa*; Kellogg oak, *Quercus kelloggii*; greenleaf manzanita, *Arctostaphylos patula* ssp. *platyphylla*;

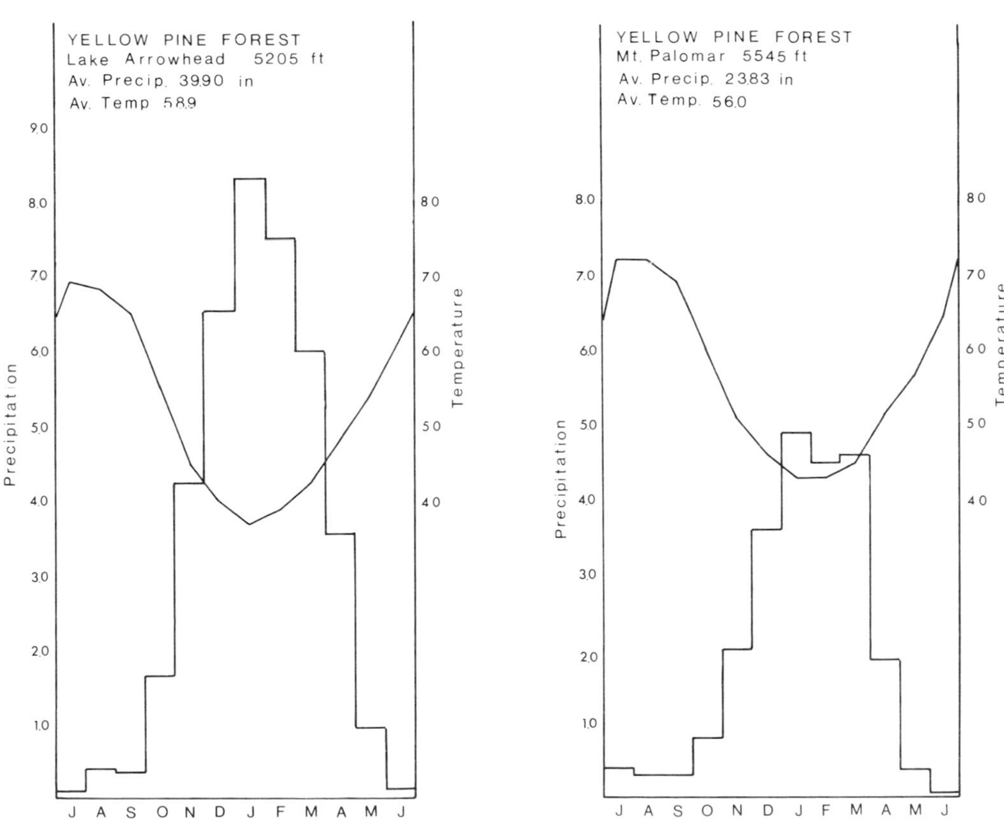

Fig. 33. Average monthly precipitation and temperature for two stations in the yellow pine forest community.

Plate 3

pinkbract manzanita, *A. pringlei* ssp. *drupacea*; deerbrush, *Ceanothus integerrimus*; hairy yerba santa, *Eriodictyon trichocalyx*; pale silktassel, *Garrya flavescens* var. *pallida*; Inyo lupine, *Lupinus excubitus*; coffeeberry, *Rhamnus californica* ssp. *cuspidata*; Sierra gooseberry, *Ribes roezlii*; and chaparral nightshade, *Solanum xantii.*

Desert Oasis. This community was not recognized by Munz and Keck and was included by Thorne in his desert oasis woodland. It is however a conspicuous community because of the presence of the California fan palm, *Washingtonia filifera,* the only palm native to California, and a species restricted to sites having a permanent water supply. It is found around the Salton Basin particularly along the San Andreas fault, on hillside seeps or in canyon washes such as in the Palm Springs area. In addition to *Washingtonia filifera,* other species include arrowweed, *Pluchea sericea*; dropseed, *Sporobolus airoides*; wiregrass, *Juncus acutus* var. *sphaerocarpus*; saltgrass, *Distichlis spicata* var. *spicata*; and screwbean, *Prosopis pubescens.*

MAJOR PLANT COMMUNITIES OF NORTHERN BAJA CALIFORNIA

In Baja California, Wiggins (1980) recognizes eight botanical regions, three of which are found in the northern portion of the peninsula. They are (1) the microphyllos desert, (2) the Californian region and (3) the Baja California coniferous forests.

The microphyllos desert lies east of the Sierra Juárez and the Sierra San Pedro Mártir. Here the dominant shrub is the creosote bush, *Larrea tridentata,* accompanied by ocotillo, *Fouquieria splendens*; littleleaf palo verde, *Cercidium microphyllum*; desert ironwood, *Olneya tesota*; elephant tree, *Bursera microphylla*; and other species. In California this assemblage of plants would be included within the creosote bush scrub community.

Wiggins' Californian region, which he does not further divide, is the same as Munz and Keck's Californian Biotic Province which extends south into northern Baja California, where it occupies the Pacific coastal strip and the foothill woodland in the Sierra Juárez and Sierra San Pedro Mártir upward to the lower edge of the Jeffrey pine belt, and continues southward to about the southern margin of the Llano Santa Maria, inland from San Quintin.

The transition from chaparral of western San Diego Co. is gradual for some distance with chamise, *Adenostoma fasciculatum,* near the coast giving way inland to redshanks, *Adenostoma sparsifolium,* and various species of *Arctostaphylos* and *Ceanothus.* According to the Munz and Keck plant community system, Wiggins' Californian region would include the coastal sage scrub and chaparral communities as well as what Mooney calls the coastal sage succulent scrub and Thorne refers to as the maritime desert scrub communities.

One shrub characteristic of this region is what Wiggins calls *Arctostaphylos oppositifolia* (= *Ornithostaphylos oppositifolia*) which in some areas forms dense, nearly pure stands. Introduced into cultivation by this botanic garden, we have found the species to be a particularly fine and dependable garden subject.

Wiggins' Baja California coniferous forest occupies the Sierra Juárez and the Sierra San Pedro Mártir above the chaparral zone. This region supports open stands of fourleaf pinyon, *Pinus quadrifolia*; and Jeffrey pine, *Pinus jeffreyi*; with the former dominant over a large portion of the Sierra Juárez and with wide stretches in the southern parts of the Sierra San Pedro Mártir. Jeffrey pine is dominant in the higher parts of the Sierra San Pedro Mártir. Other species in the coniferous forest include *Pinus murrayana*; sugar pine, *P. lambertiana*; white fir, *Abies concolor*; and incense cedar, *Calocedrus decurrens.*

←

Plate 3. a, *Arctostaphylos densiflora*; b, c, *Arctostaphylos glauca*; d, *Arctostaphylos pringlei* var. *drupacea*; e, f, *Arctostaphylos stanfordiana* var. *bakeri*; g, *Arctostaphylos* 'Trinity Ruby'; h, *Arctostaphylos 'Ophioviridis'*; i, *Arctostaphylos* 'Pt. Reyes' (front and left), *Arctostaphylos edmundsii* (right center); j, *Arctostaphylos viridissima*; k, *Arctostaphylos manzanita* (habit); l, *Arctostaphylos manzanita* (bark).

Chapter 4

A GUIDE TO SELECTED SPECIES

THE FOLLOWING LIST of species does not propose to be a complete catalogue of those plants which might be expected to thrive within the geographical limits set for this work. The plants are ones with which we have had experience in growing at Claremont or in Orange Co. and which we feel have value, as ornamentals, for soil stabilization and erosion control, as windbreaks, living barriers, for street planting, for public parks and for use along highway or freeway rights-of-way or for other specialized purposes.

In taxonomic matters we are conservative, particularly at the subspecific level. For horticulturists or others using the plants it often makes little or no difference which of several subspecies (or varieties) is used, particularly in those instances where the differences are based upon minor morphological characters.

Some of the plants listed are of hybrid origin. Unless they are the result of controlled hybridizations no attempt will be made to speculate on their parentage. Species on the rare or endangered list (California Native Plant Society, *Special Publication No. 1*, 2nd ed. 1980) are indicated by ★. One manzanita now believed to be extinct in the wild is indicated by ■.

Some common or popular species are not included often because of their susceptibility to disease or air pollution (*viz.* Monterey pine and the cypresses).

Unless otherwise indicated all species have withstood a temperature of 18 F with little or no damage.

For information on the adaptability of plants to coastal conditions we are much indebted to Dara Emery of the Santa Barbara Botanic Garden (pers. com.); for plants suitable for the Central Valley to Roberts (1979); and for information on plants suitable for southwestern Arizona to Sacamano and Jones (1975).

The botanical names used in this work are basically the ones used by Munz and Keck (1959) in *A California Flora* and by Munz (1974) in *A Flora of Southern California,* but where more recent studies have shown that other names should be applied we will use the later name with the earlier epithet given as a synonym.

In this work no attempt should be made to read any special importance into the use of either the term *variety* or *subspecies*; the two terms merely refer to subunits within the species.

ACACIA. A genus of the Fabaceae with a very large number of species native mostly to subtropical regions, especially Africa and Australia. Many species of horticultural value. One species native to California.

1. *A. greggii.* Catclaw. Plate 2a. A straggling shrub or sometimes a small tree to 12 ft., usually with prickly branches, prickles stout, curved, about ¼ in. long, young growth more or less pubescent; leaves bipinnately compound, deciduous, 1-2 in. long with 4-6 pinnae, leaflets 8-12 on each pinna, oblong-ovate, ⅛-¼ in. long, pale green, entire; flowers yellow, very small, numerous, bisexual, arranged in tight cylindrical spikes to 2 in. long; fruit pods much-flattened, 2-6 in. long and 2-11 seeded, more or less constricted between the seeds. April-July.

Catclaw is common in washes and on desert slopes in the Colorado and southern Mojave deserts, east to Texas and south into Mexico, usually at elevations below 6000 ft. where it is a member of the creosote bush scrub to pinyon-juniper woodland communities.

A. greggii presents no problems in its cultivation, is insect pest and disease free and completely drought tolerant. During the dry season it is deciduous, leafing out again with the first rains.

Regions: 1, 2, 3, 4, 5, 6. Also does well in southern Arizona and northern Sonora.

ACALYPHA. A large genus of the Euphorbiaceae with about 250 species of herbs or shrubs, mostly in tropical and subtropical areas. Some ornamental species in cultivation. One species native to extreme southern California.

1. *A. californica.* California copperleaf. An intricately branched slender shrub with thin stems, to 3 ft. and equally broad; leaves alternate, simple, ovate to cordate, to 1¼ in. long and about as broad, crenulate-dentate, prominently veined on the lower surface, glandular-pubescent to nearly glabrous in age; flowers small, without petals, unisexual but both staminate and pistillate flowers on the same plant, staminate inflorescence a short-peduncled, narrow spike ½-1¼ in. long, flowers minute, red, cream-colored anthers conspicuous, pistillate flowers inconspicuous, on short spikes or in small groups at the base of the staminate spikes; fruit a small 3-seeded capsule often surrounded by an enlarged bract. May-June, but in cultivation it flowers much of the year.

California copperleaf is locally frequent on dry granitic slopes in San Diego Co. south into Baja California, also on the islands in the Gulf of California and on Cedros Island. It is a member of the southern oak woodland and chaparral communities.

This attractive little shrub has been grown by the Rancho Santa Ana Botanic Garden since 1939. Copperleaf thrives in almost any type of soil and is at its best when it receives water every two or three weeks during the summer. Under suitable conditions it volunteers from seed but the seedlings are often eaten by rodents. The plants prefer full sun or light shade. It may be pruned to form low hedges, used as a bankcover or in planters.

This species is very sensitive to frost and plants at this botanic garden have been frozen many times. Plants are damaged by temperatures below 28 F and during the winter of 1978-79 many were killed when the temperature dropped to 18 F. If the plants are well established they generally resprout from the base after being frozen.

Fallen flower parts tend to stick to the leaves because of the glandular hairs, and the plants should be washed off occasionally to improve their appearance. Few native plants bloom over such a long period of time, and the narrow reddish catkin-like staminate inflorescences are produced in such numbers that the entire plant takes on a reddish glow.

Regions: 1, 2 and warmer areas in 3, 5.

ACER. A genus of the Aceraceae with about 115 species of deciduous or rarely evergreen trees or shrubs native to North America, Asia, Europe and North Africa. Leaves opposite, simple and usually palmately-lobed; staminate and pistillate flowers borne separately and both on the same plant or on separate plants; petals present or absent; fruits consisting of two long-winged, compressed samara. Many ornamental trees.

1. *A. macrophyllum.* Bigleaf maple. Handsome broad-topped trees to 65 ft. with trunks to 3 ft. in diameter, bark brownish-gray; leaves palmately divided into 5 parts which are then mostly 2-3 lobed or toothed, to as much as 15 in. wide and with petioles to 10 in. long, leaf blades rather thick, dark green, somewhat shiny on the upper surface, paler beneath; flowers borne in long drooping clusters, yellow, fragrant, with petals. Flowers appearing after the leaves. April-May.

Bigleaf maple is widely distributed from British Columbia to southern California, usually along streambanks and in canyons below 5000 ft. In California it occurs in the coastal sage scrub, riparian woodland and chaparral communities.

A. macrophyllum is remarkable for its large leaves and in some places for its brilliant fall coloring. It is in general of limited value in ornamental horticulture except for parks, etc., being too large for home gardens, and it should not be used as a street tree. Resistant to oak root fungus. The bigleaf maples grown at this garden in heavy soil on the mesa have all been subject to wilt caused by *Verticillum albo-atrum,* or a related organism. So far the disease has not shown up on the trees growing in sandy or gravelly soils in the plant community area. The symptoms are a sudden wilting, usually on hot summer days. Often an entire branch or one side of the tree becomes flaccid while the other side remains normal. The sapwood on the infected side will develop greenish streaks and later a slime-flux develops on the trunk.

Regions: 1, 2, 3, 6.

2. *A. negundo* ssp. *californicum.* Box elder. Round-headed trees to 50 ft., bark dark; leaves pinnately 3-foliate, terminal leaflet largest, 3-5-lobed, ovate to 6 in. long, coarsely serrate; flowers unisexual, without petals, staminate flowers fascicled, pistillate flowers in drooping racemes. March-April.

Box elder is found along streams and bottomlands from the San Bernardino and San Jacinto mts. where it is local, to northern California at elevations below 6000 ft. where it occurs in many plant communities below the yellow pine forest.

A. negundo is a fast-growing, brittle tree with many faults including harboring the box elder bug, suckering and producing a large quantity of fruits which must be cleaned from under the plants. A weedy species.

Regions: 1, 2, 3, 6.

ADENOSTOMA. A genus of the Rosaceae containing two species native to California and Baja California. Erect, evergreen shrubs or small trees with heather-like foliage; leaves simple, alternate, in clusters or scattered, linear, entire and very numerous; flowers small, white (or rarely pinkish), in terminal compound clusters; fruit an achene.

1. *A. fasciculatum.* Chamise, greasewood. Plate 2b, c. Diffusely branched shrubs to 12 ft. with slender straight branches from a well developed burl; bark reddish, becoming shreddy with age; leaves fascicled, or some single and alternate, linear, about ¼ in. long, sharp-pointed, usually channeled on one side; flowers creamy-white, almost sessile, in compact clusters 1½-4 in. long. February-July.

Chamise is a very common dominant in the chaparral community where it occurs below 5000 ft. from northern California south to Baja California. It is also found in the Sierran foothills.

Regions: 1, 2, 3, 6. Also does well in the Central Valley.

2. *A. sparsifolium.* Redshanks, ribbonwood. Plate 2d, e. Shrubs or small trees to 20 ft. with yellowish-green bark becoming reddish on old stems and peeling off in thin sheets; leaf-bearing branchlets glandular and clustered near the ends of the branches; leaves linear, ½-¾ in. long, scattered, alternate or rarely opposite, glandular-dotted; flowers white (or rarely pinkish), fragrant, in loose clusters to 2½ in. long. July-August.

Redshanks occurs mostly below 6000 ft. from San Gorgonio Pass south to Baja California and locally in Santa Barbara and San Luis Obispo cos. It is a component of the chaparral and in places where it is the dominant and sometimes almost the only species present, the community is referred to as the redshanks chaparral, considered by some to be the most attractive form of chaparral. In describing *A. sparsifolium,* Hanes (1977) said, "The foliage is feather-like and chartreuse, and open enough to allow the major branches to show through in bold relief." The seeds are very attractive to small birds.

Regions: 2, 3, 6. Also does well in the Central Valley.

Under cultivation the two species of *Adenostoma* require the same treatment; drainage must be good and once established the plants are quite drought tolerant and free from insect pests and disease. Excessive amounts of water cause the plants to sucker excessively and the foliage to turn yellow. Of the two species, *A. sparsifolium* is the more attractive and when planted to form groves, with little or no underplanting, the mature plants with their open habit, exfoliating bark and reddish branches present an especially attractive landscape. This species has not received the attention from horticulturists that it deserves.

ADOLPHIA. A genus of the Rhamnaceae containing two species, one of which is native to California.

1. *A. californica.* Plate 2f. A profusely branched, rigid, spinescent shrub forming rounded clumps to 4 ft. and twice as broad, branches bright green, ultimate twigs rigidly divaricate and spine-tipped, spine brownish or yellowish; leaves early deciduous, simple, entire, opposite, spatulate to obovate, rounded at the apex, tipped with a sharp but rather soft point, about ¼ in. long, pale green and slightly pubescent; flowers small, numerous, fragrant, fasciculate in the axils of the leaves, more or less

bowl-shaped, greenish, petals small, white; fruit a dry capsule. December-April.

Adolphia is found along arroyos and on hillsides near springs and seeps in southwestern San Diego Co. and south into Baja California where it is a component of the maritime desert scrub.

Grown at this garden in well-drained sandy and gravelly soil this species has done very well and the plants have grown together to form a solid mass some 3-4 ft. high. It has not been attacked by insects or by plant disease. Its frost tolerance is not known but it has withstood 18 F at this garden without injury. It is highly recommended for erosion control and for low barrier plantings. This species which has much to recommend it is practically unknown to horticulturists.

Regions: 1, 2, 3. Should be tried in other areas.

AESCULUS. A genus of the Hippocastanaceae with about 15 species native to the Northern Hemisphere, one species in California and one in northern Baja California.

1. *Aesculus californica.* California buckeye. Plate 2g. Deciduous trees to 40 ft. with a broad crown, or sometimes large shrubs much-branched from the base; bark smooth, whitish or gray; leaves palmately compound, opposite, usually with 5 leaflets, leaflets oblong-lanceolate, 3-6 in. long, nearly glabrous; inflorescence a many-flowered thyrse, 6-10 in. long, erect or drooping; flowers large, pinkish-white, ill-smelling, of two kinds, fertile flowers near the tip of the inflorescence with long thick styles, sterile flowers with short styles; fruit a large 3-valved capsule containing a single, large, shiny seed. April-June.

California buckeye is very common in the foothill country surrounding the Central Valley, at elevations usually below 4000 ft. where it is a dominant member of the foothill woodland community. It is also found in the coast ranges from Siskiyou Co. south to Los Angeles Co.

A. californica is one of California's most interesting trees. It comes into leaf in March and by June or early July the leaves turn brown and often drop soon thereafter. Just before the leaves turn brown the tree is covered with large multi-flowered, candle-like inflorescences but usually only a single (occasionally 2-3) fruit is set. The fruits, which are small and immature when the leaves fall, continue to develop throughout the summer and by late fall the leafless tree is attractively adorned with grayish-brown pear-shaped fruits which later split to reveal a single large, shiny brown seed.

In cultivation additional water will prevent early leaf fall; however the buckeye is one native that is nearly as attractive in its leafless state as when covered with foliage. When grown under dry conditions the plants tend to form large bushes branched from the base rather than single-trunked trees. In gardens the plants should be given plenty of room to develop.

Regions: 1, 2, 3, 6. Also does well in the Central Valley.

Aesculus parryi is a small shrubby buckeye mostly 3-12 ft. and usually branched from the base. Inflorescences are 2½-9 in. long and the fruits about 1 in. in diameter. Parry buckeye is found in the coastal desert scrub community in northern Baja California. Except for the plants at this botanic garden it is probably not in cultivation.

AGAVE. A large genus of the Agavaceae with perhaps 300 species native to the warmer parts of the Western Hemisphere. Plants stemless or short-stemmed, long lived; leaves forming basal rosettes, often very large; flower stems tall, arising from the center of the leaf-cluster, flowers usually numerous in panicles or spikes, usually yellow; fruit a many-seeded capsule.

1. *A. deserti* ssp. *deserti.* Plate 2h. Densely caespitose, acaulescent, forming large colonies; leaves gray-green, glaucous, triangular-lanceolate, to 18 in. long, edged with straight or curved prickles, terminal spine strong; inflorescence a narrow panicle to 15 ft. tall, flowers about 2 in. long, in dense clusters. May-June.

A. deserti ssp. *deserti* is found in washes and on dry slopes from the western edge of the Colorado Desert to the Providence, Old Dad, Granite and Whipple mts. of the southern Mojave Desert and south in Baja California along the western side of the Gulf of California at elevations below 5000 ft. It is a member of the creosote bush and shadscale scrub communities.

Regions: 1, 2, 3, 4, 5, 6. Also does well in the Central Valley.

2. *A. shawii* ssp. *shawii.* Shaw agave. Plants single or caespitose, rosettes small to medium with short trunks; leaves ovate to linear-ovate, green, glossy, openly concave, to 20 in. long, edges set with red, hooked prickles, terminal spine strong; flowering stems stout, to 12 ft., closely imbricated by large purple, succulent bracts, panicle congested, lateral branches stout, buds purplish-brown, flowers greenish-yellow. September-May.

Shaw agave is almost exclusively restricted to the coastal area of northwestern Baja California where it is very common on hillsides and bluffs not far removed from the ocean. Formerly it was common in southwestern San Diego Co. but the species is now nearly extinct in California. It is a member of the maritime desert scrub community.

Regions: 1, 2, 3, 6.

Agave deserti and *A. shawii* are both easily propagated from seed or by division of existing clumps, and the plants are drought tolerant and free from insect pests and disease. Great care should be taken in placing the plants since the terminal spines present hazards to children and pets.

ALNUS. A genus of the Betulaceae with about 30 species of deciduous trees or shrubs mostly in the Northern Hemisphere, in America extending south to Peru. Flowers small, borne in catkins; staminate catkins elongated, tassel-like, greenish-yellow, appearing before or with the leaves; pistillate catkins short, erect; fruit a small woody cone maturing in the fall and remaining on the trees during the winter then shed entire in early spring. The trees produce large quantities of air-borne pollen during February. Seeds attractive to birds.

1. *A. oregona.* Red or Oregon alder. Fig. 34a. A tree often 40-90 ft. tall with very white or white-mottled bark, trunk often unbranched for 20-60 ft., branches rather slim, drooping, forming a narrow dome-like crown; leaves broadly elliptic-ovate, 2-6 in. long, dark green, often rusty pubescent beneath, with coarse teeth which are in turn finely toothed, the entire margin with

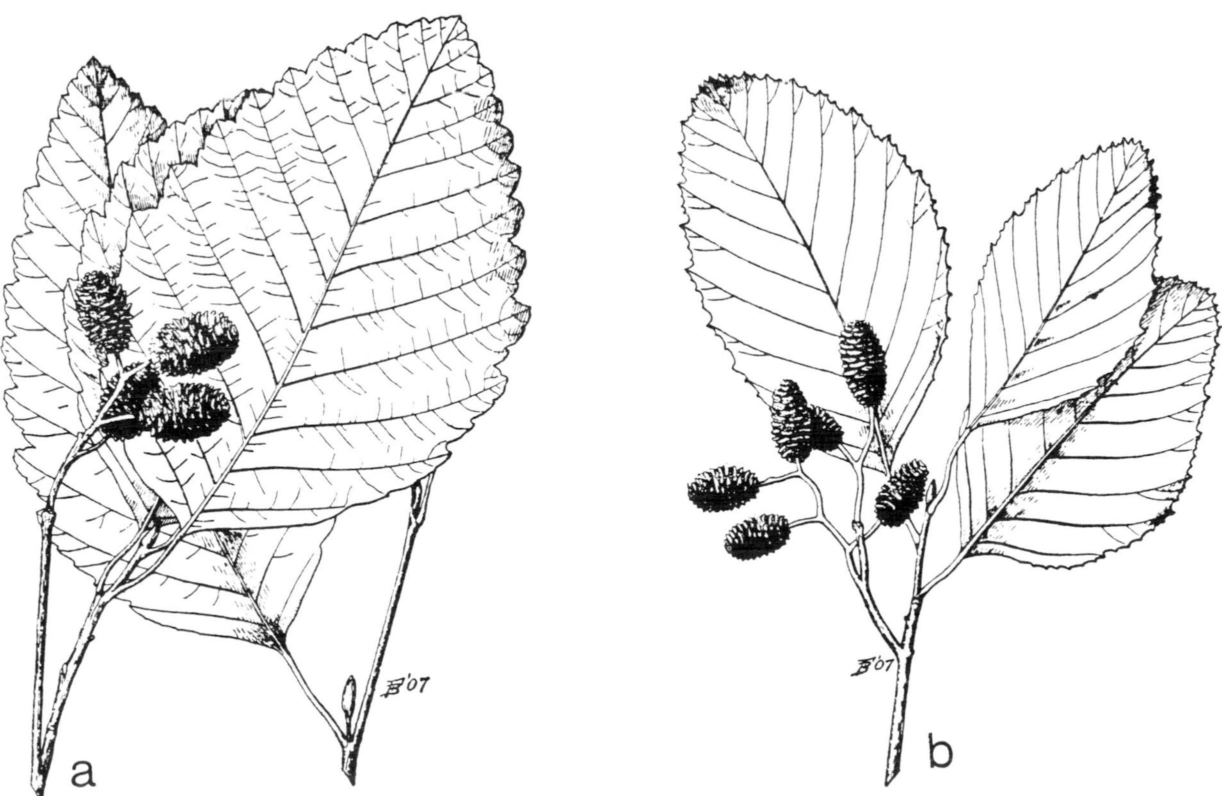

Fig. 34. a, *Alnus oregona*; b, *A. rhombifolia.* (Sudworth, 1908). Reduced.

Fig. 35. *Alnus rhombifolia.* Note mistletoe on branches.

a narrow underturned edge; staminate catkins 3-7 in. long, 2-several in a cluster, pistillate catkins to 1 in. long; fruits cone-like, to 1⅛ in. long, woody, shedding the seeds during autumn.

Oregon alder is found along streambanks and moist places from the mouth of San Carpoforo Cr., San Luis Obispo Co. north to Alaska. The trees are usually found not more than a few miles from the coast and at low elevations. It occurs in a number of plant communities being common in the mixed evergreen and redwood forests.

With adequate moisture *A. oregona* is a fast-growing tree with an invasive root system which may be made less troublesome by infrequent and deep watering. It is reported to tolerate a surprising amount of brackish water and is a useful tree where the underground water tends to be saline. Very susceptible to attack by mistletoe as is the following.

Regions: 1, 2, 3, 6.

2. *A. rhombifolia.* White alder. Figs. 34b, 35. Similar to the preceding; leaves oblong-ovate to oblong-rhomboid, 2-4 in. long, green beneath, not revolute on the margins; staminate catkins in clusters of 2-7, erect or ascending; fruit to ¾ in. long, seed shed mostly in mid-winter.

White alder is widely distributed along rivers and perennial streams from the Cuyamaca Mts., San Diego Co. north to British Columbia and east to Idaho. In California it occurs at elevations of from near sea-level to 8000 ft. in the Sierra Nevada.

A. rhombifolia is very fast-growing and heat as well as wind tolerant. It is also short lived. Native to borders of rivers and along streams, white alders require considerable water. Very susceptible to attack by mistletoe and tent caterpillars. Roots invasive and deep watering is advised.

Regions: 1, 2, 3, 6, 7.

AMORPHA. A genus of the Fabaceae with about 20 species native to the United States and northern Mexico. Deciduous shrubs with glandular-dotted, heavily scented foliage; leaves odd-pinnate; flowers many, small, borne in long, narrow terminal spikes; petals 1, folded around the stamens and pistil, more or less purplish in color; fruit a 1-2-seeded pod.

1. *A. californica.* California indigobush. Plate 2j, k. Slender shrubs to 10 ft. with prickle-like glands on branchlets and leaf-rachises; leaves 3-10 in. long, leaflets 11-27, broadly oval to elliptic, ⅜-1 in. long, pubescent on both surfaces; flowers about ¼ in. long, reddish-purple, in narrow spikes 2-8 in. long; fruit pods about ¼ in. long. May-June.

California indigobush is found in dry wooded canyons and brushy slopes of the San Gabriel, San Bernardino, San Jacinto, Santa Rosa and Santa Ana mts. north to central California at elevations below 7500 ft. where it is a component of the chaparral and yellow pine forest communities.

This species does not become leggy with age as does the following.

Regions: 1, 2, 3, 6, 7.

2. *A. fruticosa* var. *occidentalis.* Desert indigobush.

Loosely branched shrub to 12 ft. without prickle-like glands on branchlets and leaf-rachises; leaves 4-8 in. long, leaflets 11-21, ovate to oblong, ½-1½ in. long, ⅜-¾ in. wide, sparsely pubescent; flowers small, purple, numerous, borne in long narrow spikes 3-9 in. long; fruit pods about ¼ in. long. May-June.

Desert indigobush is found in canyons and along streams from San Bernardino Co. to San Diego Co., eastward to Arizona and south into Mexico. A member of the chaparral community it is usually found below 5000 ft.

Although their attractive leaves and interesting inflorescences of purple flowers make the plants useful for horticultural purposes the plants may in age become leggy with only a canopy of foliage. This condition could probably be corrected by judicious pruning.

Regions: 1, 2, 3, 4, 6.

The California species of *Amorpha* present few or no problems in cultivation and because of their spreading root systems are recommended for erosion control. *A. californica* is very common on raw roadcuts in Region 7.

ARBUTUS. A genus of the Ericaceae with about 20 species of evergreen shrubs or trees native to North and Central America, western Europe, the Mediterranean and western Asia. One species in California.

1. *A. menziesii.* Madroño, madrone. Plate 2i. Evergreen tree to 120 ft. with rounded head, bark smooth, reddish-brown, exfoliating except at the base of old trees where it becomes rough and fissured; leaves alternate, 3-6 in. long, elliptic to subovate, leathery, glossy green above, glaucous beneath, margins entire; flowers white, urn-shaped, fragrant, borne in a terminal panicle composed of 6-10 dense racemes; fruit a warty, fleshy berry about ¼ in. in diameter, red or reddish-orange, or orange, containing numerous seeds. March-May.

Madroño is found in canyons and on wooded slopes from San Luis Obispo Co. north to Vancouver Island and in the Sierra Nevada from Mariposa to Shasta cos. Rare in southern California. Usually found below 5000 ft. it is abundant in the redwood, mixed evergreen and Douglas-fir forests, less so in the foothill and oak woodland communities.

A. menziesii is one of the state's best known and most beautiful trees but its horticultural value in southern California is strictly limited. It can only be recommended for areas where the water is nonalkaline and low in total salt content. The plants prefer well-drained soil and once established should be given infrequent but deep watering. Said to be immune or highly resistant to oak root fungus.

Regions: 1, 2, 6.

ARCTOSTAPHYLOS. A large and diverse genus of the Ericaceae consisting of evergreen shrubs varying from low and prostrate to tree-like, usually with very crooked stems and dark red or chocolate-colored bark exfoliating freely, smooth and polished or sometimes fibrous and more persistent; leaves alternate, coriaceous and usually held vertically, margins mostly entire; flowers borne in terminal simple racemes or in often much-branched panicles, flowers fairly small, urn-shaped and much alike in all the species, varying in color from white through pink-tinged to pink or rose-colored; fruit berry-like with copious mealy pulp or with thin pericarp and dry, enclosing 4-10 nutlets that may be variously fused in 2s or 3s, or sometimes all fused to form a single solid stone.

The genus consists of perhaps 50 species distributed from Central America north, with one species circumpolar, but centered in California where the plants are among the best known of the native shrubs.

Nowhere in California botany is there greater disagreement among botanists than there is in the taxonomic interpretation of the manzanitas, a situation that in recent years has escalated due to the publication of large numbers of new species, many of which represent nothing more than variants of previously described polymorphic species, or often the so-called species are merely hybrids which are known to occur in abundance in nature. At present the situation can only be termed chaotic. Our treatment for the most part follows Munz and Keck (1959) and Munz (1974) and is conservative. We feel that in the absence of any consistent agreement among taxonomists that this is the best approach in writing for the horticultural public.

Some of the newly described 'species' represent horticulturally desirable subjects and since, at least in the case of hybrids, propagation must be by vegetative means if the plant's unique character is to be maintained, the

A

best way of identifying such plants is to provide them with clone names.

Biologically the genus can be separated into two distinct groups; those that produce a basal burl and stump-sprout following fires and those that do not form a burl and are killed by fires. According to Hellmers (1955), stump-sprouting species of chaparral shrubs are often those with deep penetrating roots which bind the soil and anchor it to the underlying substrate, whereas those species that do not stump-sprout are species with shallow root systems (Fig. 30).

All species of *Arctostaphylos* are reported to be relatively resistant to oak root fungus (*Armillariella*).

Manzanita branch dieback

All species of *Arctostaphylos* are subject to a disease which has been referred to variously as branch dieback or stem canker. This disease is not restricted to plants under cultivation but also occurs in native populations. In May 1947 a survey showed that nearly every plant growing in Yosemite Valley was either dead or dying from the disease (Clark, pers. com.).

The first evidence of infection is the presence of occasional blighted twigs and young branches (Plate 4b, c). The leaves on an infected stem turn brown and become dry but exfoliation does not occur for some time. On closer inspection it can be seen that the youngest infections occur as blackened spots at the base of the leaf petiole and the leaf eventually turns brown and dies. The black spots enlarge, forming elongate depressed lesions as they spread down the stem. On larger branches numerous cankers may appear secondarily which then become crateriform or fuse to form large lesions; or the initial canker may extend down the entire length of the branch to the soil line. As the cankers become older the bark becomes gray and cracked. On close examination with a hand lens, small black fruiting bodies of the fungus may be observed protruding from the cracks in the outer bark. Although many of the twigs and branches on a plant may be killed, the disease is slow in killing the entire plant.

Manzanita branch dieback is caused by a fungus which has been identified as *Botryosphaeria ribis*, an organism that attacks a wide range of species including *Sequoiadendron giganteum*. The amount of the disease present in a population and the severity of the attack is undoubtedly influenced by the environmental conditions prevailing, but detailed studies have not been reported. It is also likely though not proven, that some populations may be more resistant to the disease than others. It is known that overhead watering of the plants favors the spread of the disease. In ornamental plantings infected parts should be removed and overhead watering discontinued. At this garden experiments are under way with the use of Bordeaux Mixture in controlling the disease.

Manzanita leaf gall

Occasionally manzanitas are attacked by gall-producing aphids (*Tamalia cowenii*) which produce distinctive leaf-fold type galls (Plate 4a). Although it has been reported that all manzanitas are susceptible to attack, Russo (1979) states that he has never found galls on species with bristly hairs on the leaves and this is consistent with our observations.

When the galls, which form as a plant response to the aphid's feeding behavior, are gently unfolded the *stem mother*, a wingless female, and her parthenogenetically produced offspring can be found within a white, cottony wax which the insects secrete. The galls, usually reddish in color, are often so abundant that from a distance it appears that the plants are in flower. In most instances damage to the host is minimal and later in the season the galls turn brown and the infected leaves fall to the ground.

1. *A. andersonii* var. *andersonii.* Heartleaf manzanita. A large, erect or spreading shrub to 12 ft. with a definite trunk, or branching from the base, without a basal burl, branches often elongated, bark smooth, dark red-brown, branchlets densely short-pubescent and bristly, usually with glandular hairs; leaves ovate to oblong-ovate, to 2¾ in. long and 1¼ in. wide, pale green to dark green, heart-shaped or auriculate at the base, usually sessile, glabrous, tomentulose, or sometimes glandular-pubescent beneath; flowers white to pink, borne in large panicles; fruit depressed-globose, red-brown and viscid-pubescent, nutlets separable. January-March.

Heartleaf manzanita is found on dry, sandy and rocky ridges from the Santa Lucia Mts. north to San Francisco, usually below 2200 ft. where it is a component of the chaparral and redwood forest communities.

Fig. 36. Many species of *Arctostaphylos* of prostrate or decumbent habit spread by the rooting of stems in contact with the soil. This also occurs in some species of shrubby or arborescent habit. The two upright branches at the right (above) are connected with the main plant by means of buried branches. Below. Roots formed on branch in contact with the soil. The plant shown is *Arctostaphylos stanfordiana.*

Var. *andersonii* does very well in Claremont growing in heavy loam on the mesa or in sandy and gravelly soil in the plant community area. Although it does quite well in full sun it has been our experience that the plants are longer lived when they are shaded from the afternoon sun.

Var. *imbricata.* Differs from the preceding in having bright green, mostly glabrous leaves up to 1¼ in. long, closely imbricated along the stems. Flowers white. It occurs in the San Bruno Hills of San Mateo Co.

Var. *pallida.* This variety has pale green, glabrous leaves to 2 in. long, also closely imbricated along the stems. The fruits are bright red and glandular. Found in the hills of Alameda and Contra Costa cos. where it grows in the chaparral. All forms of *A. andersonii* have been very successful at Claremont if planted under high canopy shade or at least protected from the afternoon sun.

Regions: 1, 2, 3, 6.

2. *A. auriculata.*★ Mt. Diablo manzanita. Variable in growth habit from low and spreading to erect or even arborescent, to 12 ft., without a basal burl, bark smooth, dark red, branchlets short-pubescent and bristly-hairy; leaves ovate to oblong-ovate, to 2 in. long, gray-green with a fine canescent pubescence, sessile, subsessile or auriculate at the base; flowers white to pinkish, borne in small erect or spreading inflorescences; fruit depressed-globose, red-brown or orange-brown, pubescent, about ⅓ in. in diameter, nutlets separable. February-March.

A. auriculata, considered by some to be a form of *A. andersonii,* is found on dry sandstone slopes in Contra Costa and Alameda cos. at elevations of 500-2000 ft. where it grows in the chaparral.

This species grows very well at Claremont in either heavy soil or in alluvial sands and gravels, but the plant form is probably better when it is grown in heavy soil. It responds well to an occasional summer irrigation.

Regions: 1, 2, 3, 6.

3. *A. catalinae.*★ Catalina manzanita. A tall, erect, arborescent shrub to 18 ft. with a well defined trunk and no basal burl, bark smooth, red-brown, branchlets densely short-hispid and with longer setose hairs, most hairs at first gland-tipped; leaves ovate to oblong, acute, base from subcordate to strongly cordate, grayish-green, glandular ciliate toward the base, along the margins and on the midrib, to 1½ in. long and 1¼ in. wide; inflorescence large, openly paniculate with 5-6 branches from the main rachis, flowers white or suffused with pink; fruit oblate-spherical, glabrescent, orange-brown, nutlets separable. January-February.

Catalina manzanita is restricted to Santa Catalina Island where it grows on volcanic rocks.

This species is fairly new at this botanic garden but so far has done well when planted in either heavy loam or sandy and gravelly alluvial deposits. It tolerates drought or will take some summer irrigation and it looks better when it is irrigated. This species appears to be at least somewhat more resistant to branch dieback than some species. Recommended.

Regions: 1, 2, 3, 6.

4. *A. crustacea* var. *crustacea.* Brittleleaf manzanita. An erect shrub to 6 ft. with several spreading branches arising from a burl; bark smooth, dark purple on old stems, branchlets densely white-downy with additional long white, non-glandular hairs; leaves oblong-ovate to roundish, to broadly lanceolate, to 1½ in. long and 1¼ in. wide, obtuse, truncate or subcordate at the base, bright green, thin or thick, brittle, glabrous and rather shining above, glabrous or slightly tomentulose beneath; flowers in nearly sessile panicles, white or rosy-pink; fruit depressed-globose, reddish-brown, sparsely pubescent or glabrate. February-March.

Brittleleaf manzanita is found on brushy hills in the south coast ranges from Contra Costa and Alameda cos. south to Santa Barbara Co., also on Santa Cruz and Santa Rosa islands, usually at elevations below 3400 ft. where it is a member of the chaparral community.

Var. *rosei.* Differs from the preceding in being lower growing, with branchlets without hispid hairs and leaves that are longer and narrower. In this variety the flowers are deep pink in color. Late flowering, it is preferred to var. *crustacea* for ornamental plantings. Var. *rosei* is restricted to the hills bordering Lake Merced in San Francisco Co. where it occurs in the north coastal scrub community.

Perhaps less susceptible to branch dieback than some of the others, at this garden the plants have done well, but both varieties require good drainage. Here they are irrigated every 3-4 weeks during the summer. Should be protected from the afternoon sun.

Regions: 1, 2, 3, 6.

5. *A. densiflora.*★ Sonoma manzanita. Plate 3a. A low, spreading often nearly prostrate shrub without a

basal burl and with numerous slender, crooked, blackish branches freely rooting where touching the soil; leaves elliptic to oblong-elliptic, or rarely oblanceolate or ovate, to 1¼ in. long and ⅝ in. wide, bright green and glabrous except for minute pubescence along the margins and veins; flowers borne in short, dense many-flowered panicles, corolla white or pink; fruit flattened-globular, about ¼ in. in diameter, berry red and glossy, nutlets separable. February-March.

Fig. 37. In some species of *Arctostaphylos* flowers for the current season develop from buds formed as early as May or June of the previous year. *A. manzanita* photographed early in June.

Sonoma manzanita is known only from roadside banks near Vine Hill School, Sonoma Co. Apparently it hybridizes with *A. manzanita* and *A. stanfordiana.*

A. densiflora is an excellent garden subject and highly recommended. At this garden it grows best where planted on heavy loam and given summer irrigation every 2-3 weeks.

Regions: 1, 2, 3. Does well in the Central Valley if protected from the afternoon sun.

6. *A. edmundsii.* Little Sur manzanita. Plate 3i. A spreading, subprostrate shrub to 4 ft. without a basal burl, stems rooting when in contact with the soil, branchlets pilose; leaves broadly-ovate, elliptic or roundish, to 1 in. long and ¾ in. wide, bright green on both surfaces, margins often reddish, glabrous on both surfaces or sometimes a few hairs along the veins; flowers pinkish, borne in small compact clusters; fruit depressed-globose, brown, nutlets separable. December-February.

Little Sur manzanita is known only from the vicinity of the mouth of the Little Sur River in Monterey Co. where it grows on ocean bluffs. Experimental evidence has shown that many of the plants grown as *A. edmundsii* are of hybrid origin.

Of the low-growing manzanitas this is one of the most worthwhile and is highly recommended as a garden subject. At Claremont it has done best when planted in heavy loam and watered about every two weeks during the summer. This species responds well to shearing which makes it an ideal shrub for the formal garden. Shearing should be done after flowering.

Regions: 1, 2, 3.

7. *A. elegans.* Konocti manzanita. Erect shrubs to 6 ft. often with a 10 ft. spread, without a basal burl, bark red-brown, branches straightish, smooth, polished; leaves ovate to elliptical, to 1½ in. long, bright green, dullish, glabrous, thick, conspicuously veiny; flowers borne in ample panicles, white; fruit depressed-globose, dark red, stipitate-glandular, nutlets 5-7 with 2-3 fused.

Konocti manzanita is found on brushy or wooded slopes in the mountains of Lake Co. at elevations of 2000-4200 ft. where it is a member of the chaparral and foothill woodland communities.

This species has done well at Claremont where it has been grown in well-drained soil and given deep irrigation about every four weeks during the summer. Related to *A. stanfordiana.*

Regions: 2, 3.

8. *A. glandulosa* ssp. *glandulosa.* Eastwood manzanita. Erect spreading shrubs to 8 ft. with a basal burl, bark smooth, reddish, branchlets coarse, conspicuously glandular-hairy; leaves ovate to lanceolate, dull green, more or less glandular-hairy, to 1¾ in. long and 1 in. wide;

flowers white, borne in short, spreading, few-flowered panicles; fruit subglobose, about ⅜ in. in diameter, reddish-brown, usually viscid, nutlets separable. January-April.

Eastwood manzanita is found in dry gravelly places in the mountains of San Diego Co., north to Oregon at elevations of 1000-6000 ft. where it is a member of several plant communities including the chaparral and yellow pine forest.

Ssp. *mollis.* Differs from the preceding in having dark colored branchlets and small, compact, few-flowered inflorescences. This subspecies is found in dry places from Riverside Co. north to San Luis Obispo Co. at elevations of 2500-6000 ft. in the chaparral and yellow pine forest communities.

Ssp. *crassifolia.* Plants spreading, low-growing to 5 ft., branchlets tomentulose, sometimes with scattered longer hairs, not glandular, leaves dark green, glabrate above and more or less tomentulose beneath. It occurs on sandy mesas and bluffs in San Diego Co. where it grows in the chaparral community.

All three subspecies of *A. glandulosa* have done well at Claremont and are recommended. The subspecies *mollis* is outstanding, flowering well and having a good plant habit. There is no reason why *A. glandulosa* cannot be grown in heavy soil if care is taken in its irrigation.

Regions: 1, 2, 3, 6. The subspecies *glandulosa* does well in region 7.

9. *A. glauca.* Bigberry manzanita. Plate 3b, c. A large, erect shrub or even a small tree to 20 ft. with a distinct but short trunk and no basal burl, bark smooth, reddish-brown, branchlets pale green, glaucous and glabrous; leaves ovate or oblong-elliptic, to 1¾ in. long and 1¼ in. wide, mostly truncate or subcordate at the base, or sometimes rounded and acute, pale grayish-green, glaucous and glabrous on both surfaces, entire or on young shoots serrate; flowers in short terminal panicles, white; fruit very large, globose to ovoid, about ½ in. in diameter, light brown, very glandular-viscid, nutlets united. January-April.

Bigberry manzanita is common on the dry slopes, particularly in the cismontane and coast ranges of southern California, but extending from Contra Costa Co. south to northern Baja California where it occurs at elevations below 4500 ft. in the chaparral and Joshua tree woodland communities.

This species often forms large rounded shrubs densely branched and leafy to the ground. To be successful with *A. glauca* the plants must be provided with excellent drainage and they suffer from excessive winter rains if water stands around the plants. Many of our finest and most long-lived specimens have been grown on steep slopes. This species is very susceptible to manzanita leaf gall.

Regions: 1, 2, 3, 6.

10. *A. hookeri* ssp. *hookeri.* Hooker manzanita. A low, procumbent to erect shrub to 3 ft. without a basal burl and with lateral branches rooting where touching the soil, bark smooth, very dark reddish-brown or deep purplish, branchlets tomentose; leaves elliptical or ovate to obovate, thin, glabrous on both surfaces; flowers in short almost head-like racemes, corolla white to pink; fruit globose, slightly depressed, about ¼ in. broad, bright reddish-brown and glossy, nutlets separate or irregularly fused. February-April.

Hooker manzanita forms conspicuous mounds on sandy flats in the open pine woods around Monterey Bay where it is a member of the closed-cone pine forest.

Ssp. *franciscana.*■ Differs from the preceding in having branchlets minutely puberulent and leaves that are bright green in color. Found on serpentine outcrops on the San Francisco Peninsula.

Both subspecies of *A. hookeri* are garden tolerant plants and have done well at Claremont in both heavy loam and alluvial sand and gravel. They are perhaps best on the heavy loam and here they are irrigated about every three weeks during the summer. The subspecies *franciscana* is particularly handsome with its bright green leaves which remain attractive throughout the dry summer months when other shrubs look stressed from drought and high temperatures. Recommended.

Regions: 1, 2, 3.

11. *A. insularis.* Island manzanita. An erect, much-branched shrub or sometimes arborescent, to 12 ft., without a basal burl, bark smooth, dark reddish-brown, branchlets smooth; leaves bright green and shiny above

and glabrous or slightly puberulent beneath, ovate to elliptical, obtuse to acute at the apex, rounded or truncate at the base, to 1¾ in. long and 1 in. wide; inflorescence a large, spreading panicle, flowers white; fruit flattened-globular, bright orange-brown. December-March.

Island manzanita is confined to Santa Cruz Island.

A. insularis is conspicuous because of its bright green shiny leaves and large panicles of white flowers. At Claremont it has done very well when grown in well-drained soil in the plant community area. It is drought tolerant but probably does best when irrigated at 4-5 week intervals during the summer. This species is not as susceptible to *Botryosphaeria ribis* as are some species. Recommended.

Regions: 1, 2, 3.

12. *A. manzanita* ssp. *manzanita.* Common manzanita. Plate 3k, l; fig. 37. A tall, erect shrub with long crooked branches, or almost tree-like to 20 ft., without a basal burl, bark dark reddish-brown, branchlets usually puberulent; leaves thick and firm, elliptical to oblong, or broadly ovate to suborbicular, rarely obovate, to 1¾ in. long and 1½ in. wide, pale to dark green, usually puberulent when young becoming glabrous in age; flowers in drooping rather open panicles, white to pale pink; fruit globose, sometimes depressed, to ½ in. in diameter, deep reddish-brown, nutlets fused or distinct. February-April.

Common manzanita is widely distributed in the middle and inner coast ranges in Tehama, Trinity and Shasta cos., southward in the Sierra Nevada foothills to Mariposa Co. where it occurs on dry slopes in canyons, a member of the chaparral, foothill woodland, northern oak woodland and yellow pine forest communities. It is usually found between 300-4000 ft. elevation.

This species is extremely variable in its morphological characters. Of the larger-growing species of manzanita, *A. manzanita* is outstanding, and at Claremont it has been long-lived whether growing in heavy soil on the mesa or in the alluvial deposits in the plant community area. The species probably consists of numerous ecotypes adapted to specific habitats. The plants should be pruned to bring out the branching structure of the species.

Although quite drought tolerant, *A. manzanita* does respond well to occasional irrigations during the summer.

Regions: 1, 2, 3, 6. Also does well in the Central Valley.

13. *A. obispoensis.* Serpentine manzanita. An erect grayish shrub to 7 ft. without a basal burl and with black-purple, smooth branches and gray-tomentose branchlets; leaves ovate to ovate-lanceolate, to 1½ in. long, round or truncate to subcordate at the base, grayish-green or glaucous, tomentose on both surfaces or glabrate; inflorescences of subsessile panicles, flowers pinkish-white; fruit globose, about ⅜ in. in diameter, orange-brown to red-brown, nutlets separable. February-March.

Serpentine manzanita is found in serpentine areas in San Luis Obispo and Monterey cos. at elevations of 500-3000 ft. where it is a member of the foothill woodland and closed-cone pine forest communities.

At Claremont, *A. obispoensis* has proven to be a very fine species growing equally well in heavy loam or in sands and gravels. Growing in heavy soils the plants should be watered about every 4-5 weeks; when in well-drained soil about every 2-3 weeks.

The young growth is a striking copper color and contrasts well with the older foliage. Recommended.

Regions. 1, 2, 3, 6.

14. *A. otayensis.*★ Otay manzanita. An erect shrub to 8 ft. and nearly as broad, without a basal burl, stems with smooth, dark red bark, branchlets glandular-hairy; leaves elliptic to oblong, to 1½ in. long and ¾ in. wide, grayish- or yellowish-green, finely pubescent on both surfaces; flowers borne in open panicles or simple racemes, white or slightly pinkish; fruit globose, to ⅜ in. in diameter, pale brown. December-March.

Otay manzanita is found on dry slopes in the mountains of San Diego Co. at elevations of 1800-5000 ft. where it is a member of the chaparral community.

A. otayensis is one of the better of the manzanitas and in well-drained soil it has been long-lived, receiving one irrigation every four weeks during the summer. It is one of the first manzanitas to come into bloom. We have had no experience with this species planted in heavy soil.

Regions: 1, 2, 3, 6, 7.

15. *A. pajaroensis.* Pajaro manzanita. Erect, compact shrubs to 10 ft. without a basal burl, bark dark red, exfoliating in thin shreds with stems slowly becoming

smooth, branchlets loosely-pubescent and bristly; leaves ovate-triangular, acute at the tip and auriculate-clasping at the base, to 1¼ in. long and 1 in. wide, glabrous on both surfaces, green or slightly glaucous; flowers pink, borne in dense, sessile, many-flowered panicles; fruit globular, light red, pubescent, about ¼ in. in diameter, nutlets separable. December-January.

Another good species which has done well in both heavy soil and in alluvial deposits. When grown in well-drained soils the plants respond well to irrigation about every 2-3 weeks; in heavy soils care should be taken not to overwater.

The new growth is striking as are the flowers and fruits. Recommended.

Regions: 1, 2, 3, 6. Does well in the Central Valley if protected from the afternoon sun.

16. *A. parryana.* Parry manzanita. A diffuse, widely-spreading shrub to 6 ft. with a 9 ft. spread, without a basal burl, lateral branches often decumbent and rooting where in contact with the soil, bark smooth, dark red or reddish-brown, branchlets canescent or glabrous; leaves thick and firm, ovate, elliptic, broadly oval or rarely obovate, to 1¼ in. long and ¾ in. wide, bright green, shiny, glabrous; flowers white in simple few-flowered racemes, or few-branched panicles; fruit round-ovoid, to ⅝ in. in diameter, dark reddish-brown, nutlets usually fused. February-March.

Parry manzanita is found on dry stony slopes from the San Gabriel Mts. north to the Tehachapi Mts. and Mt. Pinos, usually at elevations of 4000-7500 ft. where it is a member of the chaparral and yellow pine forest communities.

A. parryana resembles *A. manzanita* but may be distinguished from that species by the more spreading habit, lighter green leaves and coalesced nutlets.

This species does well in filtered shade, or shaded from the afternoon sun. Grown in well-drained soil the plants respond well to 2-3 irrigations during the summer months. Parry manzanita always has a fresh green appearance. It is not as floriferous as some species.

Regions: 2, 3, 6, 7.

17. *A. pechoensis* var. *pechoensis.*★ Pecho manzanita. Bushy shrubs to about 6 ft. but sometimes taller, without a basal burl, bark smooth, dark red-brown, branchlets white-tomentose, sometimes also bristly; leaves ovate to ovate-oblong, auriculate and subsessile, or subcordate and petiolate, to 1¼ in. long and ¾ in. wide, green and glaucous with a whitish wax easily rubbed off; flowers white or pinkish, borne in few-flowered panicles; fruits depressed-globose, light brown and glabrous, to ⅜ in. in diameter, nutlets irregularly separable. January-March.

Pecho manzanita occurs on dry slopes and ridges in San Luis Obispo and Santa Barbara cos. at elevations below 2600 ft. where it is a member of the closed-cone pine forest and chaparral communities.

Regions: 1, 2, 3.

18. *A. pilosula.* Stripedberry manzanita. An erect shrub to 10 ft. with a 15 ft. spread, without a basal burl, bark dark red, branchlets pubescent also with short or long bristly hairs; leaves oblong-elliptic, to ovate, to 1¾ in. long, yellow- or glaucous-green, mature leaves glabrous or sparingly pubescent on both surfaces, young leaves densely silvery-tomentose; flowers white, or tinged pink, borne in short few-flowered panicles; fruit depressed-globose, to ⅜ in. in diameter, glabrous, light brown or red-brown usually with blue-black vertical stripes when mature. January-March.

Stripedberry manzanita is localized in San Luis Obispo Co. where it occurs on sandstone hills at elevations of 200-3600 ft., a component of the chaparral community.

This species has done well at Claremont. It requires good drainage and responds well to irrigation at 4-5 week intervals during the summer.

Regions: 1, 2, 3.

19. *A. pringlei* var. *drupacea.* Pinkbract manzanita. Plate 3d. An erect shrub to 12 ft. with spreading branches, without a basal burl, branches with smooth, dull red-brown bark, branchlets densely glandular hairy; leaves gray-green, rough to the touch with conspicuously glandular hairs, ovate to oblong, or almost orbicular, to 2 in. long and 1¾ in. wide; inflorescence a large, almost sessile panicle, flowers pink to rose-colored; fruit dark red, ovoid or spherical, nutlets fused into a single stone. February-April.

Pinkbract manzanita is found from the San Bernardino Mts. south to northern Baja California, usually on dry

slopes at elevations between 4000-7500 ft. where it is a member of the chaparral and yellow pine forest communities.

A. pringlei var. *drupacea* is one of the most distinctive and beautiful of the manzanitas and both the plants and the flowers have a very distinctive fragrance not found in other manzanitas. At Claremont the plants have been long-lived, especially when growing in poor soil and given a good irrigation every 4-5 weeks. They can be grown in heavy soil if care is taken with irrigation during the summer. Recommended.

Regions: 2, 3, 6, 7.

20. *A. pumila.*★ Sandmat manzanita. A low, decumbent or prostrate shrub, rarely more erect, without a basal burl, often forming dense mats to 4 ft. across with erect branchlets, bark reddish-brown, exfoliating, branchlets finely pubescent; leaves narrow, obovate to spatulate, to ¾ in. long and ½ in. wide, rounded to acute or mucronate at the apex, dull green, slightly pubescent on the upper surface, becoming glabrous, distinctly paler and usually permanently pubescent beneath; flowers in short, simple or few-branched racemes, corolla white or pink; fruit globose, slightly depressed, about ⅛ in. in diameter, glabrous, brown, nutlets separable. February-May.

Sandmat manzanita is known only from the sand hills and woods bordering Monterey Bay where it is a member of the northern coastal scrub community.

A. pumila has done very well at Claremont, both in heavy soil where it is watered about every three weeks during the summer, and in the plant community area on alluvial soil where it is watered about every six weeks. In the plant community area the plants tend to be lower growing than they are on the mesa, perhaps because they receive less water. Recommended.

Regions: 1, 2, 3.

21. *A. pungens* var. *pungens.* Mexican manzanita. An erect or spreading shrub to 10 ft. without a basal burl, bark reddish-brown, branchlets canescent or white-tomentulose; leaves elliptic, oblong, roundish, obovate or almost lanceolate, to 1¾ in. long and ¾ in. wide, bright green, glabrous or minutely puberulent; flowers borne in small compact, simple or branched racemes, corolla white; fruit brownish-red, globular or slightly irregular, nutlets separable or irregularly fused. February-March.

Mexican manzanita is widely distributed from Mexico City north to Texas, Nevada and southern California. In California it is found on dry slopes at elevations of 3000-7000 ft. in the chaparral and yellow pine forest communities.

A. pungens is extremely variable and in addition has hybridized with other species, all of which has led to a bewildering array of taxonomic interpretations of this and related species.

Regions: 2, 3, 6, 7.

22. *A. refugioensis.*★ Refugio manzanita. An erect shrub to 12 ft. and nearly as broad, without a basal burl, branches thick, stiff, bark dark red, smooth, branchlets with short and long gland-tipped setose hairs; leaves subsessile to mostly sessile, cordate, clasping, occasionally auriculate, to 1¾ in. long and 1¼ in. wide, glabrous, inflorescences branched, flowers white to pink-tinged; fruit large, globose, to ⅝ in. in diameter, nutlets coalesced. December-February.

Refugio manzanita is known only from a small area in the Refugio Pass area of the Santa Ynez Mts. in Santa Barbara Co. where it grows on hard and unaltered sandstone at an elevation of about 2250 ft.

A. refugioensis is a very striking and distinctive shrub characterized by its thick and stiff branches and leaves that are very closely spaced and often clasping the stem. It is a fairly recent addition to the manzanita collection at Claremont and so far we have found it to be fast-growing and very satisfactory. It probably does best in well drained soils where it has received 2-3 irrigations during the summer, but there is no reason to believe that it will not grow in heavy soil so long as care is taken in its watering. According to Roof (1967), there is nothing in the genus *Arctostaphylos* to match the color of the new spring growth with its flame-colored shoots which may be as much as 10 inches long. At Claremont the plants have not shown this characteristic to the extent reported by Roof who observed plants growing in central California under different environmental conditions than those prevailing at this botanic garden. Recommended.

Regions: 1, 2, 3, 6.

23. *A. silvicola.*★ Silverleaf manzanita. An erect, silver-gray shrub to 8 ft. without a basal burl, bark smooth, dark red, branchlets densely gray-canescent;

leaves elliptic to oblong-elliptic, to 1½ in. long, round-cuneate at the base, gray-canescent or glabrate on both surfaces; flowers white, borne in few-branched panicles; fruit globose to slightly depressed, to ⅜ in. in diameter, light brown, nutlets separable. December-March.

Silverleaf manzanita is found on marine sand deposits in the Mt. Herman region of the Santa Cruz Mts. below 1800 ft. where it grows in the yellow pine forest and chaparral communities.

A. silvicola, which has been quite long-lived at Claremont, requires good drainage and does not tolerate excess moisture. In the plant community area growing on alluvial deposits the plants are irrigated twice during the summer.

Regions: 1, 2, 3, 6.

24. *A. stanfordiana* ssp. *stanfordiana.* Stanford manzanita. An erect, much-branched shrub to 7 ft., without a basal burl and with relatively straight and slender, glabrous stems and branches and smooth reddish-brown bark, branchlets glabrous or finely pubescent; leaves elliptic or ovate to oblanceolate, acute at both ends, thick, to 1¾ in. long and 1 in. wide, bright green, glabrous and shining on both surfaces; flowers borne in profuse, loose drooping panicles, corolla pinkish; fruit depressed-globose or somewhat irregular, about ¼ in. wide, red or reddish-brown, glabrous, nutlets irregularly united, or some distinct. March-April.

Stanford manzanita is common on open ridges and slopes in Napa, Lake, Sonoma and Mendocino cos. where it is a member of the chaparral community.

Ssp. *bakeri.*★ Baker manzanita. Plate 3e, f. Differs from the preceding in having branchlets dark with close viscid puberulence and conspicuous longer hairs. It is found on dry serpentine ridges in Sonoma Co.

A. stanfordiana ssp. *bakeri* is one of the most successful of all manzanitas grown at Claremont and it is also one of the most beautiful. The plants require good drainage and if grown in heavy soil great care must be taken to see that they do not receive excessive water during the summer months. If watering is carefully controlled plants in heavy soil may last longer than those growing in alluvial deposits which are irrigated every three to four weeks. This species is unusual in showing drought stress by its flaccid leaves. In our experience ssp. *bakeri,* in addition to being more beautiful than ssp. *stanfordiana,* is also less susceptible to branch dieback. Recommended.

Regions: 1, 2, 3. Also does well in the Central Valley if protected from the afternoon sun.

25. *A. subcordata.* Santa Cruz Island manzanita. Erect shrubs to 6 ft. with an equal spread, with a basal burl, bark smooth, red-brown, branchlets cinereous-tomentose and with long spreading, gland-tipped hairs; leaves ovate-elliptic to ovate-lanceolate, dark green, to 1½ in. long and ¾ in. wide, mostly glabrous on both surfaces; inflorescence a dense terminal panicle, flowers white; fruit depressed-globose, ⅜ in. in diameter, somewhat pubescent, nutlets separable. February-March.

Santa Cruz Island manzanita occurs on stony ridges and is restricted to that island where it is a component of the chaparral community.

A. subcordata has done well at Claremont where it has been grown in well-drained soil in both full sun and partial shade. The smooth branches are particularly attractive. This species appears to be quite resistant to branch dieback. Recommended.

Regions: 1, 2, 3, 6.

26. *A. tomentosa* var. *tomentosa.* Woollyleaf manzanita. An erect, loosely-branched shrub to 12 ft. with an equal spread, bark shreddy and usually persistent, branchlets densely white-tomentose; leaves oblong-ovate to round-ovate, or broadly elliptic, to 1¾ in. long and 1 in. wide, glabrous and shining above, densely white-tomentose beneath; inflorescence a short, spreading, sessile panicle, flowers white; fruit depressed-globose, to ⅜ in. in diameter, glabrous or sparsely pubescent, nutlets variously separable. February-March.

Woollyleaf manzanita is found in sandy places from the Monterey Peninsula south to San Luis Obispo Co. at elevations below 500 ft. where it is a member of the closed-cone pine forest and chaparral communities.

Var. *tomentosiformis.* Differs from the preceding in growing to about 6 ft. and in having branchlets with white, spreading bristly hairs. Found on dry hills and ridges from San Mateo Co. south to San

Luis Obispo Co. where it is found in the closed-cone pine forest.

Var. *tomentosa* prefers well-drained soil and although drought tolerant it responds well to occasional deep watering during the summer. It has also done well in heavy soil on the mesa. Var. *tomentosiformis* also thrives in well-drained soil and at Claremont it has been watered 4-5 times during the summer; the plants would probably do better if watered about every three weeks. Although the plants withstand full sun they are better when shaded in the afternoon.

A. tomentosa is one of the few species of manzanita that has come through wet winters with no evidence of branch dieback.

Regions: 1, 2, 3 (both varieties).

27. *A. viridissima.* Lompoc manzanita. Plate 3j. An erect shrub to 9 ft. with a spread of 12 ft., without a basal burl, branchlets densely hispidulous and setose with long white, non-glandular hairs; leaves broadly ovate, rounded or acute and mucronate at the apex, to 1 in. long and 3/4 in. wide, auriculate or heart-shaped at the base, bright to dark green; flowers white, borne in dense sessile panicles; fruit globose, about 1/4 in. in diameter, reddish-brown, glabrous, nutlets separable. December-February.

Lompoc manzanita is confined to Santa Cruz Island where it is found in the closed-cone pine forest.

At this garden, *A. viridissima* has done very well in the plant community area where drainage is good. The plants, which are long-lived, bloom early and in some years the flowers have been damaged by frost. It is best when shaded from the hot afternoon sun. It has not been tried in heavy soil.

Regions: 1, 2, 3.

The following clones of *Arctostaphylos* have been introduced into the horticultural trade. Some represent selections made from species but the majority are hybrids, or suspected hybrids. Confusion often exists as to the true identity of the plants and those sold by one nursery under one name may be different from those sold under the same name by another nursery. In an effort to simplify the identification of the clones we have included leaf silhouettes (Fig. 38) which may be of some aid in correctly identifying similar introductions.

1. *A.* 'Emerald Carpet' A low-growing shrub 10-16 in. high but spreading widely, moderately fast-growing; leaves shiny emerald-green; flowers pale pink, not produced in great abundance. The most attractive feature of this clone is its foliage and its ability to retain a fresh green appearance throughout the summer. Deep irrigation every 2-3 weeks.

2. *A.* 'Festival' Appropriately named, this shrub, which may grow to 2½ ft. with an equal spread, has foliage which changes color throughout the year. During autumn the leaves may be deep green, gold, russet, or bronze. In the spring the new growth is bright scarlet. The red inflorescences, which form during the autumn and terminate the red branchlets, hang like small Christmas tree ornaments until the pink-tinged flowers open in March. It is, however, the riot of colorful foliage the plant displays between September and February that distinguish the clone. Not one of the easiest or most dependable manzanitas; it is worthwhile trying.

3. *A.* 'Greenbay' Low widely-spreading shrub to 2 ft. with very deep green, non-shining leaves and clusters of nodding pink flowers. This clone will tolerate heavy shearing in late winter. Fine.

4. *A.* 'Greensphere' A most unusual manzanita. During the first few years the plants form dense, almost perfect spheres without any pruning, the result of having a single short trunk and numerous short, thick branches well-clothed with dark green leaves. Slow-growing and during the first few years most of the flowers are borne within the sphere itself. As the plants mature they become 4-6 ft. tall with even a greater spread and more of the flowers are produced toward the tips of the branches. This clone is noted for its great uniformity of habit and it can be used successfully in formal plantings. It first appeared as a chance seedling in a large population of plants grown from open-pollinated *A. edmundsii* seed. Hardy and dependable.

5. *A.* 'Havensneck' Low-growing with long trailing red stems and bright green leaves. This clone is fast-growing, quite heat tolerant and with occasional irrigation it retains its bright green color throughout the summer.

6. *A.* 'Harmony' A selection from *A. densiflora*

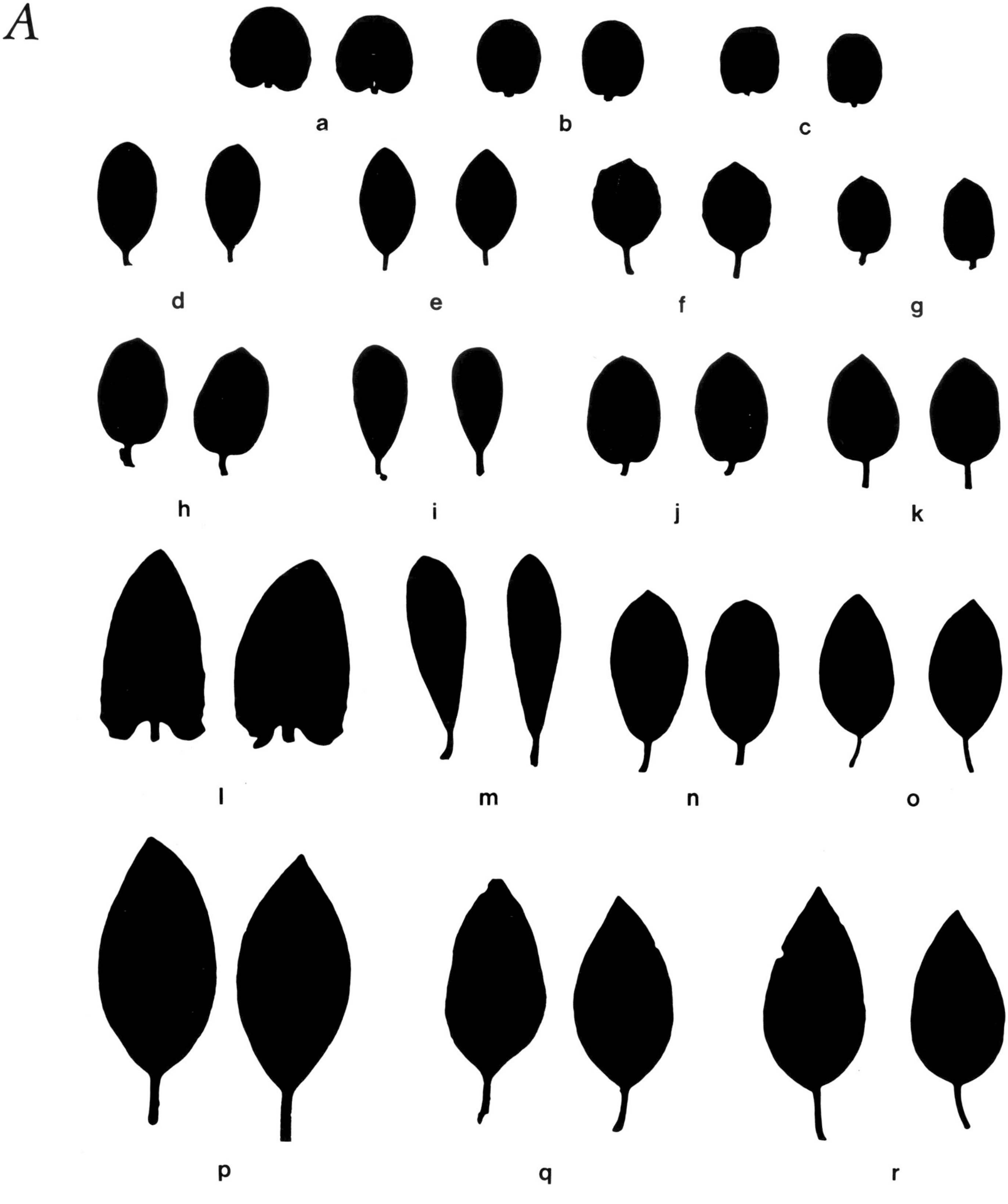

a b c

d e f g

h i j k

l m n o

p q r

growing to about 8 ft. with an equal spread; leaves bright green; flowers white, tinged pink. This and 'Sentinel' are very similar, some claim that 'Harmony' is the better.

7. *A.* 'Howard McMinn' Another selected form of *A. densiflora* eventually growing to 10 ft. and spreading widely; leaves smaller than those of 'Harmony' or 'Sentinel'; flowers white, tinged pink. Long-lived and dependable.

8. *A.* 'Indian Hill' A low-growing shrub to 1 ft. tall but spreading several feet; leaves gray-green; flowers white, tinged pink. This clone is moderately fast-growing and has a tight compact habit of growth. Its fall and winter color make it a worthwhile groundcover.

9. *A.* 'James West' Low, rather open growth to 1¼ ft. tall with small leaves and clusters of small pink flowers.

10. *A.* 'Lester Rowntree' A tall and spreading shrub to 10 ft.; leaves glaucous, gray-green, young stems red; flowers in nodding clusters, white flushed with pink; fruits shiny red to carmine. This clone is most attractive both in flower and fruit. Early.

11. *A.* 'Monterey Carpet' Compact growth to 1 ft. and spreading by rooting branches to as much as 12 ft.

12. *A.* 'Ophio-viridis' Plate 3h. An unusual groundcover to 1½ ft. tall with long spreading stems. The bright green, closely-spaced leaves are arranged in an overlapping spiral, giving the plant a scale-like or reptilian appearance which suggested the original name 'Green snake,' later changed because of some people's aversion to the association of this plant with a reptile. Beautiful in light shade on banks.

13. *A.* 'Point Reyes' Plate 3i. A low-growing shrub or groundcover to 1½ ft. tall with trailing stems from which arise short stout branchlets clothed with dark green leaves. The plants are quite compact and uniform in growth habit. The flowers, produced in March, are pinkish in color contrasting nicely with the deep green leaves. An excellent and dependable groundcover that has proven to be both heat and smog resistant. At this garden, plants growing in heavy soil are irrigated every 2-3 weeks during the summer. Highly recommended.

14. *A.* 'Sandsprite' A low-growing shrub to 1¼ ft. tall but spreading several feet, leaves light green. The rose-colored flowers, produced in March, are followed by mahogany-red fruits. In the early autumn the young emerging leaves are a brilliant red and from a distance may be mistakened for flowers. During the winter the foliage takes on attractive shade of bronze. An excellent subject for slopes where it displays to great advantage.

15. *A.* 'Seaspray' A low-growing shrub to 2 ft. forming a dense carpet. Leaves dark green and leathery. Flowers white, flushed pink. An excellent ground cover and good where it can cascade over rocks and ledges.

16. *A.* 'Sentinel' Another selection from *A. densiflora* and very similar to 'Harmony.' Upright growth to 6 ft. with at least as great a spread; leaves light green. Vigorous grower. Said to be capable of being trained to tree-form. Also reported to be sensitive to salt burn and root rots.

17. *A.* 'Trinity Ruby' Plate 3g. A selected form of *A. stanfordiana* and one of the most beautiful of all manzanitas with open, spreading panicles of purplish flowers on graceful, open plants. This variety tends to be short-lived but because of its exceptional beauty it repays frequent replanting.

18. *A.* 'Winterglow' A moderately fast-growing, low shrub to 2 ft. with pale green leaves which during March are completely hidden by masses of soft pink flowers, followed in June by bright red fruits. In early autumn the new growth displays shades of coppery-red and during the winter the entire plant appears coppery-red. Highly recommended.

With the exception of 'Emerald Carpet', the clones of manzanita listed here, all accept ordinary garden soils. At this garden they have done better and have been longer lived in the heavy loam on the mesa than when

←

Fig. 38. Leaf silhouettes of named clones of manzanita: a, 'Sandsprite'; b, 'Winterglow'; c, 'Festival'; d, 'Emerald Carpet'; e, 'James West'; f, 'Monterey Carpet'; g, 'Indian Hill'; h, 'Ophio-viridis; i, 'Pt. Reyes'; j, 'Sea Spray'; k, 'Greensphere'; l, 'Lester Rowntree'; m, 'Havensneck'; n, 'Greenbay'; o, 'Howard-McMinn'; p, 'Trinity Ruby'; q, 'Harmony'; r, 'Sentinel'. All natural size.

grown in the alluvial deposits found in the plant community area. In inland areas such as at Claremont the plants respond well during the summer to deep watering at about three-week intervals. They also respond to foliar feeding during late fall or early winter to late March. This is the period when the plants are in active growth at Claremont.

'Emerald Carpet' is best grown in neutral or slightly acid soil. Any of the clones listed here, if grown in the north, will require less frequent watering.

ARISTOLOCHIA. A genus of the Aristolochiaceae with about 180 species of mostly woody, twining vines widely distributed but very common in tropical areas.

1. *A. californica.* Dutchman's pipe. A deciduous, semi-woody vine from a woody base, to 12 ft. or more, stems slender, covered with short, silky hairs; leaves simple, alternate, ovate-cordate, to 5 in. long and 3 in. wide, bright green, pubescent on both surfaces, margins entire; flowers appearing before the leaves, axillary, solitary, pendulous, pipe-shaped, brownish to greenish-purple. January-March.

Dutchman's pipe is common in the coastal ranges from Monterey Co. north and in the foothills of the Sierra Nevada from Sacramento and Eldorado cos. north to the head of the Sacramento Valley, usually at elevations below 1500 ft., where it is a member of the foothill woodland, mixed-evergreen forest and chaparral communities.

A. californica is an interesting and attractive vine which always attracts attention with its masses of pipe-shaped flowers produced in late winter and early spring before the leaves appear. Easy to grow and not particular as to soil, it should have adequate water and protection from the afternoon sun. May be grown on fences, trellises or allowed to clamber over shrubs or into trees. Propagated by seeds or from rooted shoots from the base of the plant.

Regions: 1, 2, 3, 6. Also the Central Valley if protected from the hot afternoon sun.

ARTEMISIA. A genus of the Asteraceae with over 100 species of herbs and shrubs often with bitter and aromatic herbage, native to the Northern Hemisphere and South America, mainly in the arid regions of the Northern Hemisphere; flower-heads small, rayless, or in some with irregular ray-flowers, producing pollen in abundance which may cause hay fever. Twelve species in California.

1. *A. tridentata* ssp. *tridentata.* Great Basin sagebrush. Grayish, evergreen shrubs often with a short, thick trunk or few-branched from the base, to 15 ft., very aromatic; leaves cuneate, to 1½ in. long, usually less, and ½ in. wide, usually 3-toothed at the tip but sometimes 4-9-toothed, or entire; inflorescences mostly with erect branches, flower-heads small, disk-flowers 4-6. August-October.

Great Basin sagebrush is very widely distributed and common in the Great Basin but extending south along the western edge of the Mojave and Colorado deserts to northern Baja California, to the north it extends to British Columbia and the northern Rocky Mts. at elevations of 1500-10,600 ft. In California it is found in the sagebrush scrub, pinyon-juniper woodland, chaparral and yellow pine communities. The plants are found on deep, well-drained soils.

Ssp. *parishii.* Differs from the preceding in having leaves that are mostly linear and entire or only shallowly lobed and inflorescences with drooping branches. This subspecies is found in dry valleys from the western Mojave Desert north to the Santa Clara River Valley. Originally very abundant in the Antelope Valley but now less so due to urbanization and farming.

The sagebrushes easily grown in any well-drained soil are drought tolerant, long-lived and insect and disease free. Recommended for natural landscaping and for soil stabilization and erosion control.

Regions: 1, 2, 3, 4, 6, and selected areas in 7.

ATRIPLEX. A genus of the Chenopodiaceae with over 100 species native to many parts of the world. Especially common in alkaline places. Thirty-two species in California. Herbs or shrubs, usually grayish or whitish scurfy with inflated hairs; leaves alternate or opposite; plants monoecious or dioecious, flowers small, inconspicuous, greenish; staminate flowers with 3-5-parted calyx, stamens 3-5; pistillate flowers consisting of a naked pistil enclosed between a pair of appressed foliaceous bracts which enlarge in fruit and may be partly united; fruit a small 1-seeded inflated carpel (utricle).

The wind-blown pollen which is produced in abundance may cause hay fever in some individuals. The species listed here are fire resistant.

1. *A. canescens.* Wingscale, fourwing saltbush. Plate 21. An erect roundish gray shrub to 6 ft. with stout gray-scurfy or glabrate branches; leaves alternate, evergreen, numerous, linear, broadly elliptical or spatulate, usually obtuse at the apex, narrowed at the base, gray on both surfaces, margins entire; flower clusters borne on separate plants, staminate clusters in dense spikes forming long terminal panicles, pistillate flowers in dense leafy spike-like panicles; fruiting bracts sessile, forming a thick hard body united over the seed, projecting above into 2 free, flat wings, a second pair of wings developing from the medial line of each exposed face, the four wings entire or fringed, the whole bract ¼-¾ in. long. July-August.

Wingscale is rather abundant on dry slopes, flats and in washes in the Mojave and Colorado deserts, extending west to Los Angeles, Orange and San Diego cos., reaching the coast at Laguna Beach and San Diego. It extends south into Baja California and east to western Texas and Kansas and north to eastern Washington, Montana and Alberta, Canada. It is the most widespread of the shrubby species of *Atriplex* and consists of a number of ecotypes. Several poorly defined varieties have been described.

A. canescens tolerates a wide variety of soils including those that are alkaline or saline as well as a wide variety of climates. Since the species consists of ecotypes it is important to select a form that is adapted to the site where the plants will be grown. In most areas, after the plants are established they require no more water than they receive naturally.

Wingscale has many uses; it is valuable as a browse plant in desert and semi-desert areas, for soil stabilization, erosion control, for mass plantings, and as hedges, either clipped or unclipped.

Regions: 1, 2, 3, 4, 5, 6. Also does well in southern Arizona and in the Central Valley.

2. *A. confertifolia.* Shadscale, spiny saltbush. An erect, very rigidly branched shrub to 3 ft., compact and round in form, branches straw-colored; leaves alternate, crowded, early deciduous, branchlets then become modified into spines; leaf blades round-ovate, elliptic, rounded or cuneate at the base, grayish scurfy on both sides, margins entire; flowers borne on separate plants; staminate in globose axillary clusters on short lateral almost leafless branchlets; pistillate flowers solitary or several in the upper leaf-axils; fruiting bracts sessile, broadly elliptic or rounded, entire, surfaces smooth, convex and joined over the seed. May-June.

Shadscale is common on alkaline flats and slopes from the Mojave Desert north to Oregon, east to North Dakota and south to northern Mexico. It is usually found below 7000 ft. in the shadscale scrub and creosote bush scrub communities.

Besides being an important browse plant in desert or semi-desert areas, the plants may be used for low barrier plantings, for soil stabilization or erosion control. Cultural requirements are the same as for the preceding species. Said to tolerate high levels of boron.

Regions: 1, 2, 3, 4, 5, 6.

3. *A. hymenelytra.* Desert holly. An erect, compactly branched evergreen shrub to 3 ft., rounded in outline, branches silvery-white; leaves alternate, very numerous, roundish or round-ovate, ½-1¼ in. long and about as wide, silvery-white with a dense smooth scurf on both surfaces, margins undulate and irregularly toothed, somewhat holly-like; male and female flowers borne on separate plants; fruiting bract roundish, strongly compressed, margins smooth. February-April.

Desert holly occurs on dry alkaline washes in desert regions in California, north to Utah, west to Arizona and south into Mexico, usually below 2000 ft. where it is a member of the creosote bush scrub community.

A. hymenelytra has long been favored for Christmas decorations, and in the past vast quantities have been harvested for nationwide distribution.

The plants require a well-drained soil and watering should occur during the blooming season. Desert holly is a most attractive little shrub and highly recommended if its rather strict cultural requirements can be met. At this garden it has not been particularly successful with heavy losses occurring in the seedling and young plant stages.

Regions: 4, 5.

B

4. *A. lentiformis* ssp. *lentiformis.* Quail brush. A wide-spreading, much-branched shrub to 10 ft., and usually much broader, branches sometimes spinescent; leaves deciduous in desert forms and nearly evergreen in coastal forms, alternate, triangular or ovate to oblong to triangular-hastate, tipped with a short, soft point, ½-2 in. long, ¼-1½ in. wide, bluish-gray with a fine scurf, entire; flower clusters dioecious, crowded along the branchlets forming dense clusters 4-8 in. long; fruiting bracts sessile, roundish, ⅛-¼ in. across, surfaces smooth, united only near the base or to the middle, free portions entire or crenulate. June-August.

Quail brush occurs in alkaline depressions in the Mojave and Colorado deserts, north to Utah and south into Sonora, Mexico, usually at elevations below 2000 ft. It is a member of the alkali sink plant community.

Ssp. *breweri.* Plants dioecious or monoecious, differing only in minor ways from preceding.

In saline places from Orange Co. and western Riverside Co. north to central California. Also on San Clemente Island. It is found in the coastal salt marsh, coastal sage scrub and valley grassland communities.

Quail brush is useful for windbreaks or hedges in alkaline areas and the plants may be trimmed to improve appearance. Since this species contains numerous ecotypes it is important to select a form that is adapted to the site where the plants are to be grown. The coastal forms tend to be nearly evergreen, whereas the desert forms are deciduous.

This species is very important in providing wildlife cover, and quail are particularly attracted to it.

Although drought tolerant the plant's appearance is improved if it receives occasional deep watering. A handsome and useful species. Easily propagated by seed or cuttings.

Regions: 1, 2, 3, 4, 5, 6.

5. *A. polycarpa.* Desert saltbush. Erect, intricately-branched shrubs to 6 ft., leaves crowded on young twigs, early deciduous, oblong to spatulate, to 1¾ in. long, entire, gray-scurfy; male flowers in dense or interrupted, simple or paniculate naked spikes, female flowers crowded in small, sparsely leafy clusters arranged in paniculate spikes; fruiting bracts sessile, somewhat compressed, cuneate orbicular, shallowly to deeply lobed, plane or tuberculate. July to October.

Desert saltbush is found in alkaline soils below 5000 ft. from the Owens Valley south to Baja California and southeast to Sonora, Mexico.

A. polycarpa has been an exceptionally hardy and trouble-free plant at this botanic garden, drought tolerant and insect and disease free. The plants grow together to form large masses and are highly recommended for control of soil erosion in the desert areas, for planting at rest areas and for natural landscaping.

Regions: 3, 4, 5, 6.

BACCHARIS. A genus of the Asteraceae with about 300 species of herbaceous perennials or more often shrubs, mostly from South America; nine species in California. Flower heads rayless and either staminate or pistillate, the two borne on separate plants.

1. *B. pilularis* ssp. *pilularis.* Coyote brush. Plate 4d, e. Prostrate or decumbent evergreen shrubs to 3 ft., old stems becoming very woody, plants leafy throughout; leaves variable, numerous, cuneate-ovate, oval to obovate, with 4-6 coarse teeth, usually toward the end of the leaf, sometimes entire, to ¾ in. long, thick, resinous, dark green or gray-green; flower heads numerous, in small axillary or terminal clusters, pistillate flowers whitish, staminate flowers yellowish. August-December.

Prostrate coyote brush occurs on windswept dunes, headlands and hillsides, always near the ocean from Monterey Co. north to Sonoma Co. where it is found in the coastal strand, northern coastal sage scrub communities.

B. pilularis ssp. *pilularis* in its ability to adapt to a wide variety of soil types, amounts of moisture and temperature is one of the most remarkable of California shrubs, and from the standpoint of horticulture, one of the most valuable. Indeed there is no other native ground

→

Plate 4. a, Manzanita leaf gall; b, c, Manzanita branch dieback; d, *Baccharis pilularis* ssp. *pilularis* used as a formal clipped hedge; e, *Baccharis pilularis* ssp. *pilularis* used as a bankcover along a roadway; f, *Calliandra eriophylla*; g, *Calycanthus occidentalis*; h, *Carpenteria californica*; i, *Castela emoryi*; j, k, *Cassia armata*; l, *Ceanothus maritimus*.

Plate 4

cover that is more dependable or more widely used. It has been grown by this garden for the past 50 years.

For use on banks and hillsides the plants may be allowed to grow naturally, becoming in time perhaps 3 ft. tall and several times as broad, forming billowy masses of green across the landscape. For more formal use the plants may be kept low by regular mowing, thus forming a smooth green carpet about one foot high or the plants may be clipped to produce low formal hedges.

The plants in coastal areas, once established, can do without summer watering but inland they should be given deep irrigation every two or three weeks. Trimming should be done in late winter or early spring before new growth commences. In order to keep the plants low, old arching branches from the center of the plant should be cut out.

For large scale plantings propagation may be by seed, but for more formal uses or where the fluffy achenes produced by the female plants may be objectionable, propagation should be by cuttings taken from staminate plants. Due to the natural variation found in this species large plantings grown from seed will have more character and appear more 'natural' than plantings consisting entirely of a single clone.

B. pilularis ssp. *pilularis* is considered to be fire resistant and is approved for use in fire-prone areas.

There are several named clones of coyote brush on the market, two of them introductions of this botanic garden. *B.* 'Twin Peaks #1' has bright green leaves closely spaced along the thin branches, and normally the plants grow 1½-2 ft. tall. This clone responds well to mowing. *B.* 'Twin Peaks #2' is more compact and prostrate than the preceding, has darker and larger leaves which are often somewhat contorted. This clone does not tolerate mowing but responds well to shearing and shaping. In age it is inclined to mound or become cushion-like. *B.* 'Pigeon Point' has small light green leaves and is more prostrate than the Twin Peaks selections. It is recommended for coastal areas and it does better in the north than it does in southern California.

Prostrate coyote brush is susceptible to attacks of spider mites, lace bugs and white fly, all of which may cause considerable damage to the plants if not controlled.

Regions: 1, 2, 3, 4, 6. It also does well in the Central Valley if protected from the afternoon sun.

Ssp. *consanguinea.* Chaparral broom. An erect, compactly branched, erect or rounded evergreen shrub to 12 ft. with angular branchlets; leaves very numerous, oval or obovate, to 1½ in. long and ¾ in. wide, margins usually with 5-9 teeth, or sometimes entire, resinous, leathery, dark green; flower heads solitary or commonly several in the axils of the leaves, or in terminal clusters on the leafy branches. August-October.

Chaparral broom is found on open low hills along the coast from San Diego Co. north to Oregon, also on Santa Catalina, San Clemente, Santa Barbara, Santa Cruz and Santa Rosa islands. It belongs to the coastal strand and coastal sage scrub communities.

Easily grown and trouble free, chaparral broom may be used as a screen or windbreak in coastal areas, as well as inland at least as far as Claremont and Fallbrook. Because of its fibrous root system it is useful for erosion control. It tolerates severe pruning and may be trimmed as a hedge. Although very drought tolerant the plants look somewhat better if they receive a little summer watering.

Regions: 1, 2, 3.

2. *B. sarothroides.* Broom baccharis. A much-branched nearly leafless shrub to 6 ft. with numerous slender bright green twigs; leaves few, linear to ¾ in. long, margins entire; heads mostly solitary on naked wiry peduncles, these numerous, forming a dense broom-like panicle. June-October.

Broom baccharis is found in sandy, sometimes alkaline areas from Riverside Co. south to northern Baja California, east to New Mexico and south to Sinaloa, usually at elevations below 1200 ft. where it is a member of the coastal sage scrub and creosote bush scrub.

B. sarothroides is not particular as to soil or drainage, tolerates alkaline conditions and is free from insect pests and disease. Good for visual or traffic control, erosion control, as specimen plants and for street median plantings. The plant's appearance can be improved by occasional deep watering and pruning. Female plants are spectacular during fall and winter, but seeds may be a problem. Cutting-grown male plants may be preferred.

Regions: 3, 4, 5. Also does well in Arizona.

3. *B. sergiloides.* Squaw waterweed. An erect, rounded, glabrous, often leafless shrub to 6 ft. with slender broom-like branches; leaves obovate to spatulate, to 1¼ in. long, entire or rarely with a few marginal teeth; flower heads small, solitary on short peduncles, these numerous in dense, nearly leafless broom-like panicles. April-September.

Squaw waterweed is found in washes and canyon bottoms in both the Mojave and Colorado deserts, north to Inyo Co., east to Nevada, Utah, Arizona and south into Sonora, usually at elevations below 4500 ft. It is found in many plant communities: creosote bush scrub, pinyon-juniper woodland, chaparral, southern oak woodland, etc.

Insect pest and disease free, the plants are very drought tolerant but look better if they receive some additional summer water. Useful for erosion control, soil stabilization and natural landscaping.

Regions: 3, 4, 5, 6.

4. *B. glutinosa.* Mule fat. An erect, spreading, very leafy, loosely-branched shrub to 12 ft.; leaves resinous, narrowly lanceolate to oblong, willow-like, to 3½ in. long, pale green, margins entire to slightly denticulate; flower heads numerous, borne at the ends of short, lateral branches in close corymbose clusters. May-July, but some flowers throughout the year.

Mule fat is common throughout much of California where it occurs along stream beds, ditch banks, etc., at low elevations in many plant communities, including the valley grassland, coastal sage scrub and southern oak woodland. Also found in Utah and Arizona.

B. glutinosa is a vigorous, willow-like shrub useful for controlling erosion and for soil stabilization along waterways, natural landscaping, etc.

Perhaps the most attractive of the shrubby species of *Baccharis,* mule fat is disease and insect pest free and drought tolerant, although the plants will look better if they do receive some summer watering.

It does not tolerate pruning as well as the other species but the plant's overall appearance can be improved by thinning and shaping.

Regions: 1, 2, 3, 4.

BETULA. A genus of the Betulaceae with about 40 species of deciduous trees and shrubs in the Northern Hemisphere. Flowers borne in catkins; staminate catkins elongated, formed during the autumn and remaining on the naked trees throughout the winter; pistillate catkins oblong or cylindrical, erect, fruiting scales of the cones falling away with the small nutlets. (In *Alnus,* the entire cone falls intact.)

1. *B. fontinalis.* Water or red birch. A tall shrub or small tree to 25 ft., usually with several stems from the base, bark dark coppery-red, shining, not peeling in thin layers, twigs rough with large resinous glands; leaves broadly ovate, to 2 in. long and 1¼ in. wide, margins sharply serrate; staminate cones 1-3 in a cluster near the ends of the branchlets, cylindrical, 2 in. long when flowering, pistillate catkins usually solitary, ¾ in. long; fruiting cone to 2 in. long. April-May.

Red birch is found in moist places from northeastern Kern Co. mostly on the west side of the Owens Valley north, less common on the west side of the Sierra Nevada to Siskiyou and Humboldt cos. and north to Canada and east to New Mexico. In California it is found at elevations of 2500-9000 ft., mostly in the montane coniferous forest communities.

B. fontinalis is a fast-growing shrub requiring moisture at all times. Attractive and useful for streamside plantings providing food and cover for wildlife. Bark is very attractive.

Regions: 1, 2, 3, 6, 7.

BRICKELLIA. A genus of about 100 species of herbs or shrubs belonging to the Asteraceae. Native to the warmer parts of the United States south to South America. Ray flowers absent. Fourteen species native to California. Those listed here not of ornamental value.

1. *B. californica.* California brittlebush. Straggly, rounded, aromatic shrubs to 3½ ft., much-branched from a woody base; leaves alternate, triangular-ovate, to 2 in. long, 3-nerved, dull green, pubescent on both surfaces, margins serrate or crenulate; flower heads 10-15-flowered, borne in small clusters, terminating lateral branchlets of a leafy panicle. June-November.

California brittlebush is common in washes and on dry slopes from cismontane southern California north to Humboldt and Siskiyou cos., in the foothills of the Sierra

Nevada, occasionally on the Mojave Desert and east to Colorado, Texas and south into Mexico. It is usually found below 8000 ft. and in many plant communities.

Drought tolerant and recommended for controlling soil erosion, for roadside cuts, rest areas, etc., wherever there is good drainage. At this garden it has often been injured by frost, but the plants regenerate quickly from the woody base.

Regions: 3, 4, 5, 6.

2. *B. incana.* White brittlebush. A loosely branched, globose bush from a woody base, to 4 ft., white tomentose throughout; leaves alternate, ovate, to 1¼ in. long, densely white-tomentose, margins wavy; flower heads about 1¼ in. long, solitary, pedunculate, at the ends of the branchlets. April-October.

White brittlebush is found in sandy washes and flats in the Mojave Desert and the northern edge of the Colorado Desert, eastward to Nevada, usually below 5000 ft. where it is a member of the creosote bush scrub, shadscale scrub and Joshua tree woodland communities.

Recommended for arid areas where it can be used for soil stabilization and erosion control, roadside banks, rest areas, etc.

Regions: 3, 4, 5. Also does well in the Central Valley.

CALLIANDRA. A large genus of the Fabaceae with perhaps 150 species of herbs, shrubs or trees native to subtropical and tropical areas. One species in California.

1. *C. eriophylla.* Fairy duster, false mesquite. Plate 4f. A low, densely branched evergreen shrub to 3 ft. and equally broad, branches gray, pubescent, unarmed; leaves bluish-green, equally bipinnate, alternate with 1-4 pairs of pinnae, leaflets 5-12 pairs, oblong, about ⅛ in. long; flowers regular, showy, borne in dense heads, terminal on axillary peduncles, petals 5, dark red, stamens numerous, conspicuous, pink to deep red, ½-1 in. long; fruit a flat, straight pod 2-2½ in. long, silver-pubescent with dark red thickened margins. February-June.

Fairy duster occurs in sandy washes and gullies in eastern San Diego and Imperial cos. at elevations below 1000 ft., east to Texas and south into Mexico. In California it is a component of the creosote bush scrub.

C. eriophylla is an attractive little shrub with delicate blue-green foliage and dense spherical clusters of flowers whose long red or pink stamens give the inflorescence the appearance of fluffy balls.

The plants prefer coarse well-drained soil and once established are very drought tolerant, but their appearance is improved if they receive deep but infrequent watering, otherwise the plants will drop their leaves. Useful as accent plants or as medium-sized ground cover. There is some variation in the flower color.

An exceptionally graceful and delicate-looking plant that belies its hardy nature.

Regions: 1, 2, 3, 4, 5, 6. Also does well in southern Arizona and northern Sonora.

CALOCEDRUS. A genus of the Cupressaceae with about 20 species of evergreen trees native to both hemispheres. One species in California.

1. *C. decurrens.* Incense cedar. Fig. 39. Symmetrical trees to 150 ft. with strongly conical trunks, bark thick, red-brown, loose and fibrous, branchlets alternate, numerous, forming flat sprays; leaves scale-like, opposite pairs alternating at right angles with those above and below, giving the branchlets a jointed appearance, light or yellowish-green; cones with only a few scales, red-brown, oblong-ovate, to 1 in. long, when opened looking like a duck's bill.

Incense cedar is found on mountain slopes from northern Baja California north to Oregon and west to Nevada at elevations between 4500-8200 ft. where it is a member of the yellow pine forest community.

C. decurrens is adaptable to a wide variety of climates tolerating high summer temperatures and poor soils. If deeply and infrequently watered when young the trees will become highly drought tolerant in age. Slow-growing when young. May be used as windbreaks, high screens or as single specimens. Said to be immune or highly resistant to oak root fungus.

Regions: 1, 2, 3, 6, 7.

→

Plate 5. a, *Ceanothus* 'Concha'; b, *Ceanothus incanus*; c, *Ceanothus ramulosus*; d, *Ceanothus* 'Santa Ana'; e, *Ceanothus* 'Snowball'; f, *Ceanothus* 'Owlswood Blue'; g, *Cephalanthus occidentalis* var. *californicus*; h, i, *Cercidium floridum* (photograph taken in the field); j, k, *Cercis occidentalis* 'Claremont'; l, *Cercocarpus betuloides*.

Plate 5

C

CALYCANTHUS. A genus of the Calycanthaceae with five species, one of which is native to California.

1. *C. occidentalis.* Spicebush. Plate 4g. An erect, bushy, deciduous shrub to 12 ft. and often as broad with smooth brown bark; leaves simple, opposite, aromatic when crushed, ovate to oblong-lanceolate, 2-6 in. long, 1-3 in. wide, glabrous but with a fine sandpaper-like upper surface, smooth beneath, entire; flowers have been described as looking like small water lilies, 1½-2½ in. across, solitary at the ends of the branches, sepals and petals similar, numerous, about 1 in. long, reddish-brown, tip third soon becoming brown; fruit a large, hard, urn-shaped receptacle containing white-velvety brownish achenes. May-July.

Spicebush is found in moist places in the north coast ranges from Napa to Trinity cos. and along the western base of the Sierra Nevada from Tulare to Trinity cos. at elevations below 4000 ft. where it is a member of the foothill woodland and yellow pine forest communities.

Spicebush has long been in cultivation and is easily grown given ordinary garden care. If the plants become dry they will defoliate prematurely. In inland areas they should be given protection from afternoon sun. The plants which are free from insect pests and disease can be trained as small multi-stemmed trees. Immune or highly resistant to oak root fungus.

C. occidentalis is very similar to the eastern Carolina allspice, *C. floridus,* the former having somewhat larger flowers and leaves and the latter producing a greater abundance of flowers.

Regions: 1, 2, 3, 6.

CARPENTERIA. A monotypic genus of the Saxifragaceae.

1. *C. californica.* Carpenteria, bush anemone. Plate 4h. An evergreen shrub to 8 ft. with numerous clustered stems, branchlets opposite, and somewhat 4-angled, bark peeling in thin shreds; leaves opposite, simple, oblong-lanceolate, 2-4½ in. long, tapering at both ends, somewhat revolute, dark green and glabrous above, whitened beneath with a close pubescence, margins entire or slightly toothed; flowers large, 1½-3 in. across, fragrant, borne in terminal clusters; sepals 5-7, ovate-lanceolate, persistent; petals 5-7, white, obovate to round obovate, ¾-1½ in. long; stamens numerous; fruit a leathery capsule somewhat conical, about ½ in. long, 5-7-valved. June-July.

Carpenteria, which is one of California's rarest endemics, is confined to a few restricted areas in Fresno Co. between the San Joaquin River and the Big Creek tributary of the Kings River. There it occurs along streams, shallow gullies and on wooded slopes at elevations of 1500-4000 ft. in the foothill woodland community.

The clusters of large white flowers with their bosses of yellow stamens makes carpenteria one of California's most beautiful flowering shrubs, but its floppy habit detracts from the overall picture. For that reason it is probably best when massed rather than being planted as a

Fig. 39. *Calocedrus decurrens*; a, staminate branch; b, pistillate branch; c, seed. (Sudworth, 1908). Reduced.

single specimen. If drainage is good, the plants accept ordinary garden care and once established are drought tolerant. Carpenteria is at its best in cool areas and inland the plants should be given protection from the afternoon sun. It is reported to be resistant to oak root fungus, but it is often attacked by aphids which disfigure the plants and are difficult to eradicate by spraying due to the revolute nature of the leaves. Systematic insecticides should be tried.

Propagation is by seed, cuttings or from suckers which are produced in abundance. Because of its suckering habit carpenteria may be used in suitable areas for erosion control. Although not as variable as some other groups there are differences in flower size. Roderick (1977) reports finding a plant with small flowers produced in large clusters. *C. californica* 'Ladhams' is reported to have flowers 3¼ in. in diameter.

Regions: 1, 2, 3, 6.

CASSIA. A genus of the Fabaceae with over 450 species of perennial herbs, shrubs or trees native to the warmer parts of the world. Two species in California.

1. *C. armata.* Armed senna. Plate 4j, k. Much-branched shrubs to 6 ft. with an equal spread; leaves pinnately compound with even numbers of leaflets, leaf-rachis enlarged at the apex to form a sharp prickle or spine, leaves to 2½ in. long, leaflets 2-8, ovate to almost round, ¼ in. or less in length, glabrate or pubescent; flowers numerous, borne in terminal racemes to 6 in. long, not pea-shaped, petals yellow to salmon-colored, to ½ in. long; fruit pods linear, to 1¼ in. long, curved, not compressed between the seeds. May-July.

Armed senna is common in sandy washes and open places in the Mojave and Colorado deserts below 2000 ft. where it is a member of the creosote bush scrub.

C. armata is a most attractive shrub blooming over a long period of time. It requires a well-drained situation and once established is completely drought tolerant. It has not been successful at this botanic garden because of the susceptibility of the seedlings to damping-off. Recommended only for desert areas where it can be used as specimens or in mass plantings.

Regions: 4, 5.

CASTELA (= *Holocantha*). A genus of the Simaroubaceae with about 15 species of more or less thorny or spiny shrubs or small trees native to the West Indies, Central and South America and the Galapagos Islands. One species in California and the only native representative of the family in the state.

1. *C. emoryi* (= *Holocantha emoryi*).★ Crucifixion thorn. Plate 4i. Very thorny shrub to 10 ft.; leaves reduced to small deciduous scales, hence the plants appear leafless, consisting only of rigid greenish or greenish-gray branches each ending in a stout thorn; flowers small, dioecious, solitary or few together in the axils of the branchlets, 4-8-merous; fruit a dry drupe about ¼ in. long, borne in large, dense, long-lasting clusters. June-August.

Crucifixion thorn is rare in California, occurring in dry gravelly places in the southern Mojave Desert where it is a member of the creosote bush scrub.

C. emoryi is a wickedly thorny shrub with a distinct appearance quite unlike that of any other native plant. Completely drought tolerant when established it is recommended for barrier plantings in the desert.

Regions: 3, 4, 5, 6.

CEANOTHUS. A genus of the Rhamnaceae with 50-60 species native to temperate North America with a concentration of species in California. Shrubs or small trees with divaricate, sometimes spiny twigs and alternate or opposite, mostly evergreen leaves; flowers small but showy, white to blue, lavender, purple or pinkish, borne in terminal or lateral clusters, petals 5, stamens 5; fruit a 3-lobed capsule separating at maturity into 3 parts. The species hybridize freely both in the wild and in cultivation and many named clones are available.

Nearly forty years ago the Santa Barbara Botanic Garden published a book entitled *Ceanothus,* written by Maunsell van Rensselaer, the director of the botanic garden, and Howard E. McMinn, professor of botany at Mills College, Oakland. Our treatment follows closely that of van Rensselaer and McMinn whose work has long been out of print and rarely available.

Botanically the genus is separated into two sections, *Euceanothus* (Fig. 40a-c), with alternate leaves, fruits without horns but often with ridges or crests on the backs

Plate 6

of the locules, stipules thin and early deciduous, and leaves with stomates on the lower epidermis, never sunken into pits; and section *Cerastes* (Fig. 40d-f), with opposite leaves, fruits normally with apical or subapical horns, or these nearly obsolete in a few species, stipules usually persistent, swollen and corky, and stomates sunken in pits on the underside of the leaves. *Ceanothus verrucosus* and *C. megacarpus* have alternate leaves but otherwise possess the characters of section *Cerastes.*

California lilacs are certainly among the most beautiful and best known of California shrubs and some of them have been in cultivation for a long period of time. As a group they tend under cultivation to be short-lived often with a life span of 5-10 years. This is especially true of some of the more attractive of the cultivars, and plans should be made to replace them at rather regular intervals. Members of section *Euceanothus* are for the most part fast-growing and can be expected to produce abundant bloom the second season in the ground. Members of section *Cerastes* are slower growing and may be more difficult to establish, but once established they are, as a group, drought tolerant and very satisfactory in southern California, and most of them have done well at Claremont with some of them being very long-lived.

Most species of *Ceanothus* are susceptible to root rots and this is probably the main cause of death. To minimize such losses plants should if possible be given a well-drained location, and under no circumstances should they receive heavy watering during the summer. For this reason California lilacs should never be planted near lawns or in mixed plantings which are watered on a regular basis. Some species react well to small amounts of summer irrigation if drainage is good.

At times the plants may be attacked by aphids and white fly which can cause considerable damage to the plants unless controlled by spraying. They may also be severely attacked by the ceanothus stem gall moth (*Periploca ceanothiella*) which causes a fasciation on the branches. This insect can be controlled by the use of systemic insecticides and by carefully controlling irrigation to prevent off-season flushes of growth.

Interspecific hybridization is common in nature, and especially so under cultivation, and plants grown from seed may well turn out to be hybrids. Asexual propagation is the only safe way to maintain the species when the plants are grown in gardens.

1. *C. arboreus* (*Euceanothus*). Catalina ceanothus. A tall shrub or small tree to 25 ft., bark smooth, gray, twigs pubescent; leaves broadly ovate to elliptical, to 3 in. long and 1½ in. wide, 3-nerved from the base, dark green, dull and glabrous, or pubescent above, densely white tomentose beneath, margins serrulate to serrate; flower clusters compound, to 6 in. long, flowers pale blue; fruit triangular, to ⅓ in. in width, roughened all over, almost black when mature. February-May.

Catalina ceanothus grows on brushy slopes on Santa Cruz, Santa Rosa and Santa Catalina islands where it is a member of the chaparral community.

C. arboreus is one of the better species of the genus and a attractive and useful shrub. Plants growing in heavy soil have a better appearance than those growing in sand and gravel but the latter may be longer lived. This species is quite susceptible to ceanothus stem gall moth damage.

Regions: 1, 2, 3, 6. Also Central Valley if protected from the afternoon sun.

2. *C. cordulatus* (*Euceanothus*). Snowbush. Intricately branched, spreading shrubs to 5 ft., of a grayish or whitish aspect, bark smooth, ultimate twigs stiff, spinescent; leaves alternate, ovate to elliptic, to 1 in. long and ½ in. wide, 3-nerved from the base, green or glaucous above, grayer beneath, margins mostly entire; flower clusters short and dense, to 1½ in. long, flowers white; fruit triangular, usually distinctly lobed, about ½ in. in diameter. May-June.

Snowbush grows on dry, open flats and slopes in the San Jacinto, San Bernardino and San Gabriel mts., north to the Panamint Mts. and Oregon and south into northern Baja California where it grows at elevations mostly between 5000-10,000 ft. in the yellow pine forest community and communities above the yellow pine forest.

←

Plate 6. a, *Chamaebatiaria millefolium*; b, *Chilopsis linearis* (deeply colored form); c, *Chrysothamnus nauseosus* ssp. *bernardinus* (photograph taken in the field); d, *Clematis lasiantha*; e, *Cneoridium dumosus*; f, *Cornus stolonifera*; g, *Comarostaphylis diversifolia* var. *diversifolia*; h, *Comarostaphylis diversifolia* var. *planifolia*; i, *Cowania mexicana* var. *stansburiana*; j, k, *Dalea spinosa* (photograph taken in the field); l, *Diplacus* hybrid.

This species has not been successful at Claremont but is recommended for areas at higher elevation.

Regions: 6, 7.

3. *C. crassifolius* var. *crassifolius* (*Cerastes*). Hoary-leaf ceanothus. A tall, rather stiffly-branched, open shrub to 12 ft. with gray, brown, white or rusty tomentose branches; leaves opposite, broadly elliptic or elliptic-ovate, to 1¼ in. long and ⅜ in. wide, 1-nerved from the base, thick, leathery, usually strongly revolute, upper surface olive-green, glabrous, minutely roughened, lower surface white tomentose, margins coarsely and pungently dentate; flower clusters umbel-like, short, to ¾ in. long, flowers white; fruit globose, to ⅜ in. in diameter, viscid, with short subdorsal horns and no intermediate crests. January-April.

Hoaryleaf ceanothus is common on dry slopes and fans from Santa Barbara Co. south to northern Baja California at elevations below 3500 ft., where it is a member of the chaparral and coastal sage scrub communities.

C. crassifolius var. *crassifolius* has been satisfactory at Claremont if provided with excellent drainage. It tends to be somewhat difficult to establish as young plants probably because of the care required to see that the seedlings receive the correct amount of moisture.

Regions: 1, 2, 3, 6.

4. *C. cuneatus* (*Cerastes*). Buckbrush. An erect shrub to 8 ft. with rigid, divaricate branches and stout, short, usually opposite unequal branchlets; leaves opposite, borne on short spur-like branchlets, spatulate to obovate-cuneate, or elliptical, to 1 in. long and ¾ in. wide, 1-nerved from the base, dull gray-green, finely tomentulose beneath; flowers borne in several-flowered umbellate clusters, white, rarely pale lavender or blue; fruit sub-globose, about ¼ in. in diameter, with distinct horns but without crests or ridges. March-May.

Buckbrush is one of the most widely distributed species in the genus and an important component of the chaparral; it is found from northern Baja California north to Oregon at elevations below 6000 ft. In addition to the chaparral, it is also found in the yellow pine forest and juniper-pinyon woodland communities. Hybridizes with other species in which it comes into contact.

Good at Claremont. This species may well consist of distinct ecotypes and care should be taken to secure seed from habitats as similar as possible to those into which the species is being introduced.

Regions: 1, 2, 3, 6, 7. Also does well in the Central Valley.

5. *C. cyaneus* ★ (*Euceanothus*). San Diego ceanothus. A large shrub or tree-like to 15 ft. with angled branchlets and usually straight gray-green branches dotted with brownish tubercles; leaves alternate, ovate-elliptic, to 2 in. long and ¾ in. wide, 3-veined from the base, dark green and glabrous above, lighter and nearly glabrous beneath, margins finely glandular-denticulate, serrate or nearly entire; flower clusters large, compound, to 12 in. long, terminating the branches, flowers very deep blue in bud becoming lighter after opening; fruit subglobose, about 1/6 in. in diameter, lobed at the summit. May-June.

San Diego ceanothus is rather rare, occurring in only a few areas in the foothills of San Diego Co. where it is a member of the chaparral community.

Very striking when in flower because of its deep coloring and large inflorescences, characters which have often been transmitted to its hybrids. The plants tend to be sparsely leafed and floppy as well as being very short-lived, 5 years being a good span of life for most plants. They benefit from afternoon shade, especially in inland areas. Perhaps best in coastal areas. Recommended with qualifications.

Regions: 1, 2, 3, if given afternoon shade.

6. *C. gloriosus* var. *gloriosus* (*Cerastes*). Pt. Reyes ceanothus. Prostrate or decumbent shrubs to 1 ft. but spreading to 6 ft. or more with long, usually brownish or reddish branches and opposite branchlets; leaves opposite, broadly elliptic, roundish or broadly oblong, to 1½ in. long and 1 in. wide, 1-nerved from the base, thick, leathery, dark green and glabrous above, paler and minutely pubescent beneath, margins usually dentate or spinosely-toothed with 15-30 teeth, rarely entire; flower clusters composed of many umbels on short axillary shoots, flowers deep blue to purple; fruit globose, about ⅛ in. in diameter, viscid, 3-horned usually without intermediate crests. March-May.

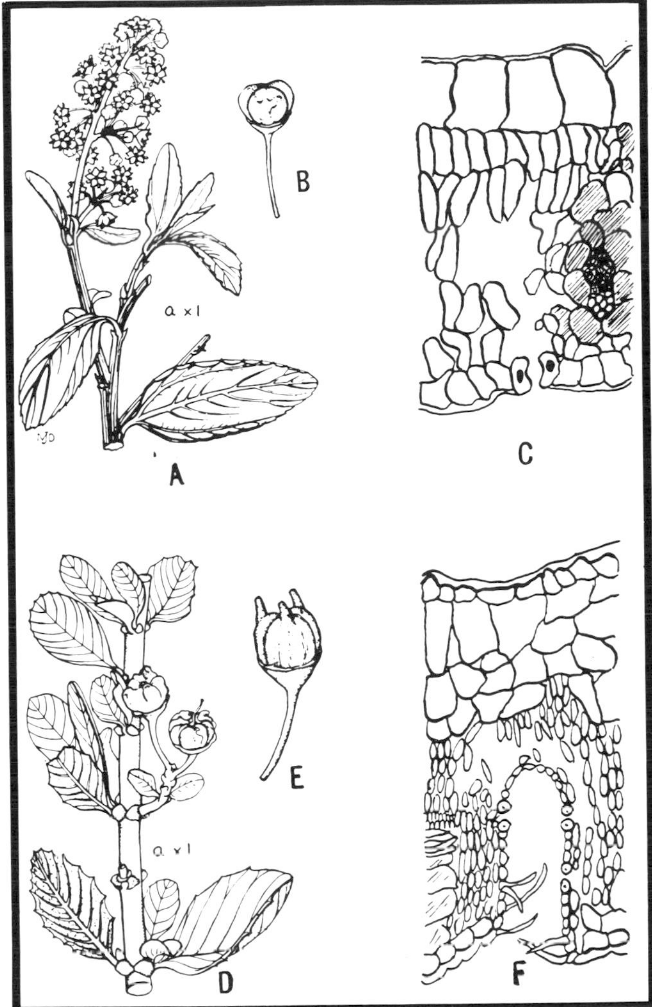

Fig. 40. *Ceanothus*, Section *Euceanothus*. A, branchlet with alternate leaves, thin, early falling stipules and umbellate clusters of flowers in a raceme; B, fruit without apical horns; C, cross section of leaf showing stomata protected by guard cells on lower surface. Section *Cerastes*. D, branchlet with opposite leaves, thick corky persistent stipules and umbel-like fruit cluster; E, fruit with apical horns; F, cross section of leaf showing stomata in sunken pits on the lower surface. (Van Rensselaer and McMinn, 1942). Reduced.

Point Reyes ceanothus occurs in sandy places in Marin and Mendocino cos. where it is a member of the coastal strand, coastal sage scrub and closed-cone pine forest communities.

Ceanothus gloriosus var. *gloriosus* does quite well in Claremont if shaded from the afternoon sun or grown under high canopy shade. It is an ideal shrub for gentle slopes.

Var. *exaltatus.* Navarro ceanothus. Differs from the preceding in being more upright with widely-spreading, fastigiate-divaricate branches. It occurs inland in the outer coast ranges from Marin to Mendocino cos. This shrub's performance in Claremont has been quite disappointing; it is short-lived, 5-7 years is about what may be expected. Apart from this it is a most attractive plant attaining a height of 10-12 ft. with a comparable spread. It is most intolerant of heavy soil on the mesa with losses occurring during winters of heavy rainfall. It should be given a location with good drainage, preferably on a gentle slope which is shaded from the afternoon sun or grown under high canopy shade.

Var. *porrectus.*★ Mt. Vision ceanothus. Sprawling shrubs to 1½ ft. with leaves to ¾ in. long. Found on Inverness Ridge and Ledum Swamp in Marin Co. where it grows in the closed-cone pine forest community. This variety does very well at Claremont if given protection from the afternoon sun.

C. gloriosus is one of the most beautiful of all ceanothus and highly recommended for coastal areas. In inland areas in full sun the plants may be short-lived, particularly var. *exaltatus,* and vars. *gloriosus* and *porrectus* should be protected from the hot afternoon sun.

Regions: 1, 2, 3 (vars. *gloriosus, porrectus*).

7. *C. greggii* var. *vestitus* (*Cerastes*). Mojave ceanothus. Erect, rigidly and intricately branched shrubs to 6 ft. with gray bark and tomentulose branchlets; leaves opposite, elliptic-ovate to oblanceolate, to ⅝ in. long, usually concave above, entire to dentate, grayish-green above, gray beneath with a close tomentum, rigid, firm; flowers borne in small, umbellate clusters, creamy-white; fruit about 1/6 in. in diameter, globose, without horns or with small spreading dorsal horns. May-June.

Mojave ceanothus is found on dry desert slopes from Mono Co. to Kern, Los Angeles and San Bernardino cos., east to Utah and Arizona at elevations of 3500-7500 ft. In California it is found in the Joshua tree woodland, pinyon-juniper woodland and sagebrush scrub communities.

Var. *perplexans.* Cupleaf ceanothus. Differs from the preceding in having leaves that are roundish to broadly elliptical or broadly obovate, to ¾ in. long, mostly conspicuously toothed, yellowish-green and glabrous above, fine tomentulose beneath. Found on dry slopes on the south face of the San Bernardino Mts., south to northern Baja California at elevations below 7000 ft. in the chaparral, pinyon-juniper woodland and yellow pine forest communities. This variety apparently grows in slightly moister habitats than var. *vestitus.*

Grown in well-drained situations, *C. greggii* has done exceptionally well at Claremont and has been very long-lived. It has not been tried on heavy soil. Useful for many landscaping purposes in southern California.

Regions: 2, 3, 4, 6, and selected areas in 7.

8. *C. griseus* var. *griseus* (*Euceanothus*). Carmel ceanothus. An erect, medium-sized shrub to 8 ft. with stout, angled, green branchlets; leaves alternate, broadly-ovate to roundish-ovate, to 1¾ in. long and 1 in. wide, dark green and glabrous above, gray-tomentose or silky beneath, veins prominent, especially beneath, margins revolute and sometimes undulate; flowers borne in dense panicles, to 2 in. long, flowers blue; fruit subglobose, about 1/6 in. in diameter, glandular-viscid when young, black and shiny when mature. March-May.

Carmel ceanothus occurs in the coast ranges from Santa Barbara to Monterey cos. and in Sonoma and Mendocino cos. where it is found at low elevations in the closed-cone pine forest and the northern coastal scrub communities.

Var. *horizontalis.* Carmel creeper. Differs from the preceding in being prostrate or low-growing in habit. Found at Yankee Point, Monterey Co.

C. griseus is one of the best. At Claremont it does very well on heavy loam and will nearly take ordinary garden conditions. There is much variation in the species and numerous forms have been selected and named.

Regions: 1, 2, 3. Also the Central Valley if protected from afternoon sun.

9. *C. impressus* var. *impressus* (*Euceanothus*). Santa Barbara ceanothus. A low, rather densely-branched shrub to 5 ft.; leaves alternate, elliptical to nearly orbicular, to ½ in. long, dark green, 1-veined from the base, upper surface deeply furrowed over the veins, margins revolute and sometimes slightly glandular; flower clusters simple, to 1 in. long, flowers blue; fruit subglobose, about 1/6 in. in diameter with prominent lateral crests. March-April.

Santa Barbara ceanothus is found only in a few areas in Santa Barbara and San Luis Obispo cos. where it grows in sand and on sandstone hills in the chaparral.

Var. *nipomensis.* Nipomo ceanothus. A poorly defined variety having leaves to 1 in. long, usually light or yellowish-green with the veins less deeply furrowed than in var. *impressus* and scarcely revolute. Nipomo Mesa, San Luis Obispo Co.

In our experience var. *impressus* is more tolerant of garden conditions than is var. *nipomensis* which we have found to be touchy, with numerous losses being reported following winters of high rainfall even if the plants were growing in well-drained situations. Here var. *nipomensis* is probably best when grown on well-drained slopes. Plants of var. *impressus* grown on alluvial sand and gravel do not have the same lush green color that they have when grown in heavy soil on the mesa.

A very beautiful species well worth the trouble required to cultivate it successfully.

Regions: 1, 2, 3.

10. *C. incanus* (*Euceanothus*). Coast whitethorn. Plate 5b. Tall shrubs to 12 ft. with numerous gray-glaucous branches and stout thorn-like branchlets; leaves alternate, broadly ovate to elliptic, to 2½ in. long and 1¾ in. wide, 3-veined from the base, minutely pubescent, or mostly glabrate on both surfaces, margins entire; flowers white, borne in short, dense axillary clusters, to 3 in. long; fruit triangular, shallowly-lobed at the summit. April-May.

Coast whitethorn is found in canyons and on open slopes from Santa Cruz Co. north to Humboldt and Siskiyou cos. usually below 3000 ft. where it occurs in the redwood and mixed evergreen forest communities.

C. incanus has done well at Claremont where it is grown in heavy soil and watered about every three weeks during the summer, and it does equally well in alluvial deposits with watering at about four-week intervals. The plants on the mesa with more water look better than those in sand and gravel, but the latter respond quickly to the autumn rains and regain their attractive appearance within a short time. Long-lived. Highly recommended.

Regions: 2, 3, 6. Also Central Valley if protected from afternoon sun.

11. *C. integerrimus* (*Euceanothus*). Deer brush. A loosely branched shrub to 12 ft. becoming untidy in age, branches green or yellowish-green, often somewhat droopy; leaves alternate, semi-deciduous or deciduous, broadly elliptical, ovate or nearly oblong, to 3 in. long and 1½ in. wide, 3-veined from the base, or less commonly pinnately-veined, light green and puberulent to almost glabrous above, paler and usually pubescent on the veins beneath, in some pubescent on both surfaces, margins entire or denticulate; flower clusters compound, rarely simple, to 6 in. long and 4 in. wide, flowers white, pale to deep blue, pinkish or rosy-lavender; fruit globose to triangular, about ¼ in. in diameter, slightly depressed at the summit, viscid, often with small lateral crests. May-July.

Deer brush and its poorly defined varieties are very widely distributed in California, extending north to Washington, east to Arizona and New Mexico.

C. integerrimus is of limited value in our area and is included here primarily because of the beautiful flower colors found within the species. In age the plants become very ragged in appearance, but this could probably be corrected with careful pruning, especially when the plants were young.

Regions: 1, 2, and marginally in 3.

12. *C. leucodermis* (*Euceanothus*). Chaparral whitethorn. A stout shrub to 12 ft. with rigid, divaricate, spinose branches which are usually gray or white in color; leaves alternate, oblong to ovate, or elliptic, to 1½ in. long and ½ in. wide, 3-veined from the base, glabrous and glaucous on both surfaces, margins variable; flower clusters usually simple, short and dense, sometimes longer and narrower, to 3 in. long, flowers blue to white; fruit globose, about ¼ in. in diameter, viscid. April-May.

Chaparral whitethorn is found on dry, rocky slopes in the mountains of cismontane southern California, south to northern Baja California and north to Alameda and Eldorado cos. at elevations below 6000 ft. where it is a member of the chaparral and southern oak woodland communities.

The flowers on this species are often a very attractive pale blue color. This species has done well at Claremont, perhaps slightly better on heavy soil on the mesa than it has in the plant community area on alluvial sand and gravel; this may be due to the fact that on the mesa they receive a little more water than do plants in the community area.

Regions: 1, 2, 3, 6, 7.

13. *C. maritimus*★ (*Cerastes*). Maritime ceanothus. Plate 41. Stiffly and intricately branched shrubs variable in height from prostrate to 3 ft. and spreading 3-8 ft.; leaves opposite, narrowly to broadly obovate or oblong, to ⅝ in. long, thick and leathery, dark green and shiny above, white-tomentose beneath, margins thickened and slightly revolute, often with 1-3 teeth on each margin; inflorescences borne on short, lateral shoots, few-flowered, flowers white to pale bluish-lavender; fruits with 3 short erect horns. January-March.

Maritime ceanothus is known only from San Luis Obispo Co. where it occurs in the vicinity of Arroyo de la Cruz, extending in places to the edge of the ocean bluffs.

C. maritimus is very un-ceanothus-like in appearance and is often mistaken for a cotoneaster. In suitable areas the prostrate forms make excellent ground covers but inland they should only be used as single plants, being very beautiful where planted so that they can drape over boulders or ledges.

Regions: 1, 2 marginal, 3 only as specimens and best with afternoon shade. Reported to do well in the Central Valley if protected from afternoon sun.

14. *C. megacarpus* ssp. *megacarpus* (*Cerastes*). Bigpod ceanothus. Rather large, compact shrubs to 12 ft. with long, slender, grayish-brown or reddish branches; leaves alternate, crowded, cuneate-obovate or elliptical, wedge-shaped at the base and usually notched at the apex, to 1 in. long and ½ in. wide, 1-veined from the base, thick, leathery, glabrous and dull green above, canescent beneath, margins entire; flowers borne in small clusters on short lateral branches, white; fruit large, globose, to ½ in. in diameter with large dorsal or subdorsal horns, viscid, wrinkled at the summit when dry. January-April.

Bigpod ceanothus occurs on dry slopes on the foothills and lower mountain slopes near the coast from Santa Barbara Co. south to northern San Diego Co. at elevations below 2000 ft. where it is a member of the chaparral community.

Ssp. *insularis.* (Sometimes known as *C. insularis.*) Differs from the preceding in having mostly opposite leaves which may be somewhat larger than those of ssp. *megacarpus.* Known only from Santa Cruz and Santa Rosa islands where it occurs in the chaparral. Intermediate forms between the two subspecies are found on Santa Catalina and San Clemente islands.

Both subspecies have done very well at this botanic garden. Ssp. *insularis* growing in heavy soil on the mesa is more tree-like than when growing in alluvial deposits. Recommended.

Regions: 1, 2, 3, 6.

15. *C. oliganthus* (*Euceanothus*). Hairy ceanothus. Medium-sized or large shrubs often with a tree-like trunk, to 10 ft., branches numerous, short and usually stiff, not spinescent but densely villous, more or less reddish and verrucose; leaves alternate, ovate-oblong to elliptic, to 1½ in. long and 1 in. wide, 3-veined from the base, upper surface dark green with scattered rather long hairs, lower surface paler with scattered hairs, especially on the veins; flower clusters simple, rather loose to 2 in. long, flowers pale to deep blue or purple, or almost white; fruit somewhat triangular, to ¼ in. in diameter, smooth or verrucosely roughened, usually viscid. February-April.

Hairy ceanothus is found on dry slopes from Los Angeles and western Riverside cos. north to San Luis Obispo Co. at elevations below 4500 ft. where it is a member of the chaparral community.

This species has done very well at Claremont either on the mesa in heavy soil or in the community area on alluvial deposits. The plants often have flowers of a pleasing shade of blue. Care should be taken in watering the plants. Recommended.

Regions: 1, 2, 3, 6. Also Central Valley if protected from afternoon sun.

16. *C. papillosus* var. *papillosus* (*Euceanothus*). Wartleaf ceanothus. A rather large, spreading, loosely-branched and often straggly shrub to 15 ft. with densely tomentose branchlets and slightly roughened branches; leaves alternate, usually crowded on the branchlets, elliptic to oblong-elliptic, to 2 in. long and ⅝ in. wide, 1-veined from the base, upper surface light to dark green, shiny, glandular-papillate, lower surface pale, usually densely pubescent with felt-like hairs, margins glandular-denticulate and revolute; flowers borne in dense, simple clusters to 2 in. long, flowers deep blue or purplish-blue; fruit triangular or subglobose, about ⅛ in. in diameter, distinctly 3-lobed. March-May.

Wartleaf ceanothus is found on open or more or less wooded slopes of the outer coast ranges from San Luis Obispo Co. north to San Mateo Co. at elevations below 3000 ft. where it is a member of the chaparral and redwood and mixed evergreen forest communities.

Var. *roweanus.* Rowe ceanothus. Differs from the preceding in having larger and lighter green, softer and less furrowed leaves but intergrades with var. *papillosus,* and a decision as to which variety a plant belongs may be purely arbitrary. It occurs mainly on dry slopes in the chaparral from San Benito and Monterey cos. south to the Santa Ana Mts. usually at elevations of 2000-4000 ft.

In our experience this species is exacting in its requirements in regard to water. We have experienced heavy losses from overhead watering during the summer, probably because the roughened leaves hold the water which is quickly heated from the sun's rays and burns the leaves. Here the plants should be protected from the hot afternoon sun. Although *C. papillosus* has very beautiful deeply-colored flowers, as a garden plant, *C. impressus* is probably to be preferred.

Regions: 1, 2, 3 with afternoon shade.

17. *C. ramulosus* var. *ramulosus* (*Cerastes*). Coast ceanothus. Plate 5c. An upright or spreading shrub to 8 ft. with slender, arching branches and slender strigose branchlets; bark grayish, smooth; leaves opposite, not crowded, roundish to elliptical, or obovate, to ¾ in. long and ⅝ in. wide, glabrous, light to dark green above, grayish canescent beneath, margins entire or with a few teeth; flowers borne in small umbels, lavender, pale blue to nearly white; fruit globose, about ¼ in. in diameter, usually prominently 3-horned. February-April.

Coast ceanothus is found in dry, rocky or sandy situations in the outer coast ranges from Monterey Co. north to Mendocino Co. at elevations below 1500 ft. where it is a member of the chaparral community.

Var. *fasciculatus.* Lompoc ceanothus. Differs from the preceding in having narrowly oblanceolate leaves that are apparently fasciculed and fruits with minute horns or no horns. Found on coastal mesas in Santa Barbara and San Luis Obispos cos.

C. ramulosus has done very well at this botanic garden, the only problem being that care must be taken in watering the plants during their early stages of growth. Mature plants receive 3-4 irrigations during the summer. Recommended.

Regions: 1, 2, 3, 6.

18. *C. rigidus* (*Cerastes*). Monterey ceanothus. A rather low-growing, intricately-branched shrub to 4 ft., branchlets slender, tomentose; leaves opposite, mostly clustered on short lateral branches, cuneate-obovate, or rounded-obovate, to ½ in. long and ⅜ in. wide, 1-veined from the base, thick, leathery, glabrous and shiny above, paler beneath, margins dentate; flower clusters composed of several few-flowered umbels, flowers bright blue to purple; fruit globose, about ¼ in. in diameter, viscid, with 3 horns and no intermediate crests. March-April.

Monterey ceanothus is found on sandy hills and flats on the Monterey Peninsula where it is a member of the coastal scrub and closed-cone pine forest communities.

Another species that has done well at this botanic garden. When growing in heavy soil care must be exercised in watering. Recommended.

Regions: 1, 2, 3. Also the Central Valley if protected from afternoon sun.

19. *C. sorediatus* (*Euceanothus*). Jim brush. A medium-sized, densely and rigidly-branched shrub to 8 ft. with smooth gray-green to purplish, somewhat villous, stiff, nearly spinescent twigs; leaves alternate, elliptic to ovate, to 1 in. long and ¾ in. wide, 3-veined from the base, dark glossy green, glabrous above and pale and usually appressed, silky beneath, margins glandular-denticulate; flower clusters short, dense, usually simple, to 1¼ in. long, flowers pale to deep blue; fruit smooth, globose, about 1/16 in. in diameter, scarcely lobed, viscid with age. February-May.

Jim brush occurs on dry slopes from Orange and western Riverside cos. north to northern California at elevations below 3500 ft. where it is a member of the chaparral community. Hybridizes with *C. oliganthus, C. papillosus* and *C. tomentosus.*

C. sorediatus has done very well at this botanic garden.

Regions: 1, 2, 3, 6.

20. *C. spinosus* (*Euceanothus*). Greenbark ceanothus, redheart. A large shrub or sometimes tree-like and 20 ft. tall, bark greenish-yellow, glabrous, twigs spinescent; leaves alternate, broadly elliptical to oblong, thick and leathery, glabrous and shiny on both surfaces, to 1¼ in. long and ¾ in. wide, usually 1-veined from the base, margins entire, rarely serrulate; flower clusters large, compound, to 6 in. long, flowers pale blue to nearly white; fruit globose, ¼ in. in diameter, viscid. February-May.

Greenbark ceanothus is found on dry slopes in the mountains of southern California, usually near the coast from San Luis Obispo Co. south to northern Baja California usually below 3000 ft. where it is a member of the coastal sage scrub and chaparral communities. Hybridizes with *C. sorediatus* and perhaps others.

C. spinosus has done exceptionally well at this garden being both dependable and long-lived. It has many uses. The plants are too large for most gardens but they can be used as tall screens or as a barrier planting. The plants should have good drainage and are perhaps better in deep soils than in sandy or gravelly deposits. They are best in coastal areas, and inland benefit from light shade.

Regions: 1, 2, 3. Also does well in the Central Valley.

21. *C. thyrsiflorus* var. *thyrsiflorus* (*Euceanothus*). Blueblossom. A large shrub or sometimes tree-like to 20 ft., with angled green branchlets; leaves alternate, oblong-ovate to elliptical, to 2 in. long and ¾ in. wide, prominently 3-veined from the base, dark green and glabrous on the upper surface and with a few coarse hairs on the prominent veins on the underside, margins finely serrate to dentate, sometimes slightly revolute; flower clusters compound, to 3 in. long, flowers light to deep blue, or almost white; fruit subglobose, about 1/6 in. in diameter, smooth and only slightly lobed, glandular-viscid and black when mature. March-June.

Blueblossom is found on wooded slopes and in can-

C

yons of the outer coast ranges from Santa Barbara Co. north to Oregon, usually below 2000 ft., a member of the chaparral and redwood and mixed evergreen forest communities.

Var. *repens.* Creeping blueblossom. Differs from the preceding in being low growing or prostrate. Found in scattered areas from Monterey to Mendocino cos.

Blueblossom has done quite well at this botanic garden but inland it is best when provided shade from the afternoon sun.

Regions: 1, 2, 3. Also the Central Valley if protected from afternoon sun.

22. *C. tomentosus* var. *olivaceous* (*Euceanothus*). Woollyleaf ceanothus. A medium-sized to large shrub with long, slender branches, bark gray or reddish, young twigs rusty-tomentose; leaves alternate, elliptic to round-ovate, to 1 in. long and ½ in. wide, 3-veined from the base or sometimes apparently 1-veined, dark green and finely pubescent above, whitish or brownish tomentose beneath, margins glandular-denticulate; flower clusters small, compound, lateral or terminal, to 2 in. long, flowers blue to nearly white; fruit subglobose, about 1/6 in. in diameter, slightly lobed, viscid when developing. April-May.

Woollyleaf ceanothus occurs on dry brush-covered slopes from Redlands and the Santa Ana Mts. south to northern Baja California at elevations below 3500 ft. where it is a member of the chaparral community.

This is a very attractive ceanothus which has done well at Claremont. Recommended.

Regions: 1, 2, 3, 6.

23. *C. verrucosus* (*Euceanothus*). Wartystem ceanothus. An erect, stiff-branched rounded shrub to 8 ft. with verrucose branches and finely pubescent branchlets; leaves alternate, crowded, round-ovate to deltoid-ovate, to ½ in. long and ⅜ in. wide, very thick and leathery, 1-veined from the base, green and glabrous above and paler and minutely pubescent beneath, margins entire, denticulate or dentate; flower clusters axillary, few-flowered, to ¾ in. long, flowers white with conspicuous dark centers; fruit globose, about ¼ in. in diameter. January-April.

Wartystem ceanothus occurs on dry hills and mesas in western San Diego Co. and adjacent Baja California where it is a member of the chaparral community.

C. verrucosus has done very well at this botanic garden when it has been grown in well-drained soil as in the plant community area. It is one of the best of the white-flowered California lilacs.

Regions: 1, 2, 3, 6.

The following clones have been introduced into the horticultural trade and are generally available. As an aid in their identification leaf silhouettes are shown in figures 41.1 and 41.2.

1. *C.* 'Blue Buttons' A handsome and fast-growing shrub or small tree to 12 ft. and often as broad, with a single trunk and rigid main branches producing a plant with a vase-like shape which has been described as taking on the form of a small live oak. Leaves very dark blackish-green; flowers pale blue in button-like clusters contrasting well with the dark leaves. March.

2. *C.* 'Blue Cascade' An erect shrub to 8 ft. with a greater spread; leaves bright green, shining; inflorescence finger-like, to 2½ in. long, flowers medium blue. April.

3. *C.* 'Blue Jeans' Erect shrub to 6 ft., branches spreading outward and upward from the base producing an open but well leaved plant; leaves dark shiny green; inflorescence small, few-flowered, on short lateral spurs, flowers lavender. Vigorous, fast-growing, does well in heavy soil.

4. *C.* 'Blue Whisp' An erect shrub to about 8 ft. with an equal spread; leaves pale green, spirally arranged in small tight rosettes on short spurbranches, pleasantly fragrant when crushed; flowers ice blue and borne in 3 in. clusters projecting from the branches like fingers. April.

5. *C.* 'Concha' (Plate 5a) A shrub to about 7 ft. with an equal spread, densely foliaged with dark green leaves;

Fig. 41-l. Leaf silhouettes of named clones of *Ceanothus*: a, 'Sierra Blue'; b, 'Snow Flurry'; c, 'Santa Ana'; d, 'Gentian Plume'; e, 'Ray Hartman'; f, 'Owlswood Blue'; g, 'Louis Edmunds'. All natural size.

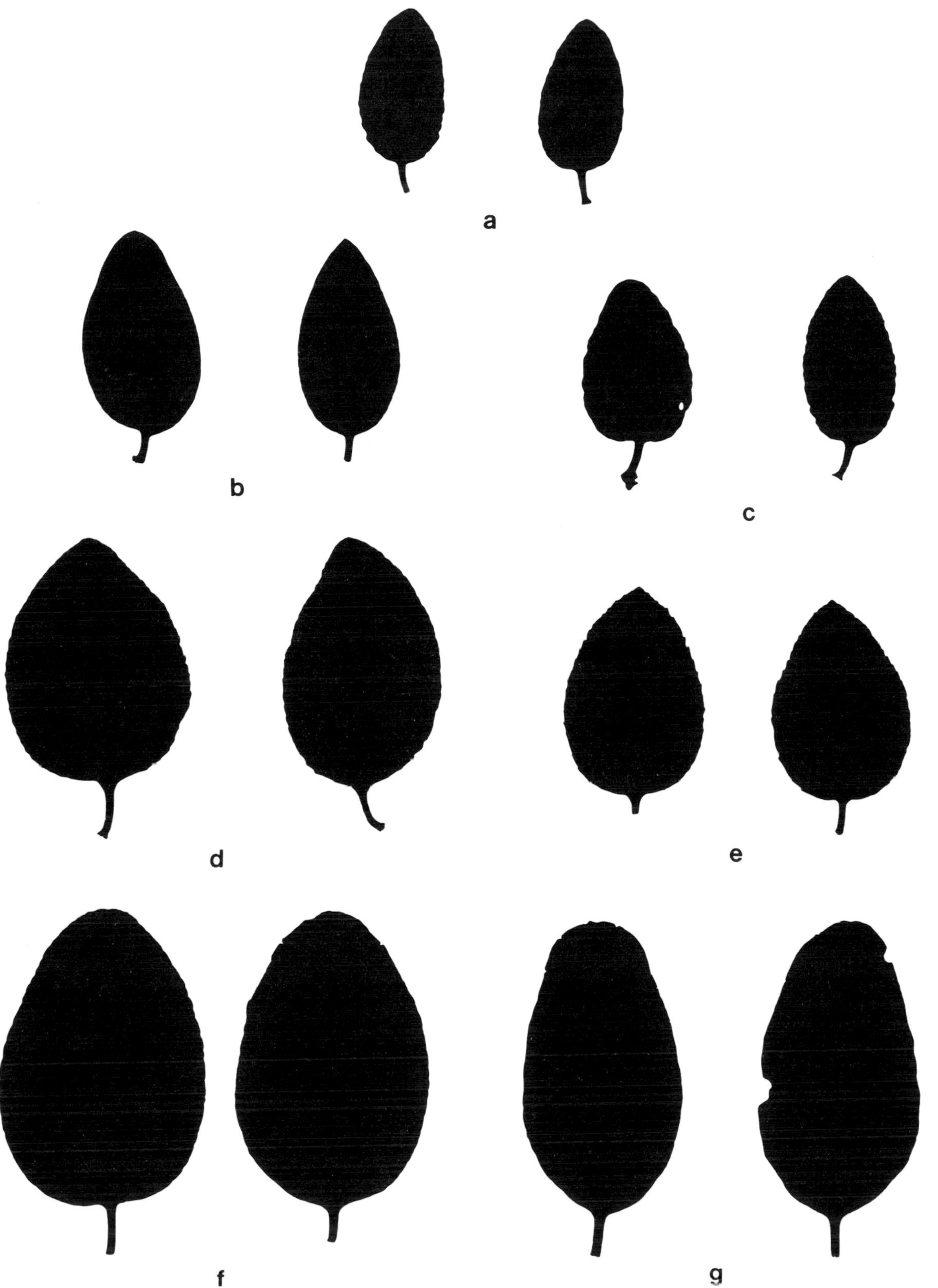
a
b
c
d
e
f
g

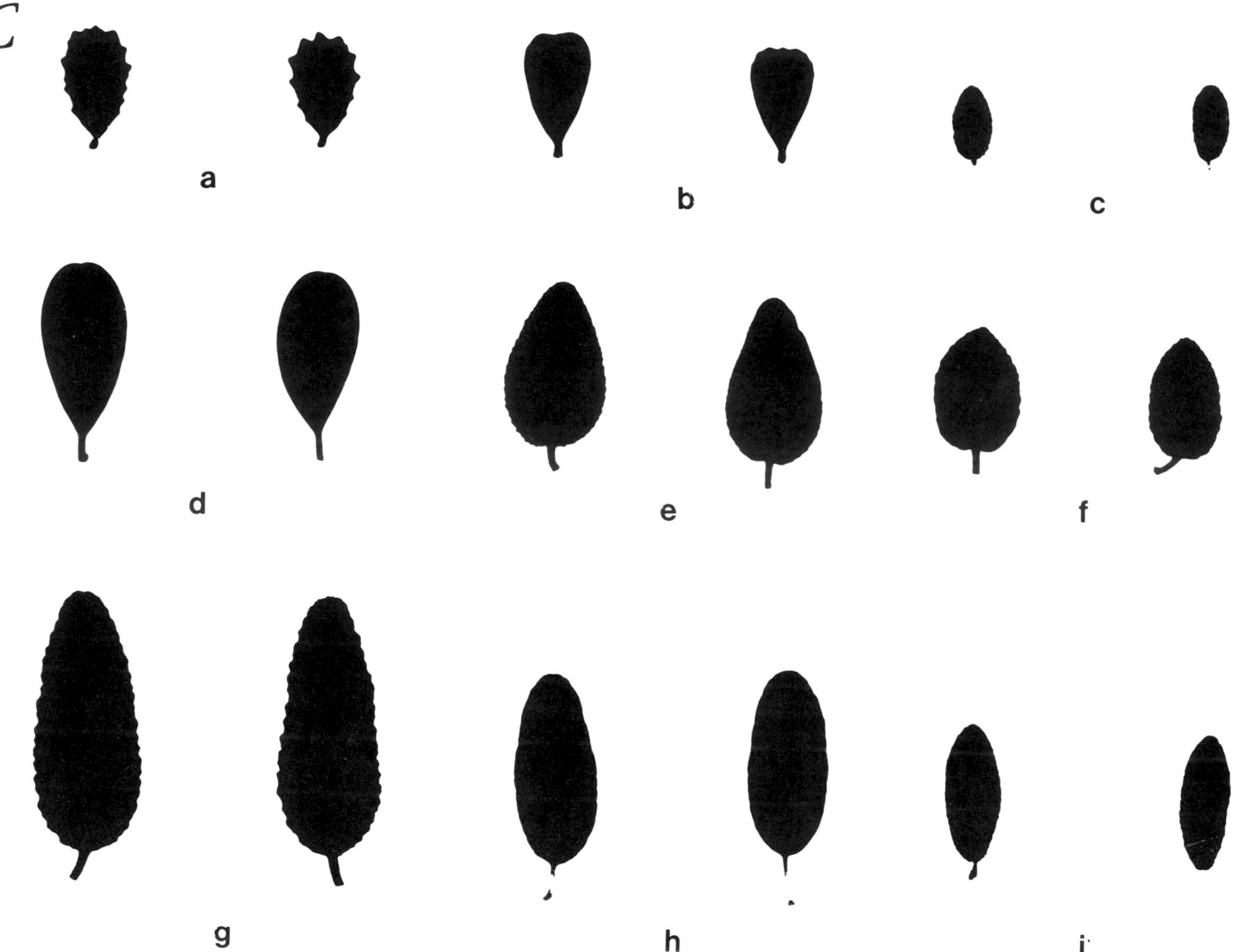

Fig. 41-2. Leaf silhouettes of named clones of *Ceanothus*. a, 'Blue Jeans'; b, 'Snowball'; c, 'Puget Blue'; d, 'Sierra Snow'; e, 'Blue Buttons'; f, 'Frosty Blue'; g, 'Joyce Coulter'; h, 'Blue Cascade'; i, 'Julia Phelps'. All natural size.

the 1 in. inflorescences are reddish in bud, flowers dark blue. Considered by some to be one of the best. April.

6. *C.* 'Consuella' Similar to the above but without the reddish buds. April.

7. *C.* 'Fallen Skies' A prostrate shrub to about 14 in. but spreading as much as 9-14 ft. in three years; leaves dark green, leathery flowers pale blue. Floriferous. A fine ground cover but it requires good drainage. March.

8. *C.* 'Frosty Blue' A shrub to 8 ft. with a 10 ft. spread; leaves dark green and deeply furrowed; inflorescences to 3½ in. long, flowers sky blue with a distinctive frosted appearance. Does well on heavy soil. Beautiful. Highly recommended. April.

9. *C.* 'Gentian Plume' A large shrub to 20 ft. with an equal spread; leaves dark green; inflorescences large and several-times branched, to 10 in. long, flowers dark gentian-blue. Tends to be leggy and requires careful pruning and pinching when young. April.

10. *C.* 'Hurricane Point' A low spreading shrub to 3 ft. with an equal spread; leaves glossy, to 2 in. long, oval; inflorescences about 1 in. long, flowers pale blue. Very fast somewhat rank grower.

11. *C.* 'Joyce Coulter' A medium-sized shrub to 5 ft. but spreading as much as 12 ft.; leaves medium green; inflorescences 3-5 in. long, flowers medium blue. Tends to mound. April.

12. *C.* 'Julia Phelps' Densely and intricately branched shrub to 6 ft. with an 8 ft. spread but often globular when young; leaves very dark green; inflorescences about 1 in. long, flowers dark indigo blue. One of the most floriferous and beautiful of the California lilacs. April.

13. *C.* 'Louis Edmunds' A shrub to 6 ft. but spreading as much as 20 ft.; leaves bright glossy green; inflorescences about 1 in. long, flowers medium blue. An old standby which tolerates heavy soil and more water than most of the California lilacs. April.

14. *C.* 'Mountain Haze' Fast-growing to about 8 ft. with a 10 ft. spread; leaves glossy, medium green; inflorescences 3-4 in. long, flowers bluish-gray. In cultivation for many years but not as beautiful as some. April.

15. *C.* 'Owlswood Blue' (Plate 5f) A spreading, rather untidy shrub to 10 ft. with an equal spread; leaves dark green; inflorescences 4-6 in. long, flowers dark blue. March.

16. *C.* 'Puget Blue' Rather openly-branched shrub to 6 ft.; leaves dark green, deeply furrowed; inflorescences small but borne in large often congested clusters, buds reddish, flowers deep blue. More tolerant of garden conditions than *C. impressus.* Blooms at the same time as *Mahonia* 'Golden Abundance'; *Dendromecon* spp. and *Fremontodendron* spp. making possible spectacular color combinations. Highly recommended. March-April.

17. *C.* 'Ray Hartman' A tall shrub to 15 ft. with an equal spread; leaves large, dark green; inflorescences 3-5 in. long, several times branched, flowers clear medium blue. An oldtimer but still one of the best as well as one of the most beautiful. April.

18. *C.* 'Santa Ana' (Plate 5d) A densely branched shrub to 14 ft. but often of much greater spread; leaves dark green; inflorescences compact, to 4 in. long, flowers very deep purplish-blue. Can be kept under control by judicious pruning. An ideal shrub for gentle slopes where it seldom gets over four feet tall but spreads widely. In Claremont, 'Santa Ana' does best when shaded from the afternoon sun. Recommended.

19. *C.* 'Sierra Blue' A shrub to 20 ft. and often nearly as broad; leaves glossy medium green; inflorescences 6-8 in. long, flowers bright medium blue. Fast grower. Good here, plants 23 years old have the largest trunks of any of the California lilacs.

20. *C.* 'Snowball' (Plate 5e) A densely and intricately branched shrub which when young may be nearly spherical in shape; leaves very dark green, thick, leathery; flowers white, borne in small tight clusters. This cultivar makes an ideal subject for hot, sunny slopes. Under such conditions it tends to spread and seldom gets over 4 ft. tall. Requires good drainage, will tolerate one good irrigation every three to four weeks. Another ceanothus that should be planted in the fall as in the young stage it is often killed by over irrigation. Distinctive; the best of the white-flowered forms.

21. *C.* 'Snow Flurry' A shrub to 10 ft. with a spread of 12 ft.; leaves rich green, to 2 in. long; flowers white, borne in profusion. This cultivar is new with us and it is too early to evaluate it for southern California. So far it has done well.

Of the above named clones 'Concha,' 'Joyce Coulter' and 'Ray Hartman' have been reported to do well in the Central Valley if protected from the hot afternoon sun.

CELTIS. A genus of the Ulmaceae with perhaps 70 species of deciduous or evergreen shrubs or trees native to both hemispheres. Related to the elms. A number planted as shade trees. One species in California.

1. *C. reticulata.* Hackberry, western hackberry. Low, densely-branched, deciduous shrubs or small spreading trees to 30 ft., often round or irregularly-crowned with short trunks, bark ashy-gray, rough, with prominent, thick, short, projecting ridges; leaves thick and leathery, prominently impressed-veined on the upper surface and with a sandpaper-like texture on both surfaces, dark green above and yellowish-green on the under surface, ovate to ovate-lanceolate, to 3½ in. long and 1½ in. wide, often slightly heart-shaped at the base with the two lobes unequal, margins serrate to almost smooth; flowers small, inconspicuous, appearing with or before

C

the leaves; fruit cherry-like, orange-red, about 1/3 in. in diameter, borne on slender stems usually singly from the axils of the leaves. April-May.

Western hackberry occurs occasionally and widely scattered in damp areas in washes and canyons in the mountain ranges bordering the California deserts, to Kern Co., northern Inyo Co. and north to Washington and east to Utah and Arizona usually at elevations of 2800-5000 ft. In California it is found in the creosote bush scrub and Joshua tree and pinyon-juniper woodlands.

C. reticulata is very adaptable, tolerating great heat and cold and although drought tolerant it benefits from additional moisture. The roots are very deep penetrating. Although sometimes symmetrical, the trees are often very irregular in shape with spreading pendulous branches. The mature fruits are much relished by birds who distribute the seeds widely with the result that young seedlings may appear at some distance from the original plant.

Although not one of California's most beautiful trees, hackberry is very useful for difficult areas taking heat, cold, drought, wind and alkaline soils.

Regions: 1, 2, 3, 4, 5, 6. Also does well in southern Arizona.

CEPHALANTHUS. A genus of the Rubiaceae with about six species of evergreen or deciduous shrubs or small trees native to Asia and Africa with one American species.

1. *C. occidentalis* var. *californicus.* Buttonbush, buttonwillow. Plate 5g. Deciduous shrub to 12 ft. or sometimes a small tree, bark gray or brown, furrowed on older stems and branches, young stems often in whorls of 3 with smooth green or yellowish bark; leaves simple, opposite or in whorls of 3, 4, 5, ovate, elliptic or oblong-elliptic, to 6 in. long and 2 in. wide, dark green, glabrous on the upper surface and with a few hairs along the veins on the lower surface, margins entire, sometimes wavy-margined; flowers numerous, small white, in dense, globose peduncled heads resembling the flower clusters of the sycamore tree, to 1½ in. in diameter, 1-6 heads terminating the branches; fruit a dry, hard capsule about ¼ in. long, tardily dehiscent into 2-4 achene-like divisions. July-September.

Buttonbush occurs along lakeshores and streams from Siskiyou Co. south in the inner coast ranges to Lake and Napa cos. and in the Sierra Nevada foothills south to Tulare Co., usually at elevations below 3000 ft. It is a member of the riparian woodland community.

C. occidentalis is an attractive and easily handled deciduous shrub which has been grown in gardens for many years. Its main requirement is that it receive adequate moisture, and because of the ultimate size of the plants care should be given in their placement. The plants can be cut back if they become too large. Valuable because of the brilliant fall coloring of the leaves.

The species has a very wide distribution extending from California east to New Brunswick, Canada and south to Florida and Cuba. It is not found in the Pacific Northwest.

Regions: 1, 2, 3, 6.

CERCIDIUM. A genus of the Fabaceae with about 10 species native to southwestern United States and extending to Mexico and northern South America. Two species native to California. Small trees or shrubs with green bark, alternate, bipinnate leaves, and yellow, nearly regular flowers borne in axillary racemes.

1. *C. floridum.* Palo verde. Plate 5h, i. Intricately branched shrubs or small multi-stemmed trees often as broad as tall, bark green, glabrous, branchlets spiny; leaves grayish, few, scattered, soon deciduous, rachis with 2 pinnae (or rarely 4-6), each pinnae with 2-4 pairs of leaflets, 1/8-3/8 in. long; flowers in axillary racemes 2-4½ in. long, flowers ½-¾ in. across, yellow, fragrant; fruit a dry pod 2-4 in. long, thickish but flattened with 1-8 seeds, pods sometimes constricted between seeds. March-June.

Palo verde is found in washes and low sandy places below 1600 ft. in the Colorado Desert, east to Arizona and south into Mexico. It is a member of the creosote bush scrub.

In bloom palo verde is a most spectacular plant producing so many flowers that they nearly hide the branches and the delightful fragrance can be detected for some distance. When out of bloom the intricately branched trees cast a light filtered shade.

The plants should be given good drainage and, al-

though very drought tolerant, they benefit from additional water which allows the plants to grow more rapidly and to become more dense and attractive. It is reported that palo verde responds well to fertilizer.

Regions: 1, 2, 3, 4 (warm areas), 5, 6. Also does well in the Central Valley and southern Arizona.

2. *C. microphyllum.* Littleleaf palo verde. A large shrub or small tree with stiff spinose-tipped branchlets; leaves with 1 (rarely 2) pairs of pinnae each ½-1¼ in. long, each pinnae with 9-17 elliptic leaflets about 1/16 in. long; flowers pale yellow, in loose racemes about 1 in. long, not fragrant; fruit pods linear-cylindric, 1-3 in. long, 1-3-seeded, constricted between the seeds. April-June.

Littleleaf palo verde occurs on gravel slopes and washes in the Whipple Mts. of eastern San Bernardino Co. east to Arizona and south into Mexico, usually at elevations below 3500 ft. It is a member of the creosote bush scrub.

The horticultural uses and requirements of the littleleaf palo verde are the same as those for *C. floridum.*

Regions: same as above.

CERCIS. A small genus of 7 species of shrubs or small trees belonging to the Fabaceae, 5 species native to the Old World, 2 species in North America.

1. *C. occidentalis.* Western redbud or Judas tree. Plate 5j, k. Large shrubs or small trees to 20 ft., usually much-branched from near the base, spreading; leaves simple, alternate, deciduous, blades round, 2-3½ in. wide, heart-shaped at the base, glabrous on both sides, entire, more or less coriaceous, bluish-green; flowers appearing before the leaves, pea-shaped, bright magenta, rosy-red or white, in simple umbel-like clusters; fruit pods oblong 1½-3 in. long, ½-⅝ in. broad, flat, few-seeded, dull red at maturity. February-April.

Western redbud is widely distributed in the coast ranges from Humboldt Co. to Solano Co., in the Sierra Nevada foothills from Shasta to Tulare cos., and on the desert slopes of the Laguna and Cuyamaca mts., eastward to Utah and Arizona. In California it is found in the foothill woodland and chaparral communities on dry slopes and in canyons and foothills usually below 3500 ft. This is one of California's most attractive flowering shrubs producing great masses of brilliant color for about two weeks in the spring. Later in the season the plants are again colorful with their attractive reddish-purple seedpods and, finally, in autumn the leaves turn yellow or red before falling.

As an ornamental, this species is highly regarded because of its early blooming and the fact that it presents few if any problems. It is drought tolerant and accepts a wide variety of soils. For heaviest bloom winter temperatures should drop to 28 F or lower. Abundant volunteer seedlings may be a problem. Western redbud is highly variable particularly in flower color and profuseness of bloom. At this botanic garden a particularly fine form of this species which is exceptionally heavy flowering and with fine deep color has been called 'Claremont' and is propagated by cuttings.

Regions: 2, 3, 4, 6. Also does well in the Central Valley if protected from the afternoon sun.

Western redbud is closely related to the more eastern *C. canadensis,* but may be distinguished from that species by the fact that the western plants have pods that average much longer and broader than those of *C. canadensis,* and the flowers of *C. occidentalis* are slightly larger and more reddish in color than those of *C. canadensis.* The eastern form grows somewhat larger and is often more tree-like than the western redbud.

C. occidentalis is reported to be immune or highly resistant to oak root fungus.

CERCOCARPUS. A genus of the Rosaceae with 8-10 species of evergreen shrubs or low trees native to western North America. Five species in California. Leaves alternate, simple, borne on short spur-like branchlets, coriaceous and straight-veined; flowers small, from winter buds, solitary or fascicled, terminal on the short branchlets, petals absent, stamens numerous; fruit a villous achene with a very much elongated, twisted soft-hairy style.

1. *C. betuloides.* Western mountain mahogany, California hardtack. Plate 51. An open shrub or small tree to 28 ft., with stiff, erect or gracefully spreading terminal branches; bark smooth, gray; leaves obovate or broadly elliptical, ½-1 in. long, to 1 in. wide, wedge-shaped and entire below the middle, serrate above the middle, con-

spicuously feather-veined, dark green and glabrous on the upper surface, paler and somewhat pubescent beneath; flowers mostly in clusters of 2-3, small, yellowish-green; fruit distinctly hairy and with a long plumose, twisted, persistent style. March-May.

Western mountain mahogany is common on dry slopes and in washes of the foothills and mountains from southern Oregon south through the coast ranges, Sierra Nevada and cismontane southern California to northern Baja California, usually at elevations below 5000 ft. The species also extends eastward to the Rocky Mts.

Cercocarpus betuloides presents no cultural problems. Plants thrive in many types of soil and, although drought tolerant, the plants appreciate some summer irrigation. Free from insect pests and disease.

Western mountain mahogany is extremely variable and numerous varieties have been described based upon leaf size, form and pubescence. All require the same cultural treatment and for horticultural purposes are interchangeable. Old plants often become coarse and unattractive and are best replaced or cut back to the base.

C. betuloides is an important component of the chaparral throughout much of the state but it rarely forms pure colonies, although in places it may be dominant (Jepson, 1936a). It crown-sprouts with marked vigor following fires and also crown-sprouts often freely upon reaching senility. It is almost never arboreous in form even if it becomes 20-28 ft. tall, but retains the aspect, form and habit of individuals much smaller.

Regions: 1, 2, 3, 6.

2. *C. intricatus.* Small-leaf mahogany. A small, intricately-branched, spinescent shrub to 4 ft.; old bark dark gray, fissured, young growth pubescent; leaves small, usually less than ½ in. long, very dark green, glabrous on the upper surface in age, white-pubescent beneath, margins strongly revolute, almost meeting over the midrib on the underside; flowers small, solitary, inconspicuous; fruit including the plumose persistent style about 1¾ in. long.

Small-leaf mahogany is found in the White, Providence and Panamint mts. of eastern San Bernardino and Inyo cos., extending eastward to Utah, southwestern Colorado, Nevada and Arizona, usually at elevations between 4000-9000 ft. In California it is found in the pinyon-juniper woodland community. It has been reported that toward the east this species intergrades with the following species.

C. intricatus is an attractive small shrub of compact form adapted to difficult situations. It prefers coarse well-drained soils and when established, the plants are very drought tolerant as well as tolerant of heat, cold and wind.

Regions: 2, 3, 4, 6.

3. *C. ledifolius.* Desert mahogany. A shrub or straggly tree usually 6-20 ft. tall but as much as 40 ft. with a trunk 2 ft. in diameter, bark gray or whitish and smooth on young branches, rough and furrowed on old stems; leaves leathery, elliptic to lanceolate, pointed at both ends, to 1 in. long and about ¼ in. wide, somewhat rolled inward, midrib very prominent; flowers small, inconspicuous, solitary or sometimes in clusters of 2-8, on short leaf-bearing branchlets; fruit including the persistent, plumose style about 2-3 in. long. April-May.

Desert mahogany is widely distributed on arid slopes, flats, plateaus and canyon walls of ranges in and bordering the deserts or arid interior, including the Santa Rosa, San Jacinto, San Bernardino and San Gabriel mts., Mt. Pinos, the easterly summits and east slopes of the Sierra Nevada from Kern Co. north to Modoc Co., west to Siskiyou Co., north to Washington and east to Colorado. It usually is found between 4000-8000 ft. In California this species is a component of several plant communities including the pinyon-juniper woodland, sagebrush scrub and bristle-cone pine forest.

C. ledifolius is one of the most important woody species in the desert ranges. As with other species of *Cercocarpus* there is a great deal of variation even among plants grown from a single collection of seed.

Desert mahogany prefers well-drained soils and is tolerant of drought, heat, cold and wind. A good accent plant, it can also be used as an informal hedge. The contrast between the dark green, almost black leaves and the white or whitish bark of the young stems is very striking.

Regions: 2, 3, 4, 6, 7.

4. *C. minutiflorus.* San Diego mountain mahogany,

mesa hardtack. A shrub usually 6-12 ft., usually with a narrow, graceful crown formed by several to many nearly vertical branches; leaves coriaceous, mostly less than 1 in. long, obovate to nearly round, serrate above the middle, glossy dark green on the upper surface, yellow-green beneath, glabrous; flowers small, several in a cluster, cream-colored. March-May.

San Diego mountain mahogany is common on the rocky hills and mesas of San Diego Co. extending south into Baja California usually at elevations between 2000-3000 ft. where it is a member of the chaparral community.

C. minutiflorus is somewhat similar to *C. betuloides,* differing mainly in having bright green leaves that are yellowish beneath and quite glabrous, whereas those of *C. betuloides* are usually dull green above and more or less whitish pubescent beneath. San Diego mountain mahogany stump-sprouts following fires, often even before the commencement of rains. As a horticultural subject it is very useful, being somewhat small in size and more refined than most forms of *C. betuloides,* and it has a rather slender growth habit. Plants can be cut back to thicken the growth or even cut to the ground and allowed to send up numerous basal suckers. It can be trimmed to form a hedge 4-10 feet in height.

Fig. 42. *Cercocarpus traskiae.* An endangered species which in its native habitat has been reduced to a single plant. (Sudworth, 1908). Reduced.

At the garden's Orange Co. site plants grew well in adobe and responded well to watering throughout the year, although once established they can do without additional water. This species has done well at Claremont where it has been drought tolerant. Highly recommended.

Regions: 1, 2, 3, 6.

5. *C. traskiae.*★ Catalina mountain mahogany. (Fig. 42). A shrub or small tree to 20 ft.; leaves coriaceous, elliptic to suborbicular or broadly ovate, serrulate toward the ends, to 2 in. long, densely white, woolly beneath, green and minutely pubescent above, margins revolute in age; flowers 4-9 in a cluster; fruit with persistent style to 2 in. long.

Catalina mountain mahogany is endemic to Santa Catalina Island and at present the entire population consists of a single plant.

Cultural requirements and uses as for *C. betuloides.*

Regions: 1, 2, 3, 6.

CHAMAEBATIARIA. A genus of the Rosaceae containing a single species native to western United States.

1. *C. millefolium.* Fern bush, desert sweet. Plate 6a. A stout, densely branched, deciduous shrub to 6 ft. with very fragrant, fern-like foliage and densely glandular branchlets; leaves bipinnately compound, alternate, with numerous minute segments, oblong-lanceolate in outline, ¾-2 in. long, more or less clustered at the ends of the branchlets, soft pubescent beneath, glabrous above; flowers white, ½-¾ in. broad, in terminal, compound, leafy panicles 2-4 in. long, petals 5, stamens numerous; fruit composed of 5 small, smooth, several-seeded pods. June-August.

Fern bush occurs on desert mountain slopes and in

rock crevices in lava beds on the eastern slope of the Sierra Nevada from Inyo to Shasta cos. and in the Panamint, White and Warner mts., north to Oregon, east to Wyoming and Arizona. In California it is found at elevations from 3400-10,000 ft. in plant communities from the sagebrush scrub to the bristle-cone pine forest. It is nowhere abundant and usually occurs in scattered clumps.

C. millefolium is a most attractive and distinctive small shrub. The drought tolerant plants do best in a coarse well-drained soil. However, the plants' appearance will be improved if they are given some additional summer irrigation. Insect pest and disease-free, the plants may be severely pruned to promote branching from the base.

Regions: 2, 3, 4, 6, 7. Also reported to do well in the Central Valley if protected from the afternoon sun.

CHILOPSIS. A genus of the Bignoneaceae with a single species native to southwestern United States and adjacent Mexico.

1. *C. linearis.* Desert willow. Plate 6b. An erect or sprawling deciduous shrub or small tree to 20 ft. with arching branches; leaves alternate, simple, willow-like or often curving and sickle-like, to 5 in. long and ¼ in. wide, sharply pointed, light green and glabrous; flowers large in short terminal clusters, catalpa-like, to 2 in. long, corolla trumpet-form, bilaterally 5-lobed, whitish, pink or purplish with purple markings; fruit a linear capsule to 10 in. long, seeds silky-hairy at both ends. April-July, but many off-season blooms.

Desert willow inhabits dry washes and streambeds of the Mojave and Colorado deserts at elevations below 5000 ft. The species extends east to Texas and south into Mexico.

C. linearis is a particularly attractive and easily grown shrub or small tree suitable for a number of purposes. When trained as a small tree it has an open and airy appearance somewhat reminiscent of the Australian willow (*Geijera parviflora*). The desert willow is fast-growing and plants may bloom the first season from 1-gallon cans.

Propagation is by seed, which germinates readily, or by cuttings or layering. The desert willow is extremely variable in flower color and only deeply colored forms should be planted. Due to the ease with which cuttings root this presents no problem. The plants should be given a location with good drainage and provided with summer irrigation. Recommended for wildlife cover, erosion control and ornamental plantings in arid regions.

Regions: 2, 3, 4, 5, 6. Also does well in southern Arizona and in the Central Valley.

CHRYSOTHAMNUS. A genus of the Asteraceae with about 13 species of shrubs native to arid regions of western North America; nine species in California.

1. *C. nauseosus.* Rabbitbrush. Shrubs 1-7 ft. with many stems from a woody base and with flexible, moderately leafy branchlets covered with gray-green or white felt-like tomentum; leaves nearly filiform to linear to 2½ in. long, more or less hairy; flower heads rayless, clustered at the ends of the branches, flowers yellow; fruit an achene.

C. nauseosus is widely distributed in western North America and consists of numerous subspecies, each restricted, more or less, to a specific habitat.

Ssp. *bernardinus.* Plate 6c. This subspecies is found on dry hillsides and benches in the San Gabriel, San Bernardino and San Jacinto mts. at elevations of 6000-9500 ft. in the yellow pine forest.

Regions: 3, 6, 7.

Ssp. *mohavensis.* Shrubs often fastigiately branched and the branches often leafless and rush-like, closely gray- or greenish-yellow tomentose. It is native to well-drained, scarcely alkaline soils from the western Mojave Desert north to central California, usually at elevations of 2500-6000 ft. in the Joshua tree woodland and creosote bush scrub communities.

This subspecies has done well at Claremont and has been long-lived, drought tolerant and free from both insect pests and disease. When in flower during the autumn, rabbitbrush is a most attractive shrub.

Regions: 3, 4, 6.

Because of their deeply penetrating roots, and the production of heavy litter as well as the ability to establish on difficult sites, rabbitbrush is valuable for controlling soil erosion and for stabilizing banks and road cuts.

Horticulturally the subspecies are rather similar and the choice of the one to use should depend upon where the plants are to be grown. Some forms do well in the Central Valley.

CLEMATIS. A genus of the Ranunculaceae with perhaps 200 species of mainly woody vines widely distributed in the temperate regions.

1. *C. lasiantha.* Pipestem clematis. Plate 6d. Deciduous, woody vines clambering over shrubs and trees by twining petioles; leaves opposite, usually composed of 3 orbicular to elliptic-ovate leaflets coarsely toothed or 3-lobed; flowers showy, about 1¼ in. in diameter, usually on 1-flowered peduncles, petals absent, sepals 4, petal-like, cream-colored; fruit an achene with long, plumose tails borne in head-like clusters. March-June.

Pipestem clematis is found mainly in canyons and near streams from northern Baja California north to Trinity and Shasta cos. usually at elevations below 6000 ft. and in many plant communities.

For a state as rich in species as is California it is remarkably lacking in woody vines. *C. lasiantha* is most attractive when in flower and can be used on trellises, fences, etc. The plants are rather rampant growers and although quite drought tolerant they respond well to some additional water during the summer. As with most species of *Clematis* they require a cool root run but are not particular as to soil, growing well in heavy loam or in alluvial deposits.

Regions: 1, 2, 3, 6.

CNEORIDIUM. A genus of the Rutaceae with a single species native to southern California and adjacent Baja California.

1. *C. dumosum.* Bushrue. Plate 6e. A strong-scented, much-branched, glabrous, evergreen shrub to 3 ft.; leaves simple, opposite, often crowded at the ends of the branchlets, linear to spatulate, rounded at the apex, tapering to the base, ½-1 in. long, ⅛ in. wide, glandular-dotted, especially on the lower surface and margins, margins entire, leaves yellowish-green; flowers small, 1-3 on short axillary or terminal peduncles, petals 4, white; fruit drupe-like, globose, about ¼ in. in diameter, reddish, drying brown-red. January-April.

Bushrue is common on mesas and bluffs in Orange and San Diego cos., extending eastward to the desert slopes of the San Jacinto and Cuyamaca mts. Also in northern Baja California. It is found in the coastal sage scrub, maritime desert scrub and chaparral communities.

C. dumosum is an attractive small shrub which presents no problems in its cultivation. In inland areas it is at its best when it receives protection from the afternoon sun. The shape of the plants can be improved by judicious pruning. Drought tolerant. Useful as a bank cover or high ground cover.

Regions: 1, 2, 3, 6 and selected areas in 4, 5.

COMAROSTAPHYLIS. A genus of the Ericaceae with about 20 species of evergreen trees or shrubs, mostly in Mexico. One species in California. Allied to *Arctostaphylos* but differences include having a warty, fleshy berry similar to those of madrone rather than a dry or mealy fruit as in *Arctostaphylos.*

1. *A. diversifolia* var. *diversifolia.* Summer holly. Plate 6g. An erect evergreen shrub or small tree to 20 ft. and often equally broad, from a basal burl (Fig. 43), bark shreddy, grayish-brown, exfoliating tardily, twigs tomentulose; leaves alternate, numerous, thick and leathery, elliptic or oblong-ovate, ¾-3½ in. long, ½-2 in. wide, serrulate, shiny green above, tomentulose beneath, strongly revolute; flowers few to many in terminal, solitary or clustered racemes, urn-shaped, white, about ¼ in. long; fruit a globular red, fleshy, drupe-like, warty berry about ¼ in. in diameter. May-June.

Summer holly is found on dry slopes at low elevations along the coast of southern California from Santa Barbara Co. south to northern Baja California where it is a component of the chaparral community.

Var. *planifolia.* Plate 6h. Differs from the preceding in having leaves that are not revolute.

Restricted to the Santa Monica and Santa Ynez mts., Santa Rosa, Santa Cruz and Catalina islands.

Summer holly is a most attractive plant, especially when trained as a small tree, and is highly recommended. The species is adaptable and tolerates a variety of soils, but it is at its best in well-drained situations and the plants should have some summer irrigation. It is very important that plants grown in containers be set out

Fig. 43. Basal burls; a, *Umbellularia californica*; b, *Comarostaphylis diversifolia*.

before they become the least bit root-bound. The fruits, which ripen in August and September, remain on the plants for several months and are much favored by birds. Somewhat slow growing at first.

According to Wolf (1943) the name summer holly was first applied to this species in 1932 by the daughter of the Clare family who then owned much of Santa Cruz Island.

Of the two varieties, *planifolia* is the more attractive and the one that should be used in ornamental horticulture.

Regions: 1, 2, 3 and selected areas in 6. Also does well in the Central Valley if protected from the afternoon sun.

CORNUS. A genus of the Cornaceae with about 25 species of deciduous or evergreen herbs, shrubs or trees native to the north temperate regions with one species in Peru. Six species in California.

1. *C. glabrata.* Brown dogwood. A deciduous shrub to 20 ft. with a broad, round crown and many flexible, drooping or pendulous branches often rooting when in contact with the soil; bark reddish-brown, smooth, nearly glabrous; leaves simple, opposite, ovate, oblong or elliptic, acute at both ends, to 4 in. long, 1½ in. wide, gray-green on both surfaces, thin, margins entire, 3-4 veins on either side of the midrib; flowers small, white, numerous, in open, flat-topped compound stalked clusters; fruit a small white or bluish drupe about ¼ in. in diameter. May-July.

Brown dogwood is widely distributed in the lower foothills of the coast ranges from southern Oregon south, infrequent in the foothills of the Sierra Nevada and in scattered localities in southern California. It occupies moist areas usually at elevations below 5000 ft. and is found in many plant communities. It is also found on Santa Catalina Island.

C. glabrata (as well as *C. stolonifera* to be treated later) are of the easiest culture, thriving in many types of soil and in many locations and with varying amounts of moisture. In nature both species are found in moist locations but when established they tolerate considerable dryness in coastal areas as well as in inland areas, if in the latter they are provided with protection from the afternoon sun. Writing of *C. glabrata,* Wolf (1944) said, "It is a pleasure to suggest the use of a California native shrub that will thrive in almost any soil and can be watered without fear of losing it from fungi attacking the roots."

C. glabrata and *C. stolonifera* are both rampant growers and spread rapidly to form large thickets by suckering and by the rooting of stem tips where they touch the soil. The plants bloom heavily in the spring and continue to produce flowers throughout much of the summer and early autumn, often showing flowers and fruit at the same time. In winter when leafless the plants with their bright reddish or purplish stems are most attractive. Ideally suited for erosion control or soil stabilization in moist areas.

Propagation is by cuttings or by removal of suckers from established plants. Highly recommended as a trouble-free deciduous shrub.

Regions: 1, 2, 3, 6, and selected areas in 7.

2. *C. nuttallii.* Mountain dogwood. A deciduous shrub or more often a tree to as much as 50 ft. with a 20 ft. spread, branches spreading horizontally; leaves opposite, elliptic to obovate, cuneate at the base, to 5 in. long, dark green, thin, veins on the leaves conspicuously impressed on the upper surface, minutely pubescent on the upper surface, softly pubescent on the underside; 'flowers' consisting of 4-7 conspicuous white or pinkish petaloid bracts surrounding a button-like cluster of small greenish-yellow flowers, appearing before the leaves and sometimes again later in the season; fruit a small, shiny, red drupe borne in dense clusters at the ends of the twigs. April-July and sometimes again in the autumn.

Mountain dogwood is widely distributed in the coast ranges from the Santa Lucia Mts. in San Luis Obispo Co. and the Sierra Nevada from Tulare Co. north to British Columbia. Rare and occasional in the mountains of southern California. It is found in several plant communities including the mixed evergreen and yellow pine forests, at elevations below 6000 ft.

It is with some hesitation that mountain dogwood is included with plants recommended for use in southern California. When in bloom, *C. nuttallii* is certainly one of the state's most spectacular trees. We have grown it at this botanic garden for many years but it has never really done well. Our plants, which were grown from seed collected in Humboldt and Siskiyou cos. may represent an ecotype that is unsuited to the Claremont area. Attempts are now being made to collect seed from plants native to southern California in order to determine whether they will be more suited to our area than those from the north. In inland areas the plants should be provided with afternoon shade. Mountain dogwood is related to the eastern *C. florida,* differing mainly in that *C. florida* has four petaloid bracts that are emarginate at the tip, whereas *C. nuttallii* usually has 4-7 bracts which are acuminate to obtuse at the tip. In cultivation *C. florida* is more easily grown than *C. nuttallii.*

Regions: 2, 6.

3. *C. stolonifera.* Red dogwood. Plate 6f. An erect but spreading deciduous shrub to 15 ft. with smooth reddish to purplish bark; leaves oval to elliptic, acute at the tip, rounded at the base, to 3½ in. long and 2 in. wide, lateral veins 4-7, very prominent on the under surface, leaves rather thick, dark green on the upper surface, lighter beneath, variously pubescent on the under surface; flowers small, numerous in branched round-topped clusters about 2 in. across; fruit almost spherical, about ¼ in. in diameter, white to bluish. April-November.

Red dogwood is found in canyons and on mountain slopes from Humboldt Co. east to the Warner Mts. in Modoc Co., southward in the Sierra Nevada to Tulare Co., occasional in the San Bernardino Mts. The species also extends northward to Alaska and eastward to the Rocky Mts., to Nova Scotia and south to Virginia, also in Mexico. It is found as high as 9000 ft. and often in the montane coniferous forest.

C. stolonifera has a very wide distribution and is extremely variable, particularly in the kind and amount of microscopic pubescence found on the undersides of the leaves, and numerous varieties have been described based on these minor morphological characters. Horticulturally the varieties are all alike, although with a species with such a wide distribution ecotypic variation might be suspected.

Cultural recommendations and uses are the same as for *C. glabrata.*

Regions: 1, 2, 3, 6, and selected areas in 7.

C. occidentalis differs only slightly from *C. stolonifera,* mainly in having twigs that are more or less hirsute. Widely distributed.

C. ×californica represents a series of hybrids between

C. occidentalis and *C. stolonifera.* Within the range of the species.

COWANIA. A genus of perhaps 6 species of the Rosaceae native to southwestern North America.

1. *C. mexicana* var. *stansburiana.* Cliffrose. Plate 6i. A much-branched evergreen shrub to 10 ft. with gray, shreddy bark and red-brown glandular twigs; leaves simple, alternate, obovate, ¼-1 in. long, pinnately 3-5-lobed, or divided, lobes revolute, glandular-dotted and green above, white-tomentose beneath; flowers cream-colored, sulfur-yellow or rarely white, pleasantly fragrant, petals 5, broadly ovate, about ¼ in. long, stamens numerous; fruit an achene with a long persistent feather-like style. April-July.

Stansbury cliffrose occurs in Inyo Co. and the eastern Mojave Desert east to Utah, Colorado, New Mexico and south into Mexico where it grows in rocky, well-drained situations at elevations of 4000-8000 ft. In California it is found in the Joshua tree woodland and pinyon-juniper woodland communities.

Stansbury cliffrose hybridizes readily with antelope bitterbrush, *Purshia tridentata*; desert bitterbrush, *P. glandulosa*; and to a limited extent with apache plume, *Fallugia paradoxa.*

When established the plants are drought tolerant and are subject to neither insect pests or plant disease but are very attractive to rabbits. The plant's appearance is improved by pruning.

Regions: 1, 2, 3, 4, 5, 6, 7.

CRATAEGUS. A very large and taxonomically difficult genus of the Rosaceae. One widely distributed species enters California.

1. *C. douglasii.* Black hawthorn, western blackhaw, Douglas hawthorn. A much-branched, deciduous shrub or small tree to 25 ft. with stout, spreading or ascending branches armed with strong thorns; leaves simple, alternate, obovate, ovate to broadly oval, 1-3 in. long, ¾-2 in. wide, 1-veined from the base, dark green and shiny above, paler beneath, doubly serrate, or often lobed; flowers in flat or convex clusters, white, fragrant, about ½ in. across, petals 5, stamens numerous; fruit drupe-like about ½ in. in diameter, black, sweet when ripe. May-July.

In California the western blackhaw is found along streams and in moist places from Marin Co. north to Siskiyou and Modoc cos. and then northward to British Columbia and east to Michigan. An inhabitant of many plant communities it is usually found at elevations of 2500-5500 ft.

Adaptable to many soils and climates the plants, which spread by suckering, require summer irrigation. They are reported to be susceptible to fire-blight. The fruits are attractive to birds.

Regions: 1, 2.

CROSSOSOMA. A genus of the Crossosomataceae with two species of summer-deciduous shrubs native to southwestern United States and offshore islands.

1. *C. californicum.* Crabapple bush. Erect shrubs or rarely tree-like, to 15 ft., branches stout, bark on older stems rough and scaly, on young stems smooth, grayish-brown; leaves pale green, alternate, entire, simple, oblong, spatulate or obovate, to 3½ in. long and ¾ in. wide; flowers solitary, about 1 in. in diameter, at the ends of short lateral branches, petals 5, white; fruit a cluster of 2-9 drooping follicles, each about 1 in. long and terminating in a curved or hooked beak. December-July.

Crabapple bush is found on dry, rocky slopes and in canyons on San Clemente, Santa Catalina and Guadalupe islands and has in the past several years been discovered on the Palos Verdes Peninsula. It is a member of the chaparral and coastal sage scrub communities.

C. californicum is notable for its long blooming season but at no one time does it produce sufficient flowers to make it a particularly attractive shrub. Added to this is the fact that the plants become dormant during summer months and the dead leaves remain hanging on the shrubs until the advent of the autumn rains, when they are shed and the plants suddenly become covered with bright green leaves and soon after with apple-blossom-like flowers as early as December. Not for the average garden but useful in environmental plantings. Botanically an interesting genus.

Regions: 1, 2, 3.

Fig. 44. *Dalea spinosa.* (Sudworth, 1908). Reduced.

DALEA. A genus of the Fabaceae with over 100 species of herbs, shrubs or small trees native from southwestern United States to South America.

1. *D. spinosa.* Smoke tree. Plate 6j, k; fig. 44. Intricately branched and spiny shrubs or small trees to 18 ft., often with a short, thickened and usually twisted and gnarled trunk, entire plant gray or whitish, branchlets minutely pubescent with white down; leaves simple, few, early deciduous, to ¾ in. long, white-downy and gland-dotted; flowers pea-shaped, small, deep indigo blue; fruit a dry, 1-seeded pod. June-July.

Smoke tree is found in sandy washes, scattered or in small groups in the southern Mojave and eastern Colorado deserts, east to Arizona and south into Sonora and Baja California usually at elevations below 1500 ft., where it is a member of the creosote bush scrub community.

D. spinosa, leafless for most of the year, is one of the best known and beloved of desert plants. It prefers deep, sandy or gravelly soils and benefits from occasional deep watering. Useful in natural landscaping, it can be used as an accent plant, a specimen shrub or planted in clumps. Seeds should be sown where the plants are wanted or in containers and the young plants set out while still small. Older plants cannot be transplanted.

Region: 5. Also southern Arizona and northern Sonora.

DENDROMECON. A genus of woody plants belonging to the Papaveraceae, variously considered as consisting of one or 20 species. Native to California and adjacent Baja California. Openly branched, glabrous, evergreen shrubs; leaves alternate, entire, coriaceous, yellowish-green to grayish-green; flowers yellow, solitary, terminal on short branchlets; sepals 2, petals 4, stamens numerous; fruit a curved, linear, 1-celled capsule, elastically 2-valved from the base upward.

1. *D. rigida* ssp. *rigida.* Tree poppy. Stems few to many from the base, plants often forming stiff rounded shrubs to 18 ft., main stems with grayish or whitish shredding bark; leaves lance-linear to lance-oblong, yellowish-green, reticulate, minutely denticulate, 1-3½ in. long, borne on very short petioles which by a twist bring the blade into a vertical position; flowers on peduncles 1-3 in. long, sepals early deciduous, flowers to 2½ in. in diameter, petals obovate to rounded; capsules 2-4 in. long. April-July, but some flowers nearly all year.

Ssp. *rigida* is rather common on dry slopes and on ridges in cismontane southern California, south into northern Baja California, north in the coast ranges and the western slopes of the Sierra Nevada to Shasta Co. It is usually found at elevations between 1000-3000 ft. and is a component of the chaparral community. It is very common after fires or in disturbed areas.

Ssp. *harfordii.* Island tree poppy. Plate 7a, b. Shrubby or sometimes tree-like to 20 ft., branches spreading or drooping; leaves crowded on the branches, elliptic to oblong-ovate, to 5½ in. long, usually not more than twice as long as wide, rounded at the tip, glaucous-green, less reticulate than in preceding; flowers to 3 in. across, golden-yellow. April-July, but some flowers nearly all year.

D

Ssp. *harfordii* is found in brushy places on Santa Cruz and Santa Rosa islands where it grows in the chaparral.

Ssp. *rhamnoides.*★ Catalina tree poppy. Leaves pale green, less crowded than preceding, mostly 2-3 times as long as broad, acute-mucronulate at tip.

This subspecies is known only from Santa Catalina and San Clemente islands where it occurs on brushy slopes.

When in bloom, the tree poppies are certainly one of California's most spectacular shrubs, especially ssp. *harfordii* with its beautiful glaucous green leaves contrasting with the deep golden-yellow flowers, but *Dendromecon* is not one of California's most dependable and easily grown shrubs. As with fremontia, if their rather demanding requirements cannot be met the horticulturist will be disappointed.

The plants should be given a coarse well-drained soil with little summer watering. Here plants grown in alluvial sand and gravel are watered about every three weeks during the summer. At this botanic garden tree poppies have never done well on the mesa where they have been grown in heavy loam. The insular forms, in well-drained soil, will tolerant more summer water than those from the mainland. The insular forms also do better if they are given protection from the afternoon sun. During summer waterings, sprinklers should not be allowed to wet the foliage when the sun is shining on the plants.

It has been our experience that tree poppies will not tolerate extensive pruning and except for the removal of dead material they should not be trimmed.

According to Jepson (1922), *Dendromecon* possesses a tap root that is very stout, fleshy and brittle and descends vertically into the ground for at least 3-5 ft. After burns, regeneration takes place from the root crown, and there may also be budding from the roots.

Tree poppies are perhaps most suited for use on well-drained hillsides and roadbanks, in rough parks or other areas where their special requirements can be met. Because of their great beauty some dedicated gardeners will accept the challenge of successfully growing one of California's most unusual and beautiful shrubs.

Propagation is by seed or cuttings.

Regions: 1, 2, 3, 4, 6. Reported to do well in the Central Valley.

DIPLACUS. A genus of about seven species of evergreen shrubs belonging to the Scrophulariaceae. A difficult genus that has suffered at the hands of taxonomists not only at the specific level but also at the generic level. Adding to the problems of natural variability is that of extensive and widespread natural hybridization between the species.

Members of this genus have often been included within *Mimulus* (Munz and Keck, 1959; Munz, 1974) while others (Jepson, 1925; McMinn, 1939; Beeks, 1961), largely on the basis that the plants are shrubs rather than herbs, have placed them in the genus *Diplacus.* Here we follow Beeks whose careful and extensive study of the southern California species has done much to clarify the situation in the southern portion of the state.

Horticulturally an interesting and important group of plants amenable to extensive manipulation at the hands of plant breeders.

1. *D. aridus.*★ San Diego bush monkeyflower. A low, usually much-branched subshrub to 1½ ft., young branches yellow, glabrous and glutinous; leaves opposite, oblong, oblanceolate or obovate, to 1¾ in. long and ⅜ in. wide, yellowish-green, crowded on the stems, entire or slightly dentate, glabrous and glutinous on both surfaces, often revolute; flowers numerous, calyx funnelform, lobes spreading abruptly at the throat, corolla pale yellow, throat rather broad, open, corolla lobes rounded, flower not strongly 2-lipped, stamens and styles exserted; fruit a capsule about ½ in. long. June-August.

San Diego bush monkeyflower is known only from a few localities in southeastern San Diego Co. and adjacent Baja California where it occurs on dry hills and ridges often growing among large boulders at elevations between 2500-3500 ft. It is a member of the chaparral community.

Regions: 1, 2, 3, 6.

2. *D. aurantiacus.* Northern bush monkeyflower. Usually erect or slightly spreading shrubs to 4 ft. with pubescent, often glandular stems; leaves opposite, oblong-lanceolate to nearly linear, to 2 in. long and ½ in. wide, dark green, shiny, glutinous and nearly glabrous above,

glandular beneath, margins serrate to entire, often revolute; flowers numerous, calyx narrow, not constricted at the middle, corolla salmon-orange, flower strongly 2-lipped, lobes spreading, unequal, stamens and styles slightly exserted; capsule linear-oblong to 1 in. April-August, but some flowers nearly all year.

Northern bush monkeyflower is found in rocky places in the coast ranges from Del Norte Co. south to Santa Barbara Co. and in the foothills of the Sierra Nevada from Placer Co. south to Tuolumne Co. Also in Oregon. Usually found below 3000 ft., this species is common in a large number of plant communities from the redwood forest to the foothill woodland.

Regions: 1, 2, 3, 6.

3. *D. calycinus.* Low, much-branched shrub with annual stems sprouting from the woody crown, branches glandular-hairy; leaves oblong-ovate or obovate, to 3½ in. long and ⅝ in. wide, yellowish-green, nearly glabrous above, pubescent beneath, entire to more or less toothed; calyx throat inflated, calyx-teeth somewhat foliaceous, densely viscid-woolly, corolla cream-colored, whitish or yellow; capsule oblong about ¾ in. long.

D. calycinus is found on rock cliffs and banks and in granitic boulders and outcrops from Fresno Co. south to the Palomar Mts. and eastward to the Little San Bernardino Mts. where it is a member of the yellow pine forest community and adjacent desert woodland.

Regions: 1, 2, 3, 4, 6.

4. *D. clevelandii.* Cleveland bush monkeyflower. An erect, freely-branching subshrub to 2 ft. with herbaceous villous-pubescent, glandular-hairy stems from a woody base; leaves oblong or lanceolate, to 4 in. long and 1 in. wide, glandular-villous but not glutinous, entire or mostly serrate along the upper half, margins sometimes revolute; calyx campanulate, constricted above the ovary, glandular-pubescent, corolla golden-yellow, about 1½ in. long, lobes rounded, flower not strongly 2-lipped, stamens and styles included; capsule to ½ in. long. May-July, but some flowers at other times.

Cleveland bush monkeyflower is a very rare plant found in the Santa Ana, Cuyamaca and Laguna mts. where it grows on dry, disturbed soil at elevations between 4000-6000 ft. in the chaparral community.

Plants of *D. clevelandii* are less woody than the other species of *Diplacus* and the plants spread by underground stems, a characteristic not otherwise found in *Diplacus* but often encountered in the perennial species of *Mimulus.* The golden-yellow color of the flowers of *D. clevelandii* is unlike that of any other species in the genus and is similar to the color sometimes seen in species of *Mimulus.* Indeed, *D. clevelandii* appears to be the connecting link between *Diplacus* and *Mimulus.*

In cultivation this species is short-lived but the plants are easily propagated from cuttings, and plantings should be replaced at about two-year intervals.

Regions: 2, 3, 6.

5. *D. longiflorus.* Southern bush monkeyflower. A profusely-branched shrub to 3 ft. with densely pubescent, glandular-hairy stems; leaves lanceolate to linear-lanceolate, or oblong, to 3½ in. long and ⅝ in. wide, yellowish-green, nearly glabrous above, pubescent beneath, entire to more or less toothed; calyx constricted above the middle and then expanding into a wide throat, corolla distinctly 2-lipped, deep red, salmon-yellow or cream colored, lobes broadly rounded and irregularly cut or wavy, stamens included, styles slightly exserted; capsule oblong, about ¾ in. long. April-May.

Southern bush monkeyflower is found on hot, dry, west-facing slopes of the inland foothills from San Luis Obispo Co. south to northern Baja California, inland to the San Jacinto Mts. and the Kern River area, at elevations below 5000 ft. where it is a member of the coastal sage scrub community.

D. longiflorus is variable particularly in flower-color, and the deep red-flowered forms with velvety corolla texture have been called *D. longiflorus* var. *rutilus* or *D. rutilus.*

Regions: 1, 2, 3, 6.

6. *D. parviflorus.* Island bush monkeyflower. A very leafy, rigid shrub to 2 ft. with nearly glabrous parts; leaves obovate or rhomboid-ovate, to 1¾ in. long and 1 in. wide, glabrous, dark green above, paler beneath, entire to irregularly serrate, sometimes revolute; flowers numerous, calyx slightly spreading at the throat, covered with sessile glands, corolla brick red, flower distinctly 2-lipped, stamens and styles exserted; capsule about ¾ in. long. January-August.

E

Island bush monkeyflower is known only from Santa Cruz, Santa Rosa, Anacapa and San Clemente islands where it occurs in rocky places in the coastal sage scrub and chaparral communities.

This species is rather distinctive because of its color and small flower size. If included in the genus *Mimulus* the correct name for this species is *Mimulus flemingii.*

Regions: 1, 2, 3, 6.

7. *D. puniceus.* Red bush monkeyflower. A freely and laxly branched shrub to 5 ft. with glabrous or puberulent, glutinous parts; leaves linear-lanceolate, thick to 2¼ in. long and ¼ in. wide, dark green and glabrous above, paler and pubescent beneath, entire or denticulate, usually revolute, partly clasping the stems; calyx tubular, flowers distinctly 2-lipped, brick red, crimson or even pink, one pair of stamens longer and slightly exserted, style exserted; capsule cylindrical, about ¾ in. long. April-July.

Red bush monkeyflower is common on temperate foothills and headlands from Orange Co. south to northern Baja California, usually at elevations below 2500 ft. and mainly in the coastal sage scrub community but also found in the chaparral.

A particularly fine form of the species has been called 'Ortega' and is now in the horticultural trade.

Regions: 1, 2, 3, 6.

All species of *Diplacus* are easily propagated from seed and the plants will bloom the first season if the seed is sown early. They may also be propagated from cuttings and this method is necessary in the case of hybrids or selected clones of the species. The plants thrive in many types of soil but should be kept dry during the summer at which time the plants are dormant. Under favorable circumstances, *D. clevelandii* spreads by underground stems, it is the only species of *Diplacus* to possess such underground parts.

The plants' overall appearance can be improved if they are severely pruned back following flowering, this will also insure vigorous new growth the following year. Writing in the *Rancho Santa Ana Botanic Garden Leaflets of Popular Information*, Wolf (1942) said, "Few other plants will yield a greater profusion of beautiful flowers for less effort than this one [*D. longiflorus* var. *rutilus*] if its rather simple requirements can be provided." These comments would also apply to the other species and the numerous hybrids.

For the amateur or hobby plant breeder, *Diplacus* presents almost unlimited opportunities for the production of new and interesting hybrids (Plate 61). All species are easily grown and bloom quickly from seed, and selected forms can be easily propagated from cuttings.

ENCELIA. A genus of the Asteraceae with about 14 species of semi-woody shrubs native to southwestern United States, Mexico, Peru and the Galapagos Islands.

1. *E. californica.* Bush sunflower. Much-branched, evergreen shrubs usually rounded in outline, to 5 ft., stems slender and woody only near the base; leaves ovate to lanceolate, to 2½ in. long and 1¼ in. wide, 3-veined from the base, entire or repand-dentate; flower heads borne on long naked peduncles, mature flower heads to 2½ in. across with about 18 clear yellow rays, each about 1 in. long and ½ in. wide, apex with 3 teeth, disk-flowers reddish-purple. March-June.

Bush sunflower is found on coastal bluffs and open or brushy slopes from Santa Barbara Co. south to Baja California; also on Santa Catalina, San Clemente and Santa Cruz islands. A member of the coastal sage scrub and chaparral communities and one of the most common shrubby plants along the coast and low foothills inland to about 25-30 miles. Frequent in disturbed soils, roadside fills and railroad embankments.

An attractive and easily grown plant whose appearance is improved with small amounts of summer water. *E. californica* crownsprouts following fires, even repeated burning. If not burned, the plants should be cut back severely before new growth commences.

Regions: 1, 2, 3. Also does well in the Central Valley.

→

Plate 7. a, b, *Dendromecon rigida* ssp. *harfordii*; c, *Epilobium canum* ssp. *mexicanum*; d, e, *Ephedra viridis*; f, *Eriodictyon crassifolium*; g, *Eriogonum crocatum* (in planter), *Arctostaphylos edmundsii* (center), *Myrica californica* (in the background used as a formal clipped hedge); h, *Eriogonum fasciculatum* ssp. *fasciculatum*; i, *Eriogonum giganteum* ssp. *giganteum*; j, *Fallugia paradoxa*; k, *Forestiera neomexicana* (male); l, *Fraxinus dipetala.*

Plate 7

E

2. *E. farinosa.* Incienso. Roundish shrubs to 3 ft. with a distinct but short trunk giving rise to numerous short branches which are very leafy towards the tips; leaves lanceolate to broadly ovate, to as much as 4 in. long and 1 in. wide, densely white- or silvery-tomentose, entire or toothed, 3-veined from the base; flowers borne in heads at the ends of long-stemmed cymes, ray-flowers 8-18, about ¾ in. long, yellow, disk-flowers yellow. April-June.

Incienso occurs on dry, stony slopes from northern Baja California north to coastal San Diego Co., inland to western San Bernardino and Riverside cos., north to Death Valley and east to Utah and Arizona usually at elevations below 3000 ft., a member of the maritime desert scrub and creosote bush scrub communities.

E. farinosa is a very floriferous shrub easily grown and recommended for natural plantings in hot interior or desert areas. Should be pruned after blooming.

Regions: 2, 3, 4, 5. Also does well in southern Arizona as well as in the Central Valley.

EPHEDRA. A genus of the Ephedraceae with about 40 species of apparently leafless shrubs with long, jointed, opposite or whorled branches and small unisexual flowers borne on separate plants. Native to arid regions of North and South America. Also in the Mediterranean region.

1. *E. californica.* California ephedra. Low, spreading or suberect shrubs to 3 ft. with numerous, opposite or whorled, pale green or glaucous branches; leafscales in whorls of 3, ¼ in. long or less, soon splitting and becoming recurved and later falling; female catkins small with 8 or more membranaceous bracts, male catkins solitary; fruits solitary, ovoid, somewhat 4-angled, about ¼ in. long. February-March.

California ephedra occurs on dry slopes and fans in the Mojave and Colorado deserts, south into Baja California, westward to the coast in San Diego Co. and northward to Santa Barbara and Merced cos. It is found in the valley grassland, creosote bush scrub and chaparral communities usually below 3000 ft.

Regions: 1, 2, 3, 4, 5, 6. Also does well in the Central Valley.

2. *E. nevadensis.* Nevada ephedra. Erect shrubs to 3 ft. with pale green or grayish, somewhat scabrous branches; leafscales usually in 2's, the tips free and deciduous in age; female catkins 2-6 at a node, male catkins solitary or several in a cluster; fruit solitary and 3-angled or in pairs with flat faces and convex-ridged backs. March-May.

Nevada ephedra is common on dry slopes and hills in the Mojave and Colorado deserts, the Owens Valley, east to Utah and Arizona. It is usually found below 4500 ft. in the creosote bush scrub and Joshua tree woodland communities.

Regions: 3, 4, 5, 6. Also does well in the Central Valley.

3. *E. viridis.* Green ephedra. Plate 7d, e. An erect shrub to 3 ft. with numerous green or yellowish-green, slender broom-like stems; leafscales usually in 2's, with free tips falling away from the brown bases; female catkins 2 at a node, male catkins 1, or sometimes 2 on each side of the node; fruit solitary and angled, or 2 in a cone with flat faces and angled convex or ridged backs, to ½ in. long. March-May.

Green ephedra is found on rocky slopes and canyon walls of desert mountains of the Mojave Desert, western slopes of the Colorado Desert, north along the eastern slopes of the Sierra Nevada to Mono and Lassen cos., east to Utah and Arizona. It is usually found between 3000-7500 ft. in the creosote bush scrub and pinyon-juniper woodland communities.

Regions: 1, 2, 3, 4, 5, 6. Also does well in the Central Valley.

The ephedras present no cultural problems but they demand excellent drainage and their appearance is improved if they are given some summer irrigation. Recommended for desert plantings. Following wet winters at Claremont many plants are lost even when planted in well-drained situations.

EPILOBIUM (section, *Zauschneria*). A genus of the Onagraceae. Clumped perennials often woody at the base, flowers zygomorphic, orange-red. A single species with six subspecies, native to western United States and northwestern Mexico.

1. *E. canum* ssp. *mexicanum* (= *Zauschneria californica* ssp. *californica* and ssp. *mexicana*). California-

fuchsia. Plate 7c. Rather low-growing much-branched, subshrubs woody at the base, to 1½ ft. tall, stems finely pubescent and with long, straight, non-glandular hairs; leaves mostly opposite, often fascicled, linear to lanceolate, to ⅜ in. wide, green with scattered, long straight non-glandular hairs; flowers spicate, large, zygomorphic, brilliant orange-red, to 1½ in. long; fruit a many-seeded capsule. August-October.

California-fuchsia occurs on dry slopes from northern Baja California north to Monterey Co. and central California at elevations below 2000 ft. where it is a member of the coastal sage scrub and chaparral communities.

It might be questioned whether California-fuchsia should be included in a book on California trees and shrubs as the plants are woody only at the base, but its exceptional beauty and usefulness in supplying color at a season when little else is in flower recommends this species very highly to the gardener. The plants, which spread by underground stems, thrive best in a well-drained soil and should be given occasional waterings during the summer. Under suitable conditions the plants reseed themselves. They may be used as ground or bank covers or in containers. Hummingbirds are much attracted to the flowers which they pollinate.

Regions: 1, 2, 3, 6.

ERIODICTYON. A genus of the Hydrophyllaceae with about eight species of aromatic, evergreen shrubs native to western United States and Mexico.

Leaves alternate, somewhat coriaceous, mostly crowded toward the ends of the branches; flowers small, many, borne in terminal one-sided, curved, compound clusters, corolla white to purple, funnelform to campanulate.

Plants often of open, weedy, invasive habit, spreading widely by woody underground rootstocks. Not suitable for ornamental horticulture but of value for soil stabilization and erosion control in difficult situations.

1. *E. californicum.* Yerba santa. Plate 7f. Shrubs to 8 ft.; leaves oblong to lanceolate, very glutinous on upper surface and shiny as if varnished, undersurface with a close, dense, felt-like pubescence, to 6½ in. long and 1 in. wide, margins dentate; flowers white or lavender. May-July.

Yerba santa grows on dry mountain slopes and ridges in the coast ranges from San Benito Co. north to Oregon and in the Sierra Nevada from Kern Co. north, usually at elevations below 5000 ft. and in many plant communities.

The plants crownsprout following fires. The leaves are often blackened by a sooty mold which grows on the resinous material on the leaf surface.

Regions: 1, 2, 3, 6.

2. *E. crassifolium.* Yerba blanco. A bluish-gray shrub with herbage covered with a dense felt-like tomentum; young leaves often snow-white, at maturity dull or greenish, oblong to oval, thick, rigid, crenate, to 4 in. long; flowers funnelform, pale lavender to rose-pink. April-June.

Yerba blanco is found in gravelly or rocky ground of streambeds and washes in cismontane southern California from Mt. Pinos to San Diego Co. at elevations below 6000 ft. in several plant communities including the chaparral and pinyon-juniper woodland.

Regions: 2, 3, 6, and selected areas in 4.

3. *E. tomentosum.* Yerba lucia. Widely branched shrubs to 8 ft., herbage covered with a thick, white felt; leaves elliptic to ovate to oblong, to 6 in. long, margins dentate or crenate; flowers white, urn-shaped.

Yerba lucia is found on dry slopes and summits of the inner coast ranges from Monterey and San Benito cos. to San Luis Obispo Co., usually below 4000 ft. where it is a member of the chaparral and foothill woodland communities.

Regions: 2, 3, 6.

4. *E. traskiae* ssp. *traskiae.* Yerba dama. Shrubs to 4 ft.; leaves ovate-elliptic to broadly oblong, crenate or subentire, greenish, to 5 in., densely pubescent above and white tomentose beneath; flowers blue. May-June.

Yerba dama is found on brushy hills in the coastal ranges from Santa Barbara to San Luis Obispo cos. and on Santa Catalina Island at elevations below 3500 ft. where it is a member of the chaparral community.

Regions: 1, 2, 3, 6.

5. *E. trichocalyx* ssp. *trichocalyx.* Yerba ynez. Shrubs to 4 ft.; leaves narrow-oblong to linear, to 3 in. long, glutinous, green, glandular and puberulent above, or gla-

E

brous, densely felt-like beneath, margins dentate; flowers narrow-campanulate, pale purplish-white.

Yerba ynez is found on dry, rocky slopes and fans in Ventura, Los Angeles and San Bernardino cos. at elevations below 8000 ft. where it is a member of the chaparral, yellow pine forest, Joshua tree woodland and pinyon-juniper woodland communities.

Regions: 2, 3, 6, and areas in 4.

The species of *Eriodictyon* are reported to do well in the Central Valley.

ERIOGONUM. A large North American genus of the Polygonaceae with over 200 species of mainly annual or perennial herbs, a few shrubby species. Well represented in California. Flowers small, without petals, borne in compound inflorescences often canopy-like above the leafy parts of the plant.

1. *E. arborescens.* Santa Cruz Island buckwheat. An erect or spreading, loosely-branched shrub to 8 ft., often with a large, thick stem, bark shreddy; leaves alternate, linear or linear-oblong, crowded at the ends of the branches, to 1½ in. long and ¼ in. wide, gray, glabrous or tomentulose above, densely white-tomentose beneath, margins strongly revolute; peduncles 2-8 in. long, bearing dense, terminal leafy-bracted cymes, calyx pink or rose-colored. May-September.

Santa Cruz Island buckwheat is found on rocky slopes and canyon walls on Santa Cruz, Santa Rosa and Anacapa islands in the coastal sage scrub and chaparral communities.

Regions: 1, 2, 3, 6. Also the Central Valley if protected from the afternoon sun.

2. *E. cinereum.* Ashyleaf buckwheat. A freely-branched shrub to 6 ft. with tomentulose branchlets; leaves alternate, ovate, to 1 in. long and ¾ in. wide, greenish-cinereous above, white-tomentose beneath; peduncles commonly formed by two's, to 18 in. long, finely tomentose; flowers in head-like clusters along the branches of the inflorescence, cream to creamy-pink. May-December.

Ashyleaf buckwheat occurs in canyons of the foothills and on beach bluffs from Santa Barbara Co. south to San Pedro, Los Angeles Co., also on Santa Rosa Island. Found in the coastal strand and coastal sage scrub communities.

Regions: 1, 2, 3, 6.

3. *E. crocatum.*★ Yellow or saffron buckwheat. Plate 7g. A small, intricately-branched, subshrub to 1½ ft. with an equal spread, bases of old stems clothed with dead leaves; leaves sheathing up the stem, ovate, to 1 in. long, ⅜ in. wide, densely white-woolly on both surfaces; flowers small, sulfur-yellow, in dense clusters forming a flat-topped cyme. April-July.

E. crocatum is restricted to a small area on the Conejo Grade along the north base of the Santa Monica Mts. where it is found in the coastal sage scrub community at an elevation of about 500 ft.

An especially attractive small subshrub ideally suited for growing in planters, large containers, rock gardens, etc. From our experience the plants thrive in rather heavy soil but should receive little or no summer watering. Easily grown from seed. Under favorable conditions the species volunteers. The contrast between the very white leaves and the sulfur-yellow flowers is most attractive. Highly recommended.

Regions: 1, 2, 3, 6.

4. *E. fasciculatum* ssp. *fasciculatum.* California buckwheat. Plate 7h. Low spreading shrubs to 3 ft., often much broader, branches somewhat decumbent, slender, flexible, bark shreddy, branches very leafy; leaves evergreen in alternate fascicles, oblong-elliptic to linear or oblong-lanceolate, tapering at both ends, to ¾ in. long, glabrous or white-tomentose above, white-tomentose beneath, margins entire, more or less revolute; leafless peduncles terminating the branches, to 10 in. long bearing simple more or less open, cymose inflorescences with many capitate clusters, calyx white or pinkish. May-October, but often much of the year.

California buckwheat is found on dry slopes and in

→

Plate 8. a-e, *Fremontodendron californicum* showing variation in plant habit and flower form; a, d, from Napa Co.; b, from Fresno Co.; c, e, from Kern Co.; f, *Fremontodendron mexicanum*; g, h, *Fremontodendron californicum* ssp. *decumbens*; i, *Fremontodendron* 'Ken Taylor', a hybrid between h and k; j, k, *Fremontodendron* 'California Glory'; l, *Fremontodendron* 'San Gabriel'; a, b, c, g, shown with yardstick for scale.

Plate 8

canyons near the immediate coast from Santa Barbara south to northern Baja California where it is an important element of the coastal sage scrub community.

Regions: 1, 2, 3, 6.

Ssp. *foliolosum* differs from the preceding in having leaves that are more revolute, green and pubescent above and white-tomentose beneath. This subspecies is common on interior cismontane slopes and mesas from Monterey Co. south to northern Baja California and is found in the coastal sage scrub and chaparral communities usually below 3000 ft.

Regions: 1, 2, 3, 6.

Ssp. *polifolium* differs mainly in having leaves that are densely canescent to hoary above and usually less revolute than the other subspecies. This subspecies is common on dry slopes in the Colorado and Mojave deserts, north to Inyo Co., the San Joaquin Valley and interior cismontane southern California, east to Utah and Arizona and south into Mexico. It is a member of the sagebrush scrub to pinyon-juniper woodland communities and is found at elevations below 7000 ft.

Regions: 2, 3, 4, 6.

5. *E. giganteum* ssp. *giganteum.* St. Catherine's Lace. Plate 7i. A freely branching shrub to 10 ft., usually broader than tall, with trunks 3-6 in. in diameter (rarely to 12 in.), bark brown, rough and shreddy; leaves alternate and borne mainly towards the ends of the branches, ovate, truncate or subcordate, or rarely tapering to the base, thick and leathery, white-tomentose on both surfaces; flowers produced in very large, branched inflorescences on long peduncles forming a canopy over the plant, flowers white to pinkish, in fruiting taking on varying shades of tan and brown. April-October.

St. Catherine's Lace is found only on Santa Catalina and San Clemente islands where it occurs on dry slopes in the coastal sage scrub and chaparral communities.

Regions: 1, 2, 3, 6. Also the Central Valley if protected from the afternoon sun.

Ssp. *compactum.*★ Differs from the preceding in being lower growing and having oblong leaves and tomentum on young growth more loose. Found only on Santa Barbara Island.

Regions: 1, 2, 3, 6.

Ssp. *formosum* with leaves oblong lanceolate. Known only from San Clemente Island.

Regions: 1, 2, 3, 6.

From the standpoint of environmental horticulture the buckwheats form a very important assemblage of plants whose usefulness has until now not been fully recognized. Except for *E. giganteum* which stands out somewhat because of its size and particularly the size of the inflorescences, the other species are of a somewhat similar aspect and size and the choice of which one to use should be dependent to a large extent upon the location where the plants are to be grown. *E. fasciculatum* is composed of three subspecies, each adapted to one or more plant communities: *E. f.* ssp. *fasciculatum* to the coastal sage scrub; *E. f.* ssp. *foliolosum* to the coastal sage scrub and chaparral communities; and *E. f.* ssp. *polifolium* to the sagebrush scrub to pinyon-juniper woodland communities at elevations as much as 7000 ft.

All the species listed here are useful for growing on banks and hillsides and as ground covers. They are highly recommended for erosion control, not only because of their spreading fibrous root systems, but also because the much-branched aerial portions of the plant help to check the force of the rain reaching the soil. The plants are attractive not only in flower when vast areas take on shades of white and pink, but also when in fruit they take on shades of tan and brown.

The plants are not particular as to soil although they are at their best in a coarse well-drained soil. Some of the species tend to be long-lived as evidenced by plants of *E. giganteum* ssp. *giganteum,* having trunk diameters of up to 12 in. Under favorable conditions the plants volunteer readily. One planting of *E. cinereum* was established at this garden in 1952, and in 1980 it was still healthy. These plants are growing in a heavy loam on a steep hillside and are watered about every two or three weeks during the summer. All the species are drought tolerant and can survive on only the water they receive from natural rainfall, but the appearance of the plants is enhanced if they receive some additional water.

Propagation is by seed, either sown in flats and the seedlings later transplanted, or by broadcasting the seed where the plants are wanted. Plants of *E. giganteum,* and probably the other species as well, develop a strong taproot at an early age and the seedlings should be transplanted to their permanent locations before they become the least bit rootbound. It is also recommended in the case of *E. giganteum* that the old inflorescences be removed to prevent the plant stems from being broken during winter storms.

When growing together the species of *Eriogonum* listed here may hybridize; a hybrid between *E. arborescens* and *E. giganteum* has been named *E. ×blissianum*. The plants were reported to be sterile. Hybrids between *E. giganteum* and *E. polifolium* have appeared at this garden.

The eriogonums are considered to be the third most valued native bee-plants (after white and black sage). The dried inflorescences are often used for home decoration. To prevent shattering of the dried flower parts the inflorescences should be sprayed with a clear acrylic.

FALLUGIA. A monotypic genus of the Rosaceae native to arid portions of Western United States.

1. *F. paradoxa.* Apache plume. Plate 7j. Much-branched deciduous shrub to 6 ft., stems straw-colored, bark flaky; leaves clustered along the branchlets, alternate, pinnately dissected into 3-7 linear divisions, margins revolute, ½-1 in. long, upper surface puberulent, rusty beneath; flowers large, white to creamy-yellow, 1-1½ in. across, solitary on 2-3 in. long peduncles at the ends of the branches, petals 5, rounded, ½-¾ in. long; fruit an achene with long plumose, persistent styles. May-June.

Apache plume is native to the mountains of eastern San Bernardino Co., east to Nevada, Texas and south into Mexico. In California it is found on dry slopes at 4000-5000 ft. in the Joshua tree and pinyon-juniper woodland communities.

This species exhibits considerable variability in morphological characters as well as physiological characters (some forms are nearly evergreen) and it occurs on many types of soil and tolerates a wide range of temperatures, both cold and hot. In addition to being a most attractive ornamental in both flower and fruit, the plants are highly useful for erosion control as well as for producing browse for animals. Natural hybrids between apache plume and Stansbury cliffrose, *Cowania mexicana* var. *stansburiana*, are known.

Propagation from seed is easy and once established the plants are drought tolerant. Free from insect pests and disease.

Regions: 1, 2, 3, 4, 5, 6, 7. Also reported to do well in the Central Valley if protected from the hot afternoon sun.

FORESTIERA. A genus of the Oleaceae containing 14 species native to North and South America.

1. *F. neo-mexicana.* Desert olive. Plate 7k. An erect, stiffly branched, deciduous shrub to 14 ft., bark smooth, light gray or yellowish; leaves often fascicled, obovate to oblong, acute, finely serrate or entire, ½-1½ in. long; flowers inconspicuous, appearing before the leaves, in sessile clusters, corolla absent, or rarely with 1-2 small petals, stamens 2-4; fruit a blue-black, ovoid drupe about ¼ in. long. March-April.

Desert olive occurs in dry ravines, along intermittent streams, on dry slopes and on lower mountain flats in the inner coastal ranges from Contra Costa Co. south to interior cismontane southern California, east to Colorado and Texas, usually below 6700 ft. In California it is found in the creosote bush scrub, chaparral, coastal sage scrub and foothill woodland communities.

F. neo-mexicana is widely adaptable, thriving in light or heavy soils, and while drought tolerant the plants are more attractive if they are given some water during the summer. The plants are free from insect pests and disease and they withstand considerable cold. The desert olive may be used as a hedge, screen or bank cover. Horticulturally little known, it is highly recommended. The fruits are much enjoyed by birds.

Regions: 1, 2, 3, 4, 6, and selected areas in 5. Also does well in the Central Valley.

FOUQUIERIA. A genus of the Fouquieriaceae with 12 species of spiny shrubs native to southwestern United States and Mexico. One species in California.

1. *F. splendens.* Ocotillo. Shrubs to 20 ft. with many slender whip-like branches from the base, gray with darker furrows and stout divaricate spines; leaves of two

F kinds, primary leaves oblong-ovate, fleshy, entire, to 1 in. long, leaf-blade soon deciduous, petiole developing into a stout spine, secondary leaves borne in the axils of the spines, blades oblanceolate, to 1¼ in. long; flowers borne in showy terminal panicles, tubular, to 1 in. long, bright red, stamens exserted; fruit a 3-valved capsule. Flowers produced following summer rains.

Ocotillo is found in dry, mostly rocky places from southeastern Mojave Desert throughout the Colorado Desert and east to Texas and south into Mexico, usually at elevations below 2500 ft. where it is a member of the creosote bush scrub community.

F. splendens is an attractive shrub with a distinctive appearance unlike that of any other California plant. It requires excellent drainage and full sun. It can be used as a specimen plant, for screening, or as an unpenetrable barrier. Easily propagated from cuttings placed directly in the soil where the plants are needed. Attractive to hummingbirds which pollinate the flowers.

Regions: 3 (marginal), 4 (warm places), 5. Also does well in southern Arizona.

FRAXINUS. A genus of deciduous shrubs or trees belonging to the Oleaceae with about 40 species in the Northern Hemisphere. Many ornamental species and some important timber trees. Four species in California.

1. *F. anomala.* Singleleaf ash, dwarf ash. A deciduous shrub or small tree to 20 ft. with 4-angled branchlets; leaves opposite, simple, or sometimes 2-3-pinnately compound, if simple, broadly ovate or round, to 2 in. long, about as wide, if compound the terminal leaflet larger, margins serrate to crenate, or entire, glabrous; flowers small, inconspicuous in axillary clusters appearing with or before the leaves, petals absent; fruit a samara, ½-¾ in. long with an oblong wing surrounding the body. April-May.

Singleleaf ash is found on the eastern slopes of the Providence Mts. in San Bernardino Co. and in the Panamint Mts. in Inyo Co., north to Utah and Colorado and east to Texas. It occurs at elevations of 4500-11,000 ft. and in California it is mostly in the pinyon-juniper woodland.

F. anomala is an interesting deciduous shrub tolerant of difficult situations of heat, cold and drought.

Regions: 2, 3, 4, 6.

2. *F. dipetala.* Flowering ash. Plate 71. A shrub or small tree with 4-angled branchlets; leaves opposite, deciduous, to 6 in. long with 3-9 leaflets, leaflets oblong-ovate, to 1½ in. long and ¾ in. wide, glabrous on both surfaces, serrate above the middle or entire; flowers small, numerous, fragrant, in compound clusters appearing with the leaves, petals creamy white; samara to 1 in. long, often notched at the tip. March-June.

Flowering ash is often common on canyon sides and mountain slopes in the Sierra Nevada from Shasta Co. south to Tulare Co. and in the coastal ranges from Siskiyou Co. south to Orange Co., usually at elevations of 400-3700 ft. where it is a member of the foothill woodland and chaparral communities.

F. dipetala is the most attractive of the native ashes and the only one with showy flowers. (*F. trifoliata* of Baja California is similar but little known in horticulture.)

Flowering ash is drought tolerant and thrives in a variety of soils. It responds favorably to additional amounts of water even when growing in heavy soil. Easily propagated by seed with the seedlings being transplanted bare-root. An attractive and useful species.

Regions: 1, 2, 3, 6. Also does well in the Central Valley.

3. *F. latifolia.* Oregon ash. Deciduous trees to as much as 80 ft., branchlets usually stout, more or less pubescent; leaves to 12 in. with 5-7 leaflets, leaflets oblong to oval, often broadest toward the tip and abruptly short-pointed, entire or toothed above the middle, to 5½ in. long; male and female flowers borne on separate trees, small, inconspicuous, without petals, appearing with or before the leaves; samara to 1¾ in. long with the wing narrowly decurrent on the body. March-May.

Oregon ash occurs along streams, in canyons and valleys in the coast ranges from Siskiyou Co. south to Santa

→

Plate 9. a, *Galvezia speciosa*; b, *Garrya veatchii*; c, *Heteromeles arbutifolia*; d, *Holodiscus discolor* var. *discolor*; e, *Isomeris arborea* var. *arborea*; f, *Isomeris arborea* var. *globosa*; g-i, *Iva hayesiana*, g, plant habit, h, recent planting, i, same planting six months later; j, *Justicia californica*; k, l, *Larrea tridentata*.

Plate 9

Clara Co. and in the Sierra Nevada from Shasta and Modoc cos. south to Tulare Co. The species extends north to British Columbia.

F. latifolia will grow in areas where the plants are subjected to standing water during the winter.

Regions: 1, 2, 3, 6.

4. *F. velutina.* Arizona ash. Deciduous trees to 30 ft. with branches often reaching the ground, young stems terete, more or less pubescent; leaves pinnately compound with 3-7 leaflets, leaflets to 4 in. long, lanceolate to narrowly ovate, oval or obovate, thick, dark green, more or less pubescent when young, or glabrous; male and female flowers borne on separate trees, flowers small, inconspicuous, appearing with or before the leaves; samara to 1½ in. long, wing decurrent barely to the middle of the body. March-April.

Arizona ash occurs along streams and in canyons from the San Gabriel, San Jacinto and Cuyamaca mts. north to Kern and Inyo cos., east to Nevada and Arizona at elevations below 5000 ft.

F. velutina is normally found in rather moist places, sometimes being associated with desert willow, *Chilopsis linearis,* and poplars, *Populus* spp., but it will tolerate drought although the trees will defoliate prematurely. *F. velutina* is extremely variable and attempts have been made to recognize varieties or subspecies within the species.

Regions: 1, 2, 3, 4, 6. Also does well in southern Arizona.

The Arizona and Oregon ashes are both fast-growing species with limited value for ornamental horticulture but useful in some environmental situations; *i.e.,* the Oregon ash will tolerate standing water during the winter. The trees of both species are brittle and easily damaged by wind and both are very susceptible to aphids. Arizona ash is reported to be immune or highly resistant to oak root fungus.

FREMONTODENDRON. A genus of the Sterculiaceae with two species of evergreen shrubs or trees native to California, Arizona and northern Baja California. Plants with stellate hairs and mucilaginous inner bark (plants often known as slippery elm); leaves simple, alternate, variable in shape and size; flowers large and showy, yellow (yellowish-orange in one subspecies), calyx petaloid, 5-lobed to below the middle, stellate pubescent on the exterior, internally with 5 nectar-bearing pits at the base of the lobes, petals absent; stamens 5, filaments joined for about ½ their length; anthers large; fruit a bristly-hairy, 4-5-valved capsule.

Fig. 45. Natural variation in leaf size and shape in *Fremontodendron californicum*, a-m; *F. mexicanum,* n. (Harvey, *Madroño* 7: 107). Reduced.

Generations of taxonomists have struggled with the problem of variability in fremontia and many attempts have been made to recognize taxonomically significant entities within the genus, but the problem remains, and there is still no satisfactory treatment of the group. According to one taxonomist, the genus most likely consists of a single polymorphic species with localized populations that might in some instances be given taxonomic recognition at the subspecific level.

Although disconcerting to the taxonomist, the variability present in fremontia offers the horticulturist almost unlimited possibilities for selecting new and interesting cultivars.

1. *F. californicum* ssp. *californicum.* Fremontia, flannel bush. Plate 8a-e. Much-branched shrubs from 4 ft. to small trees 25 ft. tall with a 30 ft. spread and with trunk diameters as much as 16 in.; leaves extremely variable (Fig. 45) from nearly entire to deeply lobed, 1-3-nerved, ½-2 in. long, usually borne on short spur-branches, mostly thick and leathery, dark green (sometimes bright green), rough above, usually densely stellate, pubescent beneath; flowers mostly on short spur-branches, yellow, flat, 1-2 in. broad, basal gland usually with hairs; capsule subglobose to 1 in. long, densely bristly-hairy; seeds brown, pubescent and usually carunculate. March-June.

F. californicum ssp. *californicum* is widely distributed in California and probably reaches its greatest development in the foothills of Kern Co. and in the San Gabriel and San Bernardino mts. It is found most often on granitic slopes at 3000-6000 ft. where it is an important component of the chaparral community, but it is also found in the yellow pine forest, foothill woodland and pinyon-juniper woodland communities.

In *F. californicum* there is a tendency for all the flowers to open at about the same time so that the plants present a spectacular show for a limited period, and this is contrasted with the following species which tends to bloom over a much longer period of time but never show the great mass of color at any one time seen in *F. californicum.* Plants from different geographical areas grown under uniform conditions may show great variation in date of flowering. Plate 8a-e shows variation in plant form and size as well as in flower form, size and color in plants grown from seed collected in different areas (Kern, Fresno, Napa cos.). The photographs illustrate individual plant variation and are not intended to suggest geographical variation.

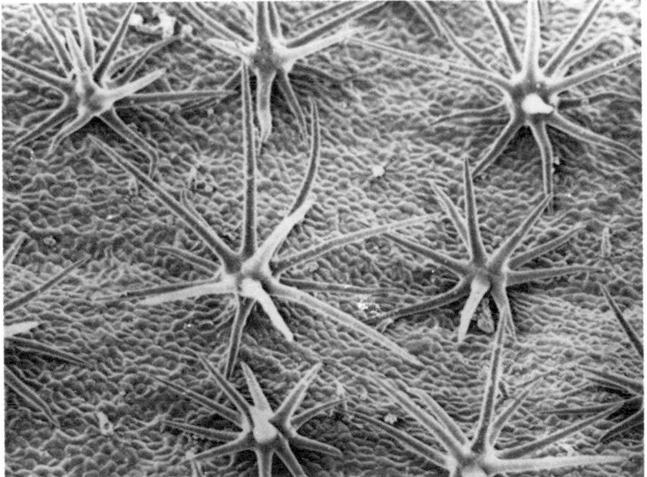

Fig. 46. Upper surface of a leaf of *Fremontodendron* showing the stellate hairs which cause skin irritation. SEM photograph by Mary O'Brien.

Plants from Pine Hill, Eldorado Co. with small yellowish-red to brownish-red flowers and a decumbent habit (Plate 8g, h) have been designated as either *F. decumbens* or *F. californicum* ssp. *decumbens.*★ In cultivation these plants have shown considerable variation in flower color and plant habit although all are somewhat decumbent. The greatest fault we have found with them is that the flowers are mostly hidden by the foliage and the plants never put on a good floral display.

Regions: 1, 2, 3, 4, 6. Also the Central Valley if protected from the afternoon sun.

2. *F. mexicanum.*★ Mexican fremontia. Plate 8f. A large shrub or tree-like to 25 ft., often equally broad, branched from the base; leaves distinctly 3-5-lobed with 5-7 veins from the base, 1¼-3 in. long, thick, dark green above with scattered dark stellate hairs, rusty pubescent beneath; flowers large, bowl-shaped, golden yellow on the inside, reddish on the outside, to 3 in. broad, solitary, disposed along the branches to appear as a long, leafy raceme, basal pits glabrous; capsules conical, 1¼-1¾ in. long; seeds black, glabrous. March-June.

Mexican fremontia is a localized, rare species found in extreme southern San Diego Co. and adjacent Baja California where it is reported to be a component of the chaparral and southern oak woodland communities. From the few collections of this species which have been made, it appears that it is never found more than about 15 miles from the ocean and at altitudes below about 2000 ft. Field-collected seed of *F. mexicanum* has been obtained by this garden on only one occasion and that was in 1936. Because many of the plants being grown today as *F. mexicanum* came from seed collected from plants grown at this garden where it has been shown that *F. californicum* and *F. mexicanum* hybridize when the two species are grown together, it seems likely that

F at least some plants labeled as *F. mexicanum* may not represent that species but rather are *mexicanum*-like derivatives of hybrids between *F. californicum* and *F. mexicanum*. In hybrids the basal glands tend to be like those of *F. mexicanum* being smooth and without hairs.

During the summer of 1980, cuttings and seedlings of *F. mexicanum* were obtained from the original Otay Mt. site and are being propagated at Claremont.

Regions: 1, 2, 3, 4, 5, 6. Also the Central Valley if protected from the afternoon sun.

In the horticultural trade today there are four named fremontia clones, one of them, 'California Glory,' a presumed hybrid between *F. californicum* and *F. mexicanum*.*

The two hybrid clones, 'Pacific Sunset' and 'San Gabriel,' are the result of controlled hybridization made at this botanic garden between the two species. 'California Glory' (Plate 8j, k) was the first to be introduced and in many respects it combines the finest characteristics of the two species; the 2½-3½ in. shallow-cupped or almost flat, lemon-yellow flowers are produced in great abundance and all tend to open at approximately the same time, similar in that respect to *F. californicum*. They are however disposed along the branches similar to the way they are in *F. mexicanum*. The flowering season is long, and in full bloom the plants appear to be a solid mass of yellow. In some respects this is probably the finest of the named clones.

From plants grown from seed produced by controlled hybridization between *F. californicum* and *F. mexicanum*, two were chosen for introduction. They differ principally in the nature of the flowers; those of 'Pacific Sunset' are a deep orange-yellow, flat or saucer-shaped and from 3½-4 in. across; those of 'San Gabriel' (Plate 8l) are buttercup-yellow, open campanulate and uniformly 3 in. across. The flowering season of 'Pacific Sunset' is somewhat longer than it is in 'San Gabriel.' In plant form and flowering habit these clones tend more to *F. mexicanum* than they do to *F. californicum*. Because of the nature of their branching, 'Pacific Sunset' and 'San Gabriel' both espalier very well.

*In plant breeding it is standard practise to list the maternal parent first followed by the name of the staminate parent. According to the rules of botanical nomenclature, when this information is not known, the parents are listed alphabetically.

A fourth clone recently introduced is called 'Ken Taylor' and is a hybrid between *F. californicum* ssp. *decumbens* and *F.* 'California Glory' (Plate 8i). It is new to our collection and as a result we cannot report upon it except to say that our plant tends to have a decumbent habit similar to that of *decumbens*. The flowers are about the same color they are in *decumbens* but are larger.

When in flower, fremontia is certainly California's most spectacular shrub (or tree) and the plants have been both the joy and despair of gardeners ever since they were first cultivated. Fremontias are not among California's most dependable shrubs and unless their rather demanding cultural requirements can be met growers will be disappointed. Since the plants are susceptible to phytophora root rot they must have no water during the summer, and they should be planted, preferably, where there is good natural drainage. The plants are also susceptible at times to attack by psyllids which must be controlled by spraying. Under favorable conditions fremontias, especially the hybrids, grow very rapidly and may be expected to bloom the first season after planting.

Because of the very irritating stellate hairs on the leaves (Fig. 46) and especially on the fruiting capsules, fremontias are not pleasant to work on or around and care should be given in their placement. They should be planted where there is no chance that old leaves and capsules will get into swimming pools or into areas subject to heavy human activity such as patios. Perhaps they are best planted where they may be enjoyed from a distance such as on hillsides where they may be grown with other colorful drought tolerant plants. If space is limited the smaller forms of *F. californicum* are recommended rather than the hybrids which may attain considerable size within a few years.

Fremontias may be propagated by seed or by cuttings. If grown from seed, plants, especially of *F. californicum*, may be expected to show variation in plant form and

→

Plate 10. a, *Keckiella antirrhinoides* ssp. *antirrhinoides*; b, *Keckiella antirrhinoides* ssp. *antirrhinoides* × *Keckiella cordifolia*; c, *Keckiella cordifolia*; d, *Keckiella ternata* ssp. *ternata*; e, *Lavatera assurgentiflora*; f, *Lonicera involucrata* var. *ledebourii*; g, *Lupinus albifrons*; h, i, *Lycium cooperi*; j, k, *Lyonothamnus floribundus* ssp. *asplenifolius*; l, *Lyonothamnus floribundus* ssp. *floribundus*.

Plate 10

size, leaf shape and color, flower size and color and flowering date. Seed should be given pretreatment to hasten germination. It has been reported (Schopmeyer, 1974) that cuttings should be made in the late summer and then grown under lights for 18 hours a day during the winter. Schopmeyer reports that seedlings and cuttings alike respond well to a long-day exposure during the winter months with about three times more growth occurring under long days than under natural day length. It has also been recommended that rooted cuttings be planted in propagating tubes of 2½-3 in. diameter by 12 in. in length with the bottom open at the time of planting so roots may readily grow into the soil below. Young plants should be staked and shaped by careful pruning and pinching.

GALVEZIA. A genus of the Scrophulariaceae containing about 4 species distributed from California to Peru.

1. *G. speciosa.* Bush snapdragon. Plate 9a. A bright green, brittle-stemmed, evergreen bush or more often vine-like to 10 ft. clambering over other shrubs; leaves in whorls of 3, glabrous or pubescent, rather thick, elliptic-ovate, entire, to 2 in. long, ¾ in. wide; flowers in terminal, leafy racemes, usually 3 at each node, flowers snapdragon-like, bright red, about 1 in. long; fruit a dry capsule. Peak flowering in spring but some flowers the year around.

Bush snapdragon is found in rocky canyons on Santa Catalina, San Clemente and Guadalupe islands.

The plants are not particular as to soil so long as drainage is good, and once established they are quite drought tolerant. In inland areas the plants require protection from the afternoon sun and they will grow in light shade although the number of flowers produced will be fewer.

G. speciosa is probably best used as a vine to cover fences and trellises, or in the absence of any supporting plant or structure, as a ground or bank cover. It should also be tried in hanging baskets.

Propagation is by seed or cuttings which root easily. It has been reported that in some areas the plants have been damaged by nematodes. Very attractive to hummingbirds.

Regions: 1, 2, 3, 6.

Fig. 47. *Garrya elliptica.* (Sudworth, 1908). Reduced.

GARRYA. A genus of about 15 species of evergreen shrubs or small trees native to the United States, Mexico and the West Indies; the only genus in the Garryaceae. The flowers are unisexual and borne in catkin-like clusters on different plants, the male inflorescences usually longer than those of the female. The species listed here are all drought tolerant (*G. elliptica* least so). Usually grown for the long attractive, pendulous male catkins which appear in late winter or early spring, these later become dried and unattractive and remain on the plants for extended periods of time unless removed.

Garryas are difficult to transplant and plants grown from seed should be set out from containers where they are wanted. They may also be grown from cuttings, and this method is necessary in the case of selected clones or to insure the sex of the plants being grown. Seed of *G. flavescens* var. *pallida* and *G. fremontii* (and perhaps others) contain dormant embryos which must be given pretreatment to hasten germination.

Taxonomically a difficult group since there is much intergradation where the species come into contact.

1. *G. congdoni.* Interior tasselbush. An erect shrub to 6 ft. with a yellowish-green aspect, young twigs silky-pubescent; leaves narrowly elliptic to oval, 1-1¾ in. long, ½-1½ in. wide, thinly puberulent or glabrous above, densely hairy beneath, margins more or less undulate, entire; staminate catkins 1¼-3 in. long, pistillate catkins ¾-2 in. long; berry roundish to broadly ovoid, densely pubescent or more or less glabrous at the base, about ¼ in. in diameter. February-April.

Interior tasselbush is found in the inner coastal ranges from San Benito Co. north to Tehama Co. and south in the foothills of the Sierra Nevada to Mariposa Co. where it occurs in dry canyons and on ridges below 2750 ft. in the chaparral and foothill woodland communities, probably mostly on serpentine.

Regions: 2, 3, 6.

2. *G. elliptica.* Coast tasselbush. Fig. 47. An erect, densely foliaged shrub or small tree, young twigs densely short-villous; leaves elliptical or oval, 1½-2½ in. long, 1-2½ in. wide, dark green, leathery, nearly glabrous above, densely felty-woolly beneath, margins strongly undulate, entire; staminate catkins 3-8 in. long (longer in selected clones), pendulous, mostly from near the ends of the branches; female catkins 2-3½ in. long becoming as long as 6 in. in fruit; berry often drupe-like, fleshy, globular, ½-⅜ in. in diameter, densely white-tomentose becoming almost glabrous in age, pulp very juicy when ripe.

Coast tasselbush is found on dry slopes and ridges in the outer coast ranges from Ventura Co. north to Oregon and on Santa Cruz Island at elevations below 2000 ft. where it is a member of several plant communities from the mixed evergreen forest to chaparral. It is nowhere abundant and usually occurs in small groups or as isolated shrubs.

G. elliptica has long been popular in cultivation because of its attractive, long pendulous greenish-yellow staminate catkins which appear from December to February. The often great masses of silky fruits, which mature from June to September, may remain on the bushes until the following year. The species is variable, particularly in the length of the inflorescence, and a fine form with long catkins has been named 'James Roof' and it is in the horticultural trade.

Coast tasselbush thrives best in coastal areas in well-drained soils in sunny or semi-shady locations. Inland the plants should be given protection from the afternoon sun. Somewhat drought tolerant, the plants are better given some summer irrigation.

Regions: 1, 2, and selected areas in 3.

3. *G. flavescens* var. *pallida.* Pale tasselbush. An erect bushy shrub to 12 ft. of grayish appearance; leaves ovate to elliptic-oblong, stiff and leathery, 1½-3 in. long, ¾-1¼ in. wide, pistillate catkins compact, densely silky, to 2 in. long; berry about ¼ in. long, silky.

Pale tasselbush is widely distributed from central California south to northern Baja California where it occurs on dry slopes at elevations between 3000-8000 ft. in many plant communities.

Regions: 1, 2, 3, 6, and selected areas in 4, 7. Said to do well in the Central Valley if protected from the afternoon sun.

4. *G. fremontii.* Fremont tasselbush. An erect shrub to 10 ft., usually yellowish-green in appearance; leaves oblong-elliptical or oblong-ovate, ¾-2½ in. long, ½-1¼ in. wide, glabrous and shiny above, usually yellowish-green, paler and glabrous or sparingly pubescent beneath, margins plane and entire; staminate catkins 3-8 in. long, yellowish, 2-5 in a cluster; pistillate catkins 1½-2 in. long (to 3½ in fruit); berry about ¼ in. in diameter, buff to purple or black, subglabrous.

Fremont tasselbush is widely distributed from southern Washington to central California at elevations between 3000-8000 ft. where it is found on dry brushy slopes in the chaparral and mixed evergreen, yellow pine and red fir forests.

Because of its abundance of glossy foliage and long pendulous staminate catkins, *G. fremontii* is probably the most desirable of all the garryas. The plants tolerate drought, heat and cold better than those of *G. elliptica.* They should be given full sun in order to prevent legginess.

Regions: 3, 6, 7, and selected areas in 4.

5. *G. veatchii.* Veatch tasselbush. Plate 9b. An erect

Plate 11

shrub to 8 ft., young twigs densely tomentulose; leaves narrowly elliptical, ovate to lanceolate, tips more or less acuminate, 1-2½ in. long, ¾-1½ in. wide, dark green and glabrous above, felty-tomentose beneath, margins plane or slightly undulate, entire; catkins 2-4 in a cluster, rarely branched, staminate 2-4 in. long, pistillate 1½-3 in. long; berry ovoid to globose, about ¼ in. in diameter, densely pubescent.

Veatch tasselbush is found on dry slopes from northern Baja California north to San Luis Obispo Co. at elevations of 750-8550 ft. where it is found in the chaparral and foothill woodland communities.

G. veatchii, which has been in the nursery trade in southern California for some time, does well in coastal and intermediate valleys.

Regions: 1, 2, 3, 6.

Of the species of *Garrya* listed above *G. elliptica, G. flavescens* var. *pallida* and *G. fremontii* are all reported to do well in the Central Valley if they are given protection from the hot afternoon sun.

GUTIERREZIA. A genus of the Asteraceae with about 25 species of perennial herbs and subshrubs; four species in California.

1. *G. microcephala.* Sticky snakeweed. Intricately branched subshrubs to 2 ft. with thin, leafy stems forming semi or completely globose masses; leaves alternate, resinous, terete, to 2 in. long, glabrous; flower heads numerous, borne in much-branched clusters, individual heads small, ray-flower 1, yellow, disk-flower 1. July-October. In cultivation it may bloom over a much longer period of time.

Sticky snakeweed is found on dry mountain slopes and in the valleys along the fringes of the Colorado and Mojave deserts, east to Colorado and Texas and south into Mexico. It is native to the shadscale scrub, creosote bush scrub and Joshua tree woodland communities and is the most xerophytic of the California species.

←

Plate 11. a, *Mahonia amplectens*; b, *Mahonia aquifolium*; c, *Mahonia pinnata* var. *insularis*; d, e, *Mahonia higginsae*; f, *Mahonia pinnata* var. *pinnata*; g-i, *Mahonia nevinii*; j, k, *Mahonia* 'Golden Abundance'; l, *Mahonia haematocarpa*.

2. *G. sarothrae.* Broom snakeweed. Similar to the preceding except that the flower heads usually contain more than four ligulate flowers and more than three disk-flowers. May-October, or longer in cultivation.

In California broom snakeweed is found on desert plains and slopes from San Diego and San Bernardino cos. east to Kansas and north to Canada, usually at elevations below 10,000 ft.

Both species of *Gutierrezia* are easily grown in well-drained soil and they are drought tolerant, but green up quickly following the first autumn rains. Both species are attractive ornamentals, particularly when closely spaced in mass plantings. Highly recommended for desert landscaping. Particularly handsome when planted with *Eriogonum fasciculatum* ssp. *polifolium.*

Regions: 1, 2, 3, 4, 5, 6.

HAPLOPAPPUS. A genus of the Asteraceae with about 150 species of annual and perennial herbs or shrubs, all American, mainly western United States, Mexico and Chile; 38 species in California.

1. *H. arborescens.* Goldenfleece. An erect, fastigiately-branched shrub to 12 ft. with slender, leafy branchlets, main stem sometimes trunk-like; leaves alternate, narrowly linear or closely revolute and becoming filiform, to 2½ in. long, straight, glabrous, impressed-punctate, usually crowded; flower heads regularly cymose, terminating the branches, rayless, disk-flowers 20-25, yellow. July-October.

Goldenfleece is found on dry open slopes of the coast ranges from Ventura Co. north to Del Norte Co. at elevations below 4000 ft. where it is a member of the chaparral community. This species is often found in disturbed areas, especially on burns where it may be associated with *Adenostoma fasciculatum, Ceanothus cuneatus* and species of *Arctostaphylos.* In cultivation the plants often develop a short, thick trunk which then branches extensively to form a most attractive shrub.

Regions: 2, 3, 6.

2. *H. berberidis.* Barberryleaf goldenbush. An erect, bushy shrub to 6 ft., main stems often 1½ in. thick, twigs brittle, plants leafy throughout; leaves alternate, oblong to obovate-oblong, spinosely-serrate, to 2 in. long and ¾ in. wide, rigid, sparsely pubescent when young,

H

soon glabrate; flower heads racemose or solitary and terminal, often 3/4 in. broad, ray-flowers 15-30, yellow, disk-flowers 20-45, yellow changing to brown or saffron.

Barberryleaf goldenbush is found on stony soil and disturbed banks on the plains and low hills of northwestern Baja California (and offshore islands), where it is a member of the maritime desert scrub community.

This species has not been grown at this botanic garden, but because of its large flower heads and attractive barberry-like foliage we believe that it would be an attractive addition to the list of *Haplopappus* species now being grown.

Regions: (probably) 1, 2, 3.

3. *H. canus.* Hoary goldenbush. A loosely-branched shrub to 4 ft. with densely white-wooly herbage, twigs stout, soft-woody; leaves alternate, obovate to oblanceolate, sometimes spatulate, to 4½ in. long and 1¼ in. wide, densely white-tomentose, or sometimes tomentum deciduous on the upper surface, margins entire to sharply serrate; flower heads usually cymose or racemose, ray-flowers 4-6 or sometimes absent, disk-flowers 40 or more. June-September.

Hoary goldenbush is found on dry slopes and rocky bluffs on San Clemente and Guadalupe islands where it is a member of the coastal sage scrub community.

H. canus is an attractive shrub that has been in cultivation in southern California for many years.

Regions: 1, 2, 3. Also the Central Valley if protected from the afternoon sun.

4. *H. detonsus.* Similar to *H. canus,* differing from that species in having leaves that are thicker and more leathery and narrowly ovate to obovate with margins coarsely serrate with only a few teeth or subentire.

This species is known only from Anacapa, Santa Rosa and Santa Cruz islands where it grows in the coastal sage scrub community.

H. detonsus is easily propagated by cuttings.

Regions: 1, 2, 3.

5. *H. ericoides.* Heather goldenbush. A compact but spreading heather-like shrub to 4 ft. with several main stems and divergent branches, the tops of the bushes often flat and densely twiggy, twigs very leafy; leaves alternate, filiform, spreading and finally reflexing, to ½ in. long, straight or curved, grooved on the back, resinous-punctate, with secondary fascicled leaves in the axils; flower heads cymose-paniculate, ray-flowers 2-6, yellow, disk-flowers 8-12. August-November.

Heather goldenbush is found on sand dunes and sandy flats, always near the ocean, from Los Angeles Co. north to Marin Co. where it is a member of the coastal strand community. Also reported on San Miguel Island.

H. ericoides is a very attractive heather-like shrub suitable for sandy areas along the coast where it may be used for soil and dune stabilization.

The lower portions of the plant endure considerable submergence by sand and the wind-pruned upper portions provide protection on the leeward side for other species.

Regions: 1, 2, 3.

6. *H. parishii.* Parish goldenbush. Fig. 48a. An erect, branched shrub to 12 ft., often arborescent, the main stems trunk-like, stems erect, densely-leafy; leaves alternate, oblanceolate to linear-oblong, to 2¼ in. long and ¼ in. wide, thick, glabrous, resinous-punctate, rather crowded but without fascicled leaves in the axils; flower heads regularly cymose, terminating the branches, ray-flowers absent, disk-flowers 9-12, yellow. August-October.

Parish goldenbush is found on dry south slopes of the San Gabriel and San Bernardino mts., south into northern Baja California where it is usually found at elevations of 1500-7500 ft., a member of the chaparral community.

H. parishii is an attractive evergreen shrub well suited for soil stabilization and erosion control on well-drained slopes.

Regions: 2, 3, 6, and selected areas in 7.

7. *H. squarrosus* ssp. *grindelioides.* Sawtooth goldenbush. Fig. 48b. Low, freely-branching subshrubs of bushy habit; leaves alternate, obovate to oblanceolate, to 2 in. long and ¾ in. wide, sharply serrate, glutinous or glandular-dotted, stiffly chartaceous, clasping at the base, some with 1-4 small leaves in the axils; flower heads racemosely paniculate or in close terminal clusters, ray-flowers absent, disk-flowers yellow, 9-30. September-October.

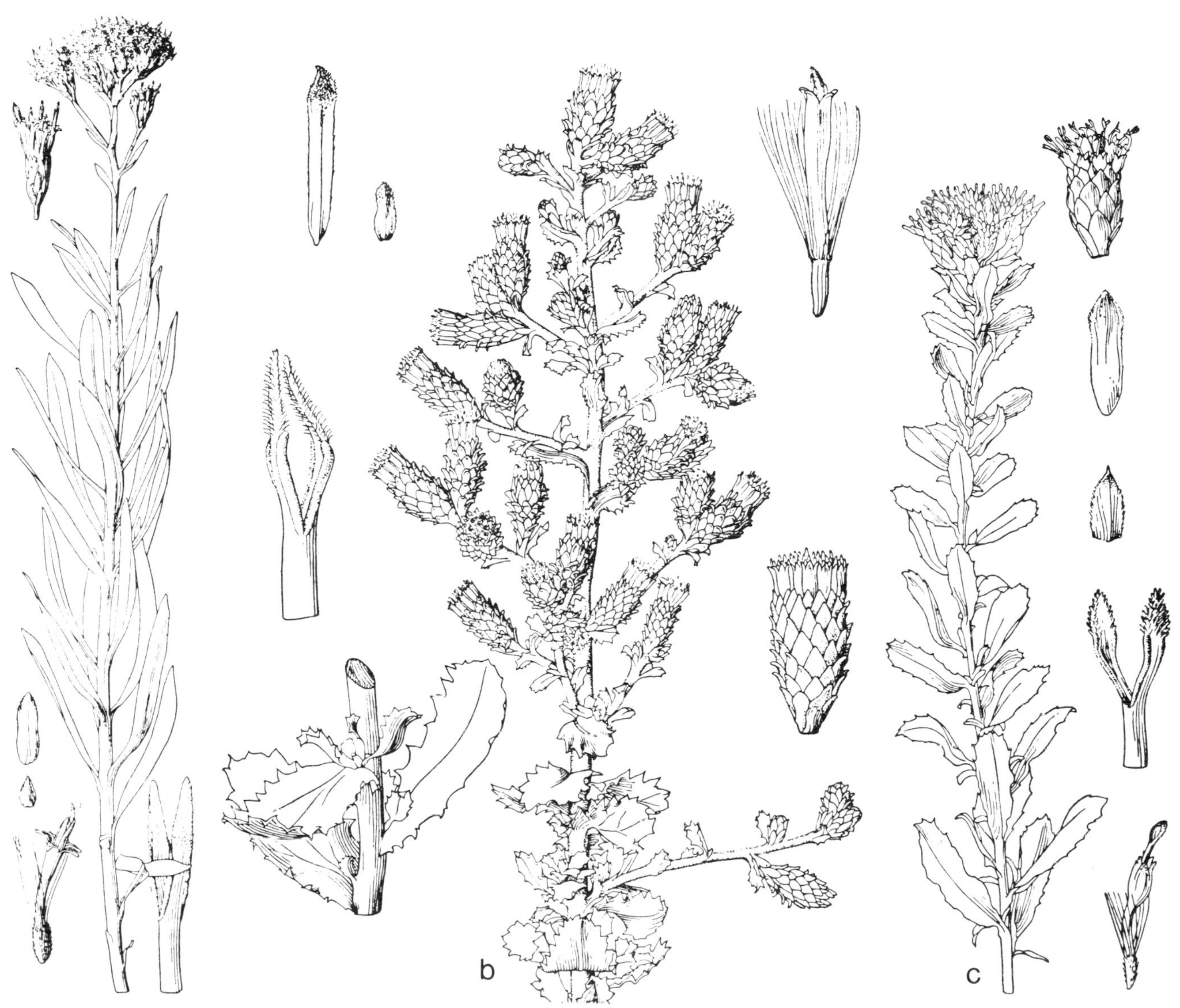

Fig. 48. a, *Haplopappus parishii*; b, *H. squarrosus*; c, *venetus* ssp. *vernonioides*. (Hall, 1928. Carnegie Inst. Wash. Publ. 389). All reduced.

Sawtooth goldenbush occurs on dry slopes in cismontane southern California from Santa Barbara Co. south to northern Baja California at elevations below 4500 ft. where it is a member of the coastal sage scrub and chaparral communities. Also on the Channel Islands.

Regions: 1, 2, 3, 6.

8. *H. venetus* ssp. *oxyphyllus.* Coast goldenbush. Erect, robust shrubs with numerous woody branches from the base, to 6 ft.; leaves alternate, oblanceolate to spatulate-oblong, rather thin, margins mostly entire; flower heads numerous, in rounded cymes, these openly paniculate, ray-flowers lacking, disk-flowers 15-25, yellow. June-November.

Coast goldenbush is found in warm valleys in San Diego Co. and northwestern Baja California where it extends inland to the base of the Sierra San Pedro Mártir Mts. often occurring in great abundance.

Regions: 1, 2, 3.

Ssp. *vernonioides.* Fig. 48c. Differs from the preceding mainly in having leaves that are dentate, lobed or incised. This species is common on dry slopes in coastal areas from central California south to San Diego Co., usually at elevations below 1200 ft. where it is a component of the coastal sage scrub community and coastal salt marsh communities. Also found on the Channel Islands.

The plants are often found growing in sand or disturbed areas sometimes forming thickets of considerable extent. Useful for sand and soil stabilization.

Regions: 1, 2, 3.

HETEROMELES. A genus of the Rosaceae with a single species native to California and northern Baja California.

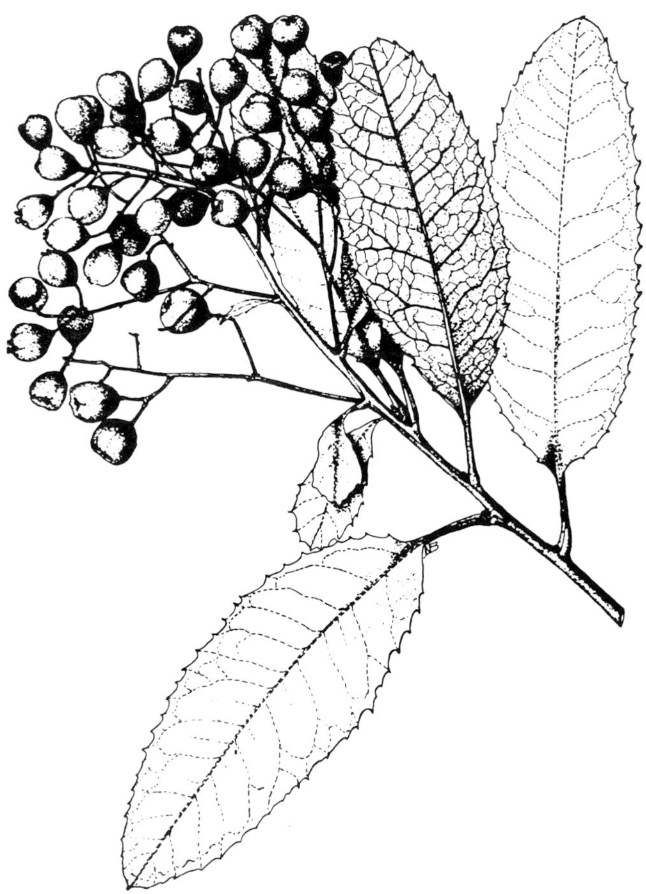

Fig. 49. *Heteromeles arbutifolia* var. *arbutifolia.* (Sudworth, 1908). Reduced.

1. *H. arbutifolia* var. *arbutifolia.* Toyon, California holly. Plate 9c; fig. 49. A large, erect, bushy, evergreen shrub or more rarely a small tree to 24 ft., bark gray and tomentulose on young branches; leaves elliptical to oblong or lance-oblong, coriaceous, plane, sharply toothed, to 4 in. long and 1½ in. wide, deep, dark green above, lighter beneath, glabrous; flowers numerous, small, white, borne in large terminal clusters on second-year wood; fruit a berry-like pome, about ¼ in. in diameter, bright red (or yellow), remaining on the plants for much of the winter. June-July.

Toyon is common on semi-dry brushy slopes and in deep soil of canyon bottoms from Humboldt and Shasta cos. south to northern Baja California, at elevations usually below 3500 ft. It is rare in Fresno and Tulare cos. *H. arbutifolia* var. *arbutifolia* is a common and important component of the chaparral community.

Toyon has long been in cultivation in California and is one of the state's most attractive shrubs as well as one of the most adaptable. Though drought tolerant it thrives with additional summer watering if the drainage is good. It may be grown in full sun or partial shade. The plants can be trimmed to form a single-trunked or multi-stemmed tree. It may also be pruned as a hedge and is valuable as a screen, for bank covers, highway planting and because of its deep penetrating root system it is highly useful for soil stabilization and erosion control. Toyon crownsprouts after fires or if the tops are removed. With careful pruning, fruiting material may be removed for holiday decorations and the plant's overall condition may be improved and the following season the fruits may be larger than from untrimmed plants. Toyon is free from insect pests and diseases except for fire-blight.

As a berry food, toyon is one of the most important for California birds and the single most important of winter foods. During late winter large flocks of waxwings and robins descend upon the garden and strip the plants within a few days.

This species consists of a number of ecotypes and for coastal areas care should be taken to use a type adapted to cool, cloudy weather.

Regions: 1, 2, 3, 6. Also does well in the Central Valley if protected from the hot afternoon sun. It has also been recommended for low desert areas.

Var. *cerina.* Differs from the preceding in having yellow fruits. A particularly large-fruited form growing at this garden has been called 'Claremont' and is being propagated vegetatively.

Var. *macrocarpa.* Differs from var. *arbutifolia* in having larger fruits which remain on the tree longer than those of var. *arbutifolia,* probably because birds prefer the smaller fruits of the latter.

Native to San Clemente and Santa Catalina islands.

HOLODISCUS. A genus of the Rosaceae, taxonomically confusing and difficult, consisting of two species native to western North America or eight species native to western North America and extending to South America.

1. *H. discolor* var. *discolor.* Creambush, oceanspray. Plate 9d. A much-branched, deciduous shrub to 20 ft., usually less, bark light brown to ashy-gray, often shreddy; leaves simple, alternate, ovate to ovate-elliptical, to 3 in. long and 2 in. wide, coarsely serrate or incised above the base, green and slightly pubescent on the upper surface, paler and soft-pubescent on the underside; flowers numerous, small, creamy-white, in large compound clusters at the ends of the drooping branches. May-July.

Creambush occurs on moist hillsides and in woods from Los Angeles Co. north to British Columbia. Also on Santa Catalina and Santa Cruz islands. Usually found below 4500 ft. elevation it is native to several plant communities, including the mixed evergreen forest and chaparral.

According to Jepson, *Holodiscus discolor* (*sensu lato*) "consists of an assemblage of forms, numerous and varied, without definite geographic segregation, and best represented as a reticulation."

In cultivation the plants are at their best in moist rich soil with partial shade, particularly in hot inland areas. In coastal areas the plants are somewhat drought tolerant. Creambush is a most attractive shrub and highly recommended. It should be pruned back after flowering. Because of its widespread distribution and the fact that it occupies such different habitats as the redwood forest and the chaparral it probably consists of a series of ecotypes.

Var. *franciscanus.* Differs only slightly in leaf characters from var. *discolor.*

Found on brushy slopes in the chaparral from Orange Co. north to British Columbia.

Regions: 1, 2, and partial shade in 3, 6.

HYPTIS. A very large genus of the Lamiaceae with perhaps 350 species of herbs and shrubs, mainly of Mexico and South America. One species in California.

1. *H. emoryi.* Desert-lavender, bee-sage. A fragrant, grayish shrub to 12 ft. and often broader with numerous straight, erect, slender branches, plant parts scurfy-tomentose; leaves simple, opposite, ovate to orbicular, truncate at the base, crenulate, to 1 in. long, and ½ in. wide, gray-green above, paler beneath; flowers in short-peduncled, axillary clusters at the ends of the branchlets, flowers small, violet to purple, bilaterally symmetrical. January-May.

Desert-lavender is found on rocky, gravelly slopes and in washes in the Colorado Desert and southern Mojave Desert, east to Arizona and south into Mexico. In California it is found in the creosote bush scrub at elevations below 3000 ft.

In a field note Jepson describes the plant, which is usually found growing singly or in small colonies, as a handsome soft gray shrub often found with *Acacia greggii, Agave deserti, Fouquieria splendens, Yucca mohavensis* (= *Y. schidigera*) and *Larrea tridentata.* When in bloom the bees work the plants constantly.

Under cultivation the plants require good drainage and, although drought tolerant, they respond very well to a little supplemental water during the summer. This species is very frost sensitive and was killed to the ground during the winter of 1977-1978 when the temperature at the botanic garden fell to 18 F, but the plants sprouted from the base and except for losing one season's flowers, the plants looked better the following year than they had for some time. In other years plants in containers in a lath house have been killed when the temperature dropped into the 20's. Pruning the plants back to ground level may have the same effect in improving the plant's overall appearance as does a freeze.

Hyptis emoryi is a very attractive gray-leaved shrub recommended for inland and desert areas where little frost is expected.

Regions: 1, 2, 3, 4 (warmer areas), 5.

I ISOMERIS. A monotypic genus of the Capparaceae native to California and Mexico with several poorly defined varieties. According to some the correct name is *Cleome*.

1. *I. arborea* var. *arborea.* Bladderpod. Plate 9e. Strongly-scented, glaucous, evergreen shrubs to 6 ft., widely branched; leaves alternate, palmately-compound with 3 leaflets, leaflets oblong to lanceolate, acute to 1½ in. long, about ¼ in. wide, greenish-yellow or grayish; flowers large, yellow, in terminal, bracteate racemes, petals 4, corolla distinctly 'bilabiate,' 2 petals erect forming an 'upper lip' and the 2 remaining petals form a 'lower lip' but are diverging; fruit an oval or broadly elliptical, inflated capsule 1-2 in. long. February-May but often throughout the year.

Bladderpod is rather common in sandy and often subalkaline valleys and on bluffs and hillsides from the western Mojave and Colorado deserts south into Mexico, west to San Diego Co., north to Ventura and western Fresno cos. Native to the coastal sage scrub, Joshua tree woodland and creosote bush scrub communities.

Var. *angustata* differs in having narrow capsules. Found in sandy washes below 4000 ft. in the Mojave and Colorado deserts west to the coast at San Diego.

Var. *globosa* (Plate 9f) has capsules that are subglobose and abruptly narrowed at both ends. With var. *arborea.*

Var. *insularis* has capsules about as thick as long, gradually narrowed at the ends. Santa Rosa, Santa Catalina and the Coronado islands.

Bladderpod presents no cultural problems, it prefers well-drained soils, and when established is drought tolerant but as with many other native shrubs, the general appearance of the plant is improved if it receives some additional water. Because the plants are found in such diverse situations as along the coast at San Diego as well as in the deserts, it is likely that the species consists of numerous ecotypes.

This species is recommended for hillsides, dry parks, roadsides, screening, etc., and not particularly for home planting.

Fig. 50. *Iva hayesiana.* The plants spread by the rooting of prostrate stems which come into contact with the soil.

Regions: 1, 2, 3, 4, 5, 6. Also does well in the Central Valley.

IVA. A genus of the Asteraceae with about 15 North American species of herbs, or low shrubs, three native to California.

1. *I. hayesiana.* Hayes iva. Plate 9g-i. Low-growing, aromatic, soft-woody evergreen shrubs to 2½ ft., freely branching from the base, branches often decumbent; leaves rather thick, spatulate-oblong to linear, to 2½ in. long and about ½ in. wide, entire, prominently 1-3-nerved on the lower surface, finely pubescent on both surfaces with short incurved hairs interspersed with short-stalked glandular hairs, giving the leaf a somewhat crystalline appearance; flowers inconspicuous, without rays, borne in small nodding heads racemosely dispersed in the axils of the upper leaves. May-July.

Hayes iva in California is restricted to southern San Diego Co. where it occurs in clayey soils that are moist during the winter and early spring months and dry the remainder of the year. It is more abundant in northwestern Baja California where it is found along margins of saline estuaries and in alkaline soil usually at elevations below 1000 ft. It is a member of the valley grassland, coastal sage scrub and maritime desert scrub communities.

Iva hayesiana is unknown to horticulturists and was first brought into cultivation at this garden in 1962. It has since been grown extensively in a wide variety of situations, in both heavy loam and alluvial sand and gravel and with varying amounts of water. It has grown well in all the situations in which it has been tried. Cuttings root easily and very rapidly and the young plants grow rapidly and strongly when placed in the ground. The plants spread by layering (Fig. 50). The oldest plantings at this garden are in alluvial sand and gravel and are watered two or three times during the summer. The cold tolerance of this species is unknown, it has withstood a temperature of 18 F without damage. It has not been damaged by rabbits even as young plants.

We believe that Hayes iva is the most promising native ground cover to have come along in recent years. It is highly recommended for use on banks and hillsides for erosion control. The woody portions of the plant are soft and, when cut, quickly send out new growth; it seems likely that plantings can be kept low by regularly trimming the plants.

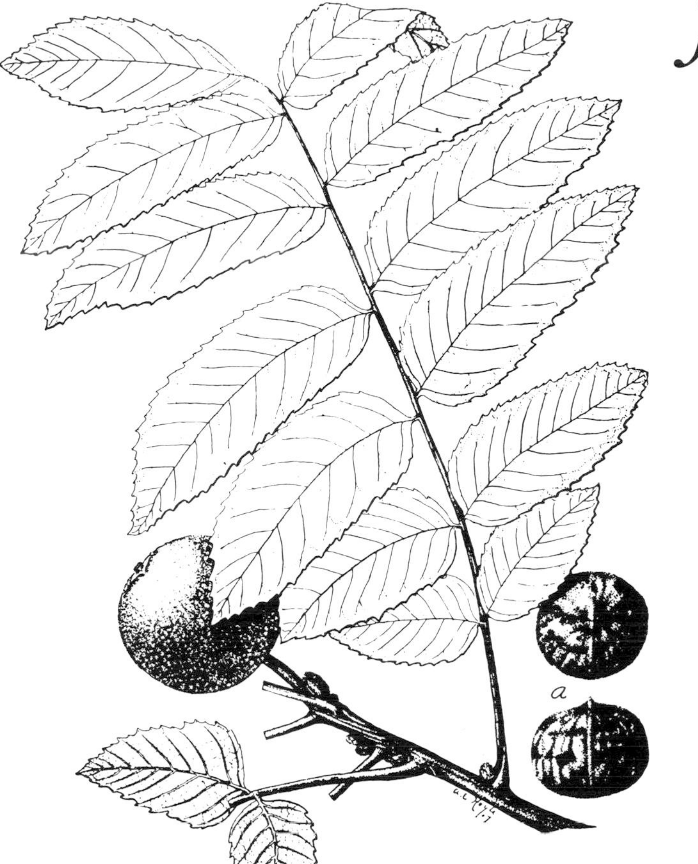

Fig. 51. *Juglans californica.* (Sudworth, 1908). Reduced.

To date it has not been possible to propagate this species by seed.

Regions: 1, 2, 3, 6. Should be tried in other areas.

JUGLANS. A genus of the Juglandaceae with about 15 species native to the North Temperate Zone and the northern Andes. Deciduous trees or shrubs with strong-scented bark, hollow branchlets divided into small chambers by pithy partitions; flowers appearing after the leaves, pistillate and staminate flowers borne separately but on the same plant; fruit a large indehiscent drupe. Two species in California.

1. *J. californica.* California walnut. Fig. 51. Trees, or often shrub-like in habit although imposing in size, bark rough, nearly black; leaves pinnately compound, 6-13 in. long, leaflets 11-19, oblong-lanceolate, serrate to 4

Fig. 52. a, *Juniperus osteosperma,* 40 years old, 18 feet tall with a 12 foot spread; b, *J. occidentalis* ssp. *australis*, 25 years old, 34 feet tall with a 15 foot spread; c, *J. californica*, 29 years old, 27 feet tall with a 20 foot spread.

in. long; staminate catkins pendant, to 4 in. long, borne on wood of the previous season, pistillate flowers borne in small clusters at the ends of the new growth, usually on the same branch with the staminate flowers; fruit a spherical, hard nut covered by a dry, brown, or black in age, husk which separates from the shell irregularly or in a partial manner, shell almost smooth. April-May.

California walnut is locally common from Ventura Co. south to San Bernardino Co. where it is a member of the southern oak woodland community usually below 4500 ft. elevation.

The walnut of central California has been described as *J. hindsii* but it is probably not specifically distinct. *J. californica* hybridizes with the English walnut, *J. regiae.*

California walnut is a hardy, drought tolerant species which thrives in poor soil and is reported to be resistant to oak root fungus. Of limited value for home use it is important in environmental landscaping. Some of the finest remaining stands in southern California are now being destroyed by developers.

Regions: 1, 2, 3, 6. Also does well in the Central Valley if protected from the afternoon sun.

JUNIPERUS. A genus of about 60 species of trees or shrubs native to the Northern Hemisphere with leaves opposite or in whorls of 3, needle-like and spreading or scale-like and appressed, the cones berry-like and fleshy, Many species are in cultivation and large numbers of cultivars are listed, especially of low-growing, ground cover types. Four species in California.

1. *J. californica.* California juniper. Fig. 52c. Usually a shrub, much-branched from the base, or sometimes tree-like; bark thin, brown or ashen-gray; leaves in 3's, scale-like, glandular-pitted on the back, about ¼ in. long; berries at first bluish with a dense whitish bloom, later reddish-brown beneath the bloom.

California juniper is rather widely distributed in California extending from the western edge of the Lower Colorado Valley and the Joshua Tree National Monument north along the western slopes of the southern Sierra Nevada and the inner coast ranges north to Tehama Co. It is found in the pinyon-juniper woodland, Joshua tree woodland and foothill woodland communities, mostly below 5000 ft.

A valuable shrub or small tree for use in difficult situations of poor but well drained soil and limited water availability. The plants withstand wind as well as air pollution and are not attacked by insect pests or disease. May be used as a windbreak or for screening. At Claremont a 26-year-old specimen measured 24 ft. in height with an 18 ft. spread.

Regions: 1, 2, 3, 4, 6. Also reported to do well in the Central Valley if protected from the afternoon sun.

2. *J. occidentalis* ssp. *australis.* Western or Sierra juniper. Fig. 52b. A small to medium-sized tree often with large spreading branches; bark deeply furrowed, reddish-brown; leaves in 3's, scale-like, bluish-green, glandular; berry bluish-black with a white bloom.

Western juniper occurs in the San Bernardino and San Gabriel mts. and northward through the Sierra Nevada to Lassen Co. where it is found on dry slopes and flats to 10,000 ft. in the montane coniferous forest.

Western juniper is the handsomest of the California junipers and at Claremont has done well in poor but well-drained soil with limited summer irrigation. Here it has not been attacked by insect pests or disease and withstands air pollution. A 20-year-old tree growing in the botanic garden measured 25 ft. in height with an 8 ft. spread.

Regions: 2, 3, 6, 7. Also does well in the Central Valley if protected from the afternoon sun.

3. *J. osteosperma.* Utah or desert juniper. Fig. 52a. A shrub or small tree of bushy habit, branches stiff; leaves in 2's or 3's, scale-like, usually glandless; berry when mature dark brown with a whitish bloom.

Utah juniper is found in the mountains of the eastern Mojave Desert, north to Idaho, Wyoming and east to New Mexico. In California it is a component of the pinyon-juniper woodland where it occurs on dry slopes and flats between 4800-8500 ft. elevation.

In habit, Utah juniper is intermediate between the two preceding species, larger and more tree-like in form than California juniper and smaller than western juniper. A 37-year-old specimen which had been brought to Claremont from the garden's original site in Orange Co. measured 25 ft. in height with an 18 ft. spread.

Regions: 1, 2, 3, 4, 6, 7.

K

JUSTICIA. In Munz and Keck (1959) and Munz (1974) this genus is treated as *Beloperone.* A genus of subtropical shrubs belonging to the Acanthaceae, one species of which is native to California.

1. *J. californica.* Chuperosa. Plate 9j. A low, intricately-branched, gray-green, canescent, soft-woody shrub to 5 ft. with an equal spread; leaves ovate, to ½ in. long, yellowish-green, soon deciduous, the plant remaining for most of the year a leafless shrub; flowers tubular, straight, bright red, or occasionally yellow, borne at the ends of the branches. Most abundant bloom March-May but some flowers nearly all year.

Chuperosa is common in sandy places along water courses or on alluvial fans from the west and north edges of the Colorado Desert south to Baja California, east to Arizona and south to Sonora, Mexico, usually at elevations below 2500 ft.

J. californica is an unusual looking and most attractive shrub which in cultivation produces some flowers nearly all year. It is frost sensitive and may be frozen to the ground, but established plants soon recover. Very drought tolerant, it is highly recommended for warmer desert areas where it may be used in a variety of ways. There is much flower color variation in this species and deeply colored forms should be selected and propagated asexually. The yellow form is rare in cultivation.

The California species is related to the common shrimp plant, *J. brandegeana,* frequently seen in California homes and gardens.

Regions: 1, 2, 3, 5. Also does well in the Central Valley.

KECKIELLA. Formerly included in the genus *Penstemon, Keckiella* consists of five shrubby species of Scrophulariaceae native to California and adjacent Baja California.

1. *K. antirrhinoides* ssp. *antirrhinoides.* Yellow penstemon, snapdragon penstemon. Plate 10a. A soft-woody, bushy shrub to 8 ft. with spreading stems and branches, young stems finely puberulent; leaves ovate to broadly elliptical, linear or oblanceolate, ½-¾ in. long, ⅛-¼ in. wide, glabrous or glandular-puberulent, entire to slightly denticulate; flowers yellow, ½-¾ in. long, in large leafy clusters, corolla gaping, upper lip erect, lower lip deeply parted, lobes turned downward. April-June.

Yellow penstemon occurs on dry, often rocky slopes, usually at elevations below 4000 ft. in Los Angeles, San Bernardino, Riverside, Orange and San Diego cos. south into Baja California. It is a member of the chaparral community.

K. antirrhinoides is a most attractive yellow-flowered shrub highly recommended for horticultural use. It should be given a well-drained soil with little summer water since the plants become dormant during the hot summer months. They should be heavily pruned in order to improve their overall shape. Easily propagated by seed or cuttings, the plants tend to be short-lived.

Regions: 1, 2, 3, 6.

2. *K. cordifolia.* Climbing penstemon. Plate 10c. A bushy shrub clambering over other shrubs with long straggling stems to 12 ft. long; leaves fuchsia-like, ¾-2¼ in. long, ½-1¼ in. broad, ovate to nearly round, truncate to heart-shaped at the base, glabrous, margins serrate to nearly entire; flowers red to scarlet, strongly two-lipped, 1¼-1½ in. long, in long arching, leafy panicles. May-July.

Climbing penstemon occurs in canyons and on hillsides in the coastal mountains from San Luis Obispo Co. south to northern Baja California and eastward as far as the San Jacinto and San Bernardino mts. at elevations from 500-2000 ft. Found primarily in the chaparral it also occurs in the coastal sage scrub community. This species is found on the Channel Islands.

K. cordifolia does well in cultivation in many types of soil but does best in a well-drained soil with some summer irrigation and it prefers a cool root run. Propagation is by seed or cuttings.

Regions: 1, 2, 3, 6.

K. antirrhinoides and *K. cordifolia* have hybridized in gardens and the hybrids are fertile, and segregation occurs in advanced generations. Many of the plants seen at the Rancho Santa Ana Botanic Garden are of hybrid origin, some of them being most attractive and worthy of asexual propagation (Plate 10b).

3. *K. ternata* ssp. *ternata.* Whorlleaf penstemon. Plate 10d. A straggling shrub to 6 ft. with long, slender,

straight flowering stems from a woody base; leaves in whorls of 3, lanceolate or linear-lanceolate, or ovate, ¾-1½ in. long, ¼-⅜ in. wide, glabrous on both sides, serrulate; flowers scarlet to terra-cotta color, about 1 in. long and in an elongated terminal cluster. June-July.

Whorlleaf penstemon is found on dry slopes and in canyons from the San Gabriel and San Bernardino mts. south to Baja California, usually at elevations below 6000 ft. where it is a member of the chaparral community.

K. ternata is an attractive, small, soft-wooded shrub notable for its many straight stems from the base each terminating in an elongated cluster of brightly colored flowers. Drought tolerant and free from insect pests and disease. Easily propagated by seeds or cuttings, the plants should be pruned back to the base to encourage new growth.

Regions: 2, 3, and selected areas in 6.

LARREA. A genus of the Zygophyllaceae with five species, four native to arid and semi-arid regions of Argentina, Chile, Bolivia, Peru and Mexico and one species in southwestern United States and northwestern Mexico.

1. *L. tridentata.* Creosote bush. Plate 9k, l. An erect, diffusely branched, strong-scented, evergreen shrub to 12 ft. and often equally broad, branches brittle, grayish, densely leafy toward the tips, young growth with dark, glutinous or corky rings at the nodes; leaves opposite, consisting of a pair of sessile, crescent-shaped leaflets, leathery, dark olive green, about ⅜ in. long; flowers yellow, solitary and terminal on short, lateral branchlets, petals 5, often more or less twisted; fruit a small globose capsule about ¼ in. long covered with dense white or rusty-colored hair.

Creosote bush is the dominant shrub (of the creosote bush scrub community) over vast areas of the desert in southwestern United States and northwestern Mexico, occupying well-drained, sandy flats, bajadas and upland slopes usually below 5000 ft.

Due to the monotony of vast stretches of the desert, often uniformly and evenly covered with creosote bush, one often fails to appreciate the attractiveness of the individual plants. As a horticultural subject, creosote bush is a handsome, dark olive green-leaved shrub, usually well branched from the base and rather open and airy in appearance, well adapted to withstand harsh and difficult situations in the desert. Good as a filler or as a windbreak.

L. tridentata does best in a coarse well-drained soil, and its appearance is improved if it receives some additional water as evidenced by the more attractive plants seen along the edges of highways in the desert where the plants receive road runoff. Although growing in well-drained alluvial sand and gravel the plants at this botanic garden were severely damaged during the past two years of high winter rainfall.

Creosote bush stems and leaves are covered with a sticky resinous secretion, and in the presence of high humidity the leaves may be attacked by a sooty-mold fungus which causes the plants to defoliate.

L. tridentata is recommended for use as specimen plants, as a windbreak, for screening and as a hedge in desert areas. The plant's form may be improved by moderate pruning.

Recent work in the Mojave Desert (Vasek, 1980) has shown that some of the creosote bushes there are of great age and have spread to form clones, usually circular in outline with a hollow center. The largest one, known as King Clone, described as similar in structure to a hollow tree but stem crown segmentation obscures a direct comparison, is estimated to be 11,700 years old, or about twice the age of the oldest known bristlecone pine.

Regions: 3, 4, 5. Also does well in the Central Valley and in Arizona.

LAVATERA. A genus of the Malvaceae with about 20 species mostly native to the Mediterranean region. One species on the California offshore islands.

1. *L. assurgentiflora.*★ California tree mallow. Plate 10e. An erect, somewhat brittle, bushy, evergreen shrub or sometimes tree-like to 15 ft.; leaves alternate, maple-like, palmately 5-7-lobed, 3-5 in. long and about as wide, lobes coarsely toothed or lobed, grayish-green above, glabrous, silvery-green beneath, not shiny; flowers reddish-lavender, white- and purplish-striped, about 2 in. in diameter, petals 5, reflexed. April-August, but some blooms nearly all year.

California tree mallow is known only from the Santa Barbara islands and on Santa Catalina Island where it occurs on sandy flats and rocky places in the coastal sage scrub community.

L. assurgentiflora has long been in cultivation in California particularly in coastal areas where it resists drought, wind and salt spray. The plants grow rapidly and may bloom the first season from seed. It is especially useful in coastal areas as a fast-growing hedge resistant to both wind and salt spray. It should be sheared to make it dense. Attractive to rodents.

Regions: 1, 2, 3.

LEPTODACTYLON. A genus of the Polemoniaceae with 13 species of perennial herbs and shrubs native to western North America. Three species in California.

1. *L. californicum* ssp. *californicum.* Prickly phlox. An erect, widely branched shrub to 3½ ft., stems tomentose; clusters of prickly leaves in the axils of the main leaves, main leaves usually alternate, falling the second season, palmately-divided into 5-9 linear divisions about ⅛-½ in. long; flowers in terminal, congested clusters, corolla salverform, bright rose to rose-purple, or almost white; fruit a dry capsule.

Prickly phlox occurs on dry slopes and banks from San Luis Obispo Co. south to the San Gabriel and Santa Ana mts. at elevations below 500 ft. where it is a member of the chaparral and foothill woodland communities.

L. californicum is one of California's most beautiful small shrubs and worthy of a place in any native planting scheme where its requirements can be met. The plants require excellent drainage and once established they should receive no water during the summer. Even under ideal circumstances the plants may be short-lived.

Regions: 1, 2, 3, 6.

LITHOCARPUS. A genus of about 100 species of evergreen trees or shrubs belonging to the Fagaceae native to southeast Asia with one species in western North America.

1. *Lithocarpus densiflora.* Tanbark oak. An evergreen tree as much as 150 ft. tall with thick furrowed bark, ascending or spreading branches and tomentose young branchlets; leaves simple, alternate, thick and leathery, oblong or oblong-ovate, 2-5 in. long, ¾-2½ in. wide, strongly straight-veined, glabrous or with scattered pubescence on the upper surface, and white to rusty tomentose on the underside, becoming glabrous in age, edges of leaves dentate with thick teeth; flowers inconspicuous, staminate, numerous in ill-smelling catkins, 2½ in. long, pistillate flowers solitary in the axils of the involucral bracts; fruit an ovoid nut ⅔-1 in. long surrounded at the base by an involucral cup covered with numerous scales, nut maturing the second year.

Tanbark oak is found in the coast ranges from Ventura Co. north to southern Oregon where it occurs on wooded slopes in the redwood and mixed evergreen forests at elevations below 4500 ft.

This species is remarkably shade tolerant and may be found as low bushes or small trees in dense redwood or Douglas fir forests. Under garden conditions the tanbark oak often carries branches to the ground forming exceptionally handsome columnar or broadly conical, densely foliaged trees. Adaptable to various soil types it does best in good soil with some summer irrigation. In cultivation the growth rate is moderate. In sandy or gravelly soil with but little summer water the plants tend to be more shrub-like. Although native to areas farther north, this species has done well at Claremont and has been free from disease and the only insect pests attacking the plants have been aphids. Propagation is by seed which should be sown soon after collecting. Plants may show some variation in leaf characters.

Regions: 1, 2, 3.

LONICERA. A genus of the Caprifoliaceae with about 100 species of mainly deciduous shrubs or twining vines native to the Northern Hemisphere. Many ornamental species in cultivation.

1. *L. involucrata* var. *ledebourii.* Twinberry. Plate 10f. Densely foliaged deciduous shrub to 6 ft., often spreading to 12 ft., branches slender and wand-like; leaves opposite, ovate, 2-5 in. long, 1-2½ in. wide, dark green above, somewhat lighter beneath and slightly hairy; flowers borne in pairs on slender peduncles 1-2 in. long, longer in fruit, corolla tubular ½-¾ in. long, rich orange-yellow to reddish tinged, subtended by two or more somewhat glandular-hairy bright red ovate bracts ½-1½ in. across; fruit a berry about ⅜ in. in diameter, lustrous black, juicy. March-July.

Ledebour's twinberry occurs in moist ravines and on slopes along the coast from Santa Barbara Co. north to Oregon.

L. involucrata var. *ledebourii* is an attractive shrub both in flower and fruit, easily propagated by seed or cuttings, and the plants are not particular as to soil but should have adequate moisture, and in inland areas they are best if protected from the afternoon sun. Flowers and fruit are produced on young shoots, so pruning should consist of the removal of old branches to ground level. Free from insect pests and disease.

Regions: 1, 2, 3, 6.

LUPINUS. A genus of the Fabaceae with over 100 species of herbs or, more rarely, shrubs native to all continents except Australia but most well developed in North America. Over 50 species in California, mostly herbs.

1. *L. albifrons* var. *albifrons.* Silver lupine. Plate 10g. Rounded, very leafy shrubs usually not more than 5 ft. tall but often much broader with distinctly woody trunks; herbage silky-pubescent, leaves pinnately compound with 7-11 leaflets, leaflets oblanceolate to 1¼ in. long, silvery-silky on both surfaces; flowers borne in narrow racemes to 12 in. long, flowers in whorls, pea-shaped, fragrant, blue, purple, white or pink; fruit a dry, flat pod. March-July.

Silver lupine is found on dry hillsides and in canyons from Ventura Co. north to Humboldt Co. and in the Sierra Nevada from Shasta Co. south to Tulare Co. It is found in many plant communities.

Var. *eminens.* Differs from the preceding in having slightly larger flowers. Native to San Diego Co. and northern Baja California where it is a member of the coastal sage scrub community.

Silver lupine is a most attractive shrub which does well here in heavy soil with summer irrigation about every 2-3 weeks. The plants are subject to attacks by loopers which may be controlled by regular treatment with an insecticide. In our experience *L. albifrons* is the most resistant to air pollution of the lupines but during very smoggy periods the leaves may show some damage; however the plants recover quickly with the return to normal conditions.

Regions: 1, 2, 3, 6.

LYCIUM. A genus of the Solanaceae with about 100 species of spiny shrubs native to arid regions of all continents.

L

1. *L. brevipes* var. *brevipes.* Desert thorn. A leafy, erect, spreading or somewhat clambering much-branched spiny shrub to 8 ft., sometimes forming dense thickets, branches often flexuous, spines thick, sharp; leaves elliptic, obovate to spatulate, about ¾ in. long and ⅜ in. wide, thick and fleshy, glabrate to densely pubescent; flowers campanulate to tubular, about ⅜ in. long, pink to violet, often with black stripes in the throat; fruit a bright red berry, ovoid, many-seeded. March-April.

Desert thorn occurs in sandy washes and on hillsides in the western Colorado Desert, south throughout the length of Baja California, in Sonora and on San Clemente Island, usually at low elevations.

For some reason *L. brevipes* has never been grown at this garden and our attention was first directed to the species when we saw plants growing at the botanic garden of the University of California, Riverside. Covered with bright red berries they made a most attractive display. There the plants, growing in heavy soil, receive no special care. From the size of the plants it is apparent that some of them are quite old. Useful for screening and windbreaks, as specimens or for barrier plantings.

Regions: (probably) 1, 2, 3, 5. Does well in the Central Valley.

2. *L. cooperi.* Cooper desert thorn. Plate 10h, i. Densely leafy, intricately branched, spiny shrubs to 7 ft. and often as broad, spines short, stout; leaves 3-10 in a fascicle, oblanceolate, oblong or spatulate, to 1¼ in. long and ⅜ in. wide, glabrate to densely glandular-pubescent; flowers numerous, 1-3 in fascicles, pendant, greenish-white with lavender veins; fruit ovoid, laterally constricted above the middle. March-April.

Cooper desert thorn is found on dry mesas and slopes in the Mojave and Colorado deserts, north to the upper San Joaquin Valley, east to Utah and Arizona, usually at elevations below 5000 ft. where it is a member of the creosote bush scrub and pinyon-juniper woodland communities.

L. cooperi requires good drainage and once established needs no special care. When in flower the plants are most attractive. Because of their leafy habit the plants can be used for screening and windbreaks and they are admirably suited for barrier plantings.

Regions: 3 (marginal), 4, 5.

LYONOTHAMNUS. A genus of the Rosaceae with a single species native to the southern California offshore islands.

Slender evergreen trees to 60 ft., or sometimes shrubby, bark red-brown to grayish, exfoliating in long thin strips; leaves opposite, polymorphic; flowers small, white, ill-smelling, borne in large flat-topped clusters to 12 in. across, petals 5, stamens numerous; fruit a pair of small woody follicles, seeds minute.

1. *L. floribundus* ssp. *floribundus.* Catalina ironwood. Plate 10l. Leaves variable even on a single plant, oleander-like, 3-5 in. long, ½-¾ in. wide, or variously lobed or pinnately divided, dark glossy green above, more or less pubescent beneath. June.

Catalina ironwood occurs on dry slopes and in canyons on Catalina Island where it is a member of the island chaparral community.

Ssp. *asplenifolius.* Plate 10j, k. Leaves broadly ovate in outline, pinnately divided into 2-7 divisions, these in turn broadly lobed forming a leaf like an asplenium fern. June.

The subspecies *asplenifolius* is found on San Clemente, Santa Rosa and Santa Cruz islands.

Catalina ironwood has long been in cultivation and is one of California's most unusual and well-known trees. Growth is rapid, especially if additional water is provided, but the plants must have excellent drainage. Inland they do better if given some protection from hot drying winds. This species is particularly attractive when planted in groves. In coastal areas the species is very drought tolerant. Propagation is by seed and the percentage of germination at times is very low.

Regions: 1, 2, 3. Does well in the Central Valley if protected from the afternoon sun.

MAHONIA. A genus of the Berberidaceae with 104 species native to eastern Asia and North and South America. Thirteen species in California.

A word of explanation should be given regarding the use of the name *Mahonia* rather than of *Berberis.* In *A California Flora* (1959), Munz and Keck list as the latest taxonomic treatment of the group, the work of Abrams (1934) who placed all the Pacific Coast barberries in the genus *Mahonia.* Munz and Keck, however, did not follow Abrams but rather, with no explanation, placed the California species in *Berberis.* In *A Flora of Southern California* (1974), Munz cited a more recent monograph of the barberries of the world (Ahrendt, 1961) in which two genera are recognized, *Berberis* and *Mahonia.* According to Ahrendt, the leaves in *Mahonia* are always pinnate whereas those of *Berberis* are simple, and there are no spines on the stems of *Mahonia* whereas they are nearly always present in *Berberis.* On the basis of these (and other) distinctions he placed all the Pacific Coast species in *Mahonia.* Although citing Ahrendt's monograph, Munz, again with no explanation, continued to use the name *Berberis.* Here we follow Ahrendt and use the name *Mahonia.*

Evergreen shrubs with yellow wood and inner bark; leaves imparipinnate (paired leaflets + a terminal leaflet), alternate, leaflets 3 to many, prickly, glabrous, sessile; rachis jointed at the insertion of the leaflets; petioles somewhat clasping; flowers perfect, yellow, borne in racemes which are often densely crowded; sepals 6, in two series, petal-like, deciduous; petals 6 in two series, with a pair of glands at the base; fruit a few-seeded berry sometimes becoming dry.

1. *M. amplectens.* Plate 11a. A much-branched shrub to 5 ft.; leaves to 6 in. long, leaflets 5-7, suborbicular, to 3½ in. long and often nearly as broad, stiff-coriaceous, dull green to bluish-green, glaucous, paler and dull beneath, each leaflet with 8-20 spinose teeth on each margin; flowers in fascicled racemes; berries bluish-purple with a white bloom. April-May.

M. amplectens occurs on rocky slopes in the chaparral and yellow pine forest communities at elevations of 3000-5000 ft. in the Santa Rosa Mts. of Riverside Co., south to the Laguna Mts. in San Diego Co. In its better forms *M. amplectens* is a most distinctive species with beautiful bluish-green, glaucous leaves which are very leathery in texture. Adaptable to many types of soil and drought tolerant. It should be given full sun.

Regions: 1, 2, 3, 4, 6. Also reported to do well in the Central Valley if given protection from the afternoon sun.

2. *M. aquifolium.* Oregon grape, hollyleaf mahonia. Plate 11b. An erect, freely branching shrub to 6 ft.; leaves 4-10 in. long, leaflets 5-9, oblong-ovate, 1-2½ in. long, ¾-1½ in. wide, thin, glossy, dark green above and beneath, plane to somewhat undulate with 10-20 slender spines on each margin; berries bluish with a bloom. March-May.

Oregon grape is found from southern British Columbia and Idaho south to Humboldt, Trinity and Modoc cos. In California it is found on wooded slopes in the Douglas fir and red fir forest communities at elevations below 7000 ft.

The Oregon grape has been widely grown for many years in gardens in the United States as well as in Europe and is the species gardeners most often think of when the name *Mahonia* is mentioned. It is also the parent of numerous hybrids.

This species is adaptable to many soil types but in southern California it should be provided with some shade as well as an occasional irrigation during the summer. Because of its thin leaves it is often attacked by leaf-eating caterpillars, often called loopers (*Coryphista meadii*), which may quickly disfigure the plants. This species spreads by underground stems and the size and shape of the plants may be controlled by judicious pruning and the removal of unwanted suckers. Long bare stems should be cut back nearly to the ground, after which the plants will quickly fill in with new growth. This species is reported to be resistant to oak root fungus and is recommended for gardens where the parasite is prevalent.

Due to the natural variation inherent in this species as well as by chance unrecognized hybridization, plants grown from seed may show great dissimilarity in overall size, leaf form and color as well as in flower and fruit production. For that reason, if uniformity is desired, plantings should be by cuttings taken from a single plant displaying the characteristics desired.

Regions: 1, 2, 3, 6, and selected areas in 7.

3. *M. haematocarpa.* Plate 11l. Shrub to 6 ft. with stiff, thick, spreading branches bearing numerous short spurbranches; leaves stiff-rigid, crowded and strongly crisped, to 2 in. long, 1½ in. wide, leaflets mostly 5 with about 3 sharp spines on each margin, gray-green on the upper surface, lighter beneath; inflorescences few-flowered; berries reddish-purple with a bloom. May-June.

M. haematocarpa occurs from the eastern Mojave Desert east to western Texas and northern Mexico. In California it is found in the pinyon-juniper woodland in dry, rocky places at elevations of 4500-5000 ft.

This species is very drought tolerant and not particular as to soil but it should be given full sun. Probably because of the small size of the leaves, as well as the fact that they are very coriaceous and provided with sharp spines, this species is not attacked by loopers.

Regions: 2, 3, 4, 6, selected areas in 5; also in the Central Valley if protected from the afternoon sun.

4. *M. higginsae.*★ Higgins barberry. Plate 11d, e. Similar to the preceding species differing in the proportion of the terminal leaflets (*M. haematocarpa* being 2-5 times as long as wide, *M. higginsae* not more than twice as long as wide) as well as in fruit color, *M. haematocarpa* having plum-colored berries, whereas those in *M. higginsae* are yellowish-red. May-June.

M. higginsae occurs in southeastern San Diego Co. and adjacent northern Baja California where it is found on dry, rocky slopes in the chaparral and pinyon-juniper forest communities at 2500-4000 ft. elevation.

Tolerant of adverse conditions, it is a most attractive species and is highly recommended. It should be given full sun.

Regions: 1, 2, 3, 4, 6, and selected areas in 5. Also reported to do well in the Central Valley if protected from the afternoon sun.

5. *M. nevinii.*★ Nevin barberry. Plate 11g-i. Large, rounded, much-branched shrub to 12 ft.; leaves to 3 in. long, leaflets 3-5, lanceolate to lance-ovate, somewhat coriaceous, 1-1½ in. long, ¼-½ in. wide, blue-green on the upper surface, paler beneath, dull, mostly plane, with 5-16 bristle-like teeth on each margin; racemes loosely flowered; berries juicy, yellow-red to red.

Nevin barberry occurs in sandy and gravelly places at elevations of less than 2000 ft. in the coastal sage scrub and chaparral communities, mostly in Los Angeles Co.

but occasionally in Riverside Co. Extinct in the San Fernando Valley, this species is on the endangered list. Nevin barberry is an exceptionally handsome species both in flower and in fruit and is highly recommended for use in southern California. It adapts well to most soils, even adobe, and while it is drought tolerant it will also accept copious amounts of water without apparent damage. Free from insect pests and disease.

Regions: 1, 2, 3, 4, 6, and selected areas in 5. Does well in the Central Valley.

6. *M. pinnata* ssp. *pinnata.* Shinyleaf or California barberry. Plate 11c, f. Shrub to 10 ft. with long, stiff, upright branches; leaves to 10 in. long, leaflets 5-11, ovate, ovate-elliptical or oblong, 1-3½ in. long, ¾-1¾ in. wide, thick, glossy-green above, paler but glossy beneath, margins plane to undulate and sinuate, with 10-20 bristle-tipped teeth on each margin; racemes many-flowered; berries blue-glaucous. February-May.

M. pinnata is found in rocky, exposed places in the Douglas fir, redwood and mixed evergreen forests of the northern coastal ranges where in morphological characters, some specimens approach *M. aquifolium.* The species is also found in the chaparral from Santa Barbara Co. south to northern Baja California at elevations below 4000 ft. The latter may represent an undescribed species. A form found in the closed-cone pine forest on Santa Cruz and Santa Rosa islands has been described as ssp. *insularis.*★

Because of its thin leaves this species may be attacked by loopers. For its best development the plants should be given full sun. *Mahonia pinnata* has done particularly well at Claremont and is highly recommended.

Regions: 1, 2, 3, 6. Also the Central Valley if protected from the afternoon sun.

7. *M. piperiana.* Piper barberry. An erect shrub to 12 ft. tall; leaves 4-8 in. long, leaflets 5-9, ovate, 1-2½ in. long, upper surface glossy-green and finely reticulate, lower surface gray-green, and papillate, usually with 7-10 spinescent teeth on each margin; racemes loosely to densely fascicled; berries blue-black. March-April.

Piper barberry occurs on dry, open, or wooded slopes in the mixed evergreen, Douglas fir and yellow pine forests of the coast ranges from Del Norte and Siskiyou cos. south to Lake Co. and less commonly in the chaparral and yellow pine forest of the San Gabriel Mts. south to northern Baja California. This species is usually found at elevations of from 3000-5000 ft.

Regions: 1, 2, 3, 6.

It is likely that this and the preceding species both contain ecotypes adapted to such widely divergent plant communities as the mesic redwood, mixed evergreen and Douglas fir forests on the one hand and to the dry and hot chaparral on the other. For southern California, propagating material should be sought in the southern populations.

8. *M.* 'Golden Abundance' (Pl. Patent 3233). Plate 11j, k. A densely and intricately branched shrub to 8 ft.; leaves to 10 in. long, 5 in. wide, leaflets 7-9, orbicular to ovate, with 6-10 slender spines on each margin, nearly plane, dark, bright green above, lighter on the lower surface, rachis of leaf dark, shiny red; flowers in large, densely compacted clusters; fruit deep purple with a white bloom.

M. 'Golden Abundance' which at Seattle, Washington has withstood a minimum temperature of 10 F, and five days with maximum temperatures of less than 29 F with damage only to the flower buds, is of hybrid origin, and with its very heavy flower and fruit set it is the most spectacular of the mahonias. It is highly recommended for many areas in southern California and from the results reported from Seattle it should be tried in other areas as well.

The handsome evergreen foliage, masses of yellow flowers followed by attractively colored berries, much relished by birds, makes the species of *Mahonia* highly desirable garden subjects. The plants can be propagated by seed, suckers, cuttings or by layering. Because of their stiff and very spinescent leaves some of the species can be used to advantage as barrier plantings.

MALOSMA. A monotypic genus of the Anacardiaceae native to southwestern California and northwestern Baja California.

1. *M. laurina* (= *Rhus laurina*). Laurel sumac. Plate 12a, b. A leafy, aromatic, evergreen shrub usually forming a dense rounded mound to 20 ft.; leaves alternate, ovate or lanceolate, entire, rounded at the base, 2-4 in.

long, ¾-1½ in. wide, leathery dark green and somewhat glossy on the upper surface, lighter on the under surface, reddish when young, midrib conspicuous, the leaf often folded upward, trough-like; flowers in dense intricately branched panicles, small, white; fruit a small whitish, oily drupe attractive to birds. May-July.

Laurel sumac occurs in the valleys and on dry slopes from Santa Barbara Co. south to northern Baja California and inland to the edge of the desert. It is a member of the coastal sage scrub and chaparral communities where it is found at elevations of less than 3000 ft.

M. laurina is not particular as to soil provided that drainage is good, and the plants are free from both diseases and insect pests. The ability of this species to remain green and attractive throughout the year makes it a valuable ornamental. The plants may be used as clipped hedges, or espaliered, and because of its extensive root system it is useful in erosion control. Laurel sumac however is very sensitive to frost and it should not be planted in areas where temperatures regularly drop to 25 F. Indeed it is one of the very few natives that, when growing in its native habitat, is often damaged by frost.

Regions: 1, 2, 3. Also does well in the Central Valley if protected from the hot afternoon sun.

MYRICA. One of two genera of the Myricaceae containing about 30 species of mainly tropical, deciduous or evergreen, shrubs or small trees. One species in California.

1. *M. californica.* Wax myrtle, California wax myrtle. Plate 7g. Densely branched, evergreen shrub or small tree to 25 ft. with smooth gray or light brown bark; leaves thick, dark green above, slightly paler beneath, glossy, oblong or oblanceolate-oblong, to 5 in. long, ½-¾ in. broad, serrate to nearly entire, glabrous; flowers inconspicuous, monoecious, borne in catkins in the axils of the upper leaves, staminate catkins to 1 in. long, pistillate catkins ¼-½ in. long, in the leaf axils terminal to the staminate catkins; fruit globose, 1/6-1/4 in. in diameter, brownish-purple, covered with a thin covering of wax. March-April.

Wax myrtle is found on sand dunes, moist hillsides and canyons along the coast from near Santa Monica north to Washington. In California it occurs in the coastal sage scrub and chaparral communities.

M. californica has long been grown in cultivation and according to the *New Western Garden Book,* in gardens it is one of the best-looking of the native plants, its great virtue being the clean-looking foliage which is attractive throughout the year. In coastal areas it is drought tolerant, in hot inland areas it requires some summer irrigation and is probably best if given protection from the afternoon sun. The plants can be grown as specimens either as bushes or trained as trees, or they may be used for hedges, either unclipped or clipped.

Regions: 1, 2, 3.

NOLINA. A genus of the Agavaceae with about 25 species native to southwestern North America. Long-lived perennials with a yucca-like aspect; trunks thick, woody, in one species forming an underground platform; leaves very numerous, narrow with dilated bases; flowers very numerous, in congested panicles, small, whitish, pistillate and staminate flowers borne on different plants; fruit a 3-winged capsule.

1. *N. bigelovii.* Bigelow nolina. Stems stout much-branched, to 3 ft. tall; leaves flat, to 4 ft. long and ¾ in. wide above the dilated base, margins entire or scarcely serrulate, shredding into brown fibers; flower stalk to 9 ft., panicle dense, many-flowered; capsule about ¼ in. in diameter. May-June.

Bigelow nolina is found on dry slopes on the western and northern edges of the Colorado Desert, east to Arizona and south to Baja California usually at elevations below 3000 ft. where it is a member of the creosote bush scrub community.

2. *N. wolfii.* Wolf nolina. Plate 12c. Stems very stout, to 15 ft. tall usually unbranched; leaves flat, stiff, conspicuously hispid-serrulate on the margins, to 5 ft. long, 1½ in. wide above the base; inflorescence massive on a stalk to 12 ft. tall, base bare for about 3-6 ft.; capsule about ½ in. in diameter. May-June.

Wolf nolina is found on dry slopes in the Kingston, Eagle, Little San Bernardino and southeastern San Jacinto mts. and on the Kern Plateau growing at elevations of 3500-6000 ft., a member of the pinyon-juniper woodland and Joshua tree woodland communities.

The nolinas are highly desirable plants for landscape use in desert areas, very drought tolerant and insect pest

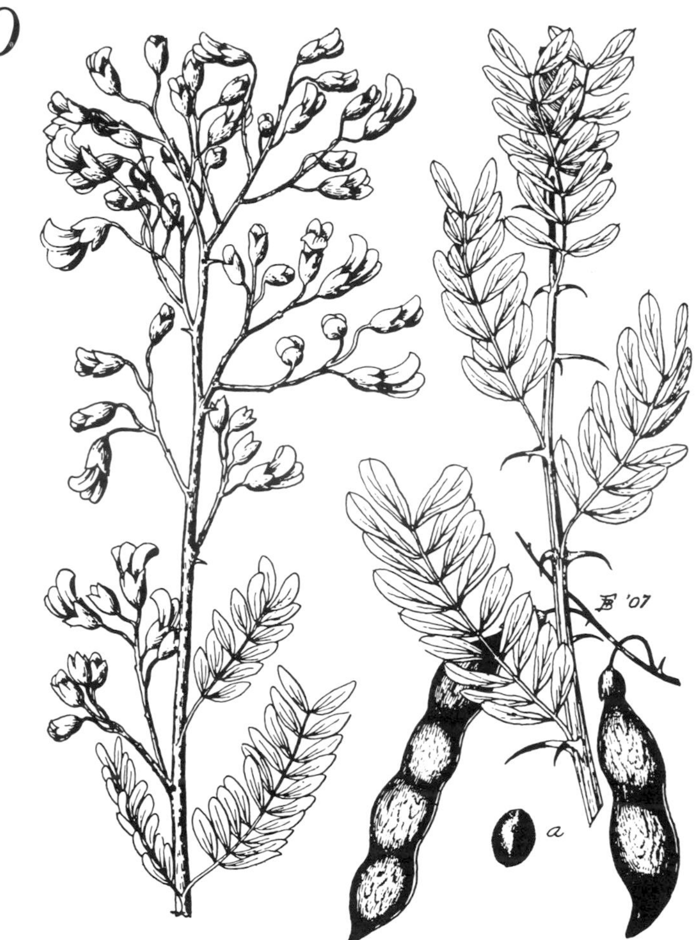

Fig. 53. *Olynea tesota.* (Sudworth, 1908). Reduced.

and disease free. They should be given good drainage. Unlike agaves and yuccas, where the rosette usually dies following blooming, nolinas continue to grow and flower once they reach blooming size. Wolf nolina is a spectacular plant when in flower and because of its size care should be given to its placement. Ideal for desert parks, rest areas, etc. *N. wolfii* is slow to reach maturity, but because of its fountain-like habit it is attractive without flowers.

Regions: (both species) 1, 2, 3, 4, 5, 6. Also do well in the Central Valley.

OLNEYA. A genus of the Fabaceae with a single species native to Southwestern United States, Sonora and Baja California.

1. *O. tesota.* Desert ironwood. Fig. 53. Broad-crowned evergreen tree to 30 ft. with a single trunk or branched from the base, bark thin, flaky, deep red-brown, branchlets spiny, entire plant gray-green; leaves pinnately-compound, to 2 in. long, leaflets 7-15, white-pubescent, leaves falling when the new leaves appear; flowers pea-shaped, pale rose-purple, borne in small, showy clusters; fruit a dry russet brown pod densely coated with gland-tipped hairs. April-May.

Desert ironwood occurs in desert washes in the Colorado Desert, east to Arizona and south into Sonora and Baja California at elevations usually below 2000 ft. where it is a member of the creosote bush scrub.

O. tesota is a handsome, slow-growing tree which has been described as, from a distance, resembling an olive tree. In warm areas it can be used as a specimen tree, for shade, as a large hedge or for barrier plantings.

The plants require good drainage and the rate of growth can be hastened by regular watering.

Region: 5. Also southern Arizona and northern Sonora.

ORNITHOSTAPHYLOS. A genus of the Ericaceae containing a single species native to northern Baja California and southwesternmost San Diego Co. Allied to *Arctostaphylos.*

1. *O. oppositifolia.* Palo blanco. Plate 12d-f. Erect evergreen shrub to 12 ft., much-branched from the base, bark reddish-brown, exfoliating in thin sheets to expose the satiny-smooth white or greenish-white stems; leaves usually in whorls of 3, linear, to 2¾ in. long, ¼ in. wide, thick, shiny green above, pale green and finely puberulent beneath; flowers in partially drooping panicles, urn-shaped, greenish-white; fruit a dry drupe with thin pulp, very attractive to birds. February-April.

Palo blanco is known in California from a single collection made near San Ysidro, San Diego Co., in 1973. The species is common in northwestern Baja California at elevations below 4000 ft. where it is a member of the chaparral community.

O. oppositifolia was first brought into cultivation in 1963 from seed collected by the senior author south of Ensenada. In 1980 the original plantation is still extant. The few plants which have lost their top growth have resprouted from the basal burl. Palo blanco is remarkably free from insect pests and disease. The plants prefer a well-drained soil and are drought tolerant, but the plant's appearance is improved if some additional water is given during the summer. At this botanic garden they have done well in heavy soil but care has been taken in watering.

There is some plant to plant variation in habit, and selected forms may be propagated by cuttings. Also easily grown from seed. Attractive throughout the year, easily propagated and very adaptable, palo blanco, which is lighter and more open in appearance than *Arctostaphylos,* is highly recommended.

Regions: 1, 2, 3, 6. Also does well in the Central Valley if protected from the afternoon sun.

OSMARONIA. A monotypic genus of the Rosaceae native to the coast ranges of the three western states.

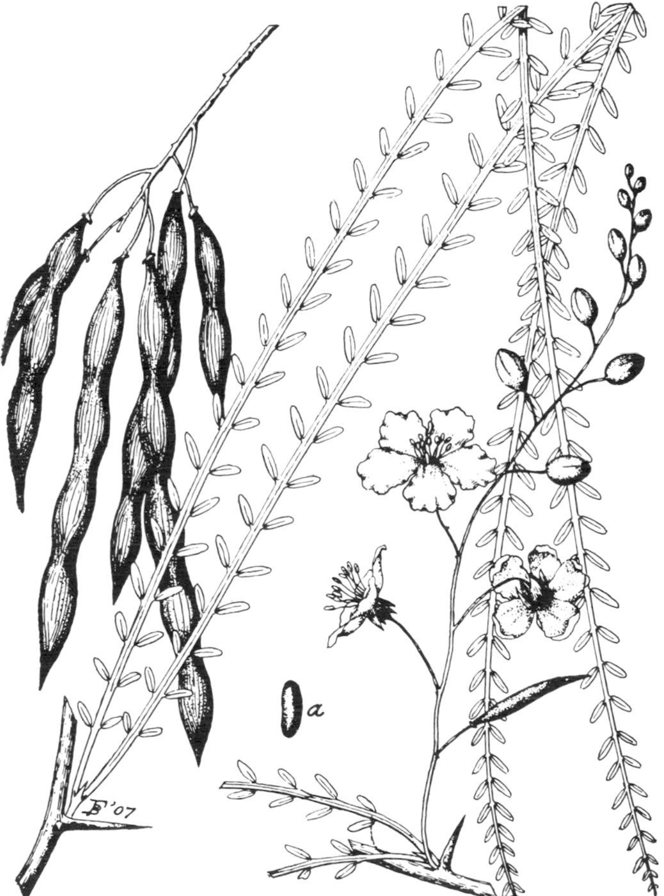

Fig. 54. *Parkinsonia aculeata.* (Sudworth, 1908). Reduced.

1. *O. cerasiformis.* Osoberry. Plate 12g, h. An erect, deciduous shrub to 12 ft. with mostly straight slender stems and smooth gray or reddish bark; leaves simple, alternate, scattered along the branches, oblong-ovate, narrowed at each end, 1½-3½ in. long, ¾-1 in. wide, light yellowish-green, paler beneath, entire; flowers dioecious, small, white, borne in terminal clusters; fruit consisting of 1-5 ovoid blue-black drupes about ⅜ in. long. January-April.

Osoberry is found on moist to fairly dry slopes in the coast ranges from Santa Barbara Co. north to British Columbia and in the Sierra Nevada from Tulare to Shasta cos. at elevations below 5600 ft. It is found in several plant communities including the redwood, mixed evergreen and yellow pine forests and in the chaparral.

In cultivation it is free from diseases and insect pests and easily propagated by seeds, which must be pretreated to hasten germination, or by cuttings. It does best in moist habitats, but when established it is quite drought tolerant. In inland areas it should have afternoon shade to prevent early defoliation.

As an ornamental, osoberry is of interest as an early-leafing and blooming shrub with light green leaves and later for its interesting fruit, much relished by birds. Since the plants may produce suckers some distance from the original plant this species should be useful in erosion control.

Regions: 1, 2, 3, 6, and selected areas in 7.

PARKINSONIA. A genus of the Fabaceae with two species, one widespread in southwest and southern United States.

1. *P. aculeata.* Mexican palo verde. Fig. 54. Shrubs or trees to 30 ft., often equally broad, branches green, spines present in pairs at each node and a spine terminating the primary leaf-rachis; leaves sparse, bipinnate, 6-9 in. long, primary leaflets in 1-3 pairs, crowded, secondary leaflets many, early deciduous; flowers in long racemes, yellow with red spotting; seedpod bulging, to 4 in. long, strongly constricted between seeds.

Mexican palo verde is native to a wide area of southern United States to the West Indies and northern South America. It is widely cultivated and now naturalized in southern California. Mexican palo verde is highly recommended for planting in hot areas in California where it tolerates alkaline soil and drought. It should be provided with good drainage and it does not do well with lawn watering. The plants, which have a long blooming season in the spring and bloom intermittently throughout the year, should be staked and carefully pruned when young. As with other palo verdes, *P. aculeata* casts light shade.

Regions: 1, 2, 3, 5, and warm areas in 4, 6. Also does well in southern Arizona.

PHILADELPHUS. A genus of the Saxifragaceae with perhaps 40 species widely distributed throughout the Northern Hemisphere. Taxonomically a difficult group and the California representatives have been afforded different treatments by systematists. One native species of horticultural importance.

1. *P. lewisii.* Wild mock orange. Plate 12i. A loosely branched, deciduous shrub to 10 ft. with reddish, glabrous bark on young growth and brown exfoliating bark on old stems; leaves opposite, ovate to lanceolate-ovate, 1¼-3¼ in. long, ¾-1½ in. wide, entire, minutely denticulate or sometimes toothed, mostly glabrous above, variously pubescent beneath; flowers white, fragrant, ¾-1¾ in. across, usually numerous in terminal clusters, petals 4-5, stamens numerous; fruit a dry, many-seeded capsule. May-July.

Wild mock orange is widely distributed in western United States. In California a form often referred to as *P. lewisii* ssp. *californicus* occurs on rocky slopes and in canyons in the Sierra Nevada from Tulare Co. north to Humboldt, Siskiyou and Trinity cos. at elevations of 1000-4500 ft., where it is found in the yellow pine forest and foothill woodland communities. Another form often referred to as *P. lewisii* ssp. *gordonianus* occurs in the coast ranges from Lake Co. north to Del Norte and Siskiyou cos. at elevations below 4000 ft. in the mixed evergreen and yellow pine forest communities.

P. lewisii is extremely variable in nature and the two forms referred to, which differ only slightly, do not cover the total diversity to be found in the species. Horticulturally the differences are not significant.

Wild mock orange is a most attractive and dependable shrub, easily propagated by seed, root-sections or by cuttings. It is somewhat drought tolerant, but in hot inland areas it is better if protected from the hot afternoon sun and given some summer irrigation.

Regions: 1, 2, 3.

PHYSOCARPUS. A genus of the Rosaceae with about 10 species of deciduous shrubs native to North America and Asia. Two species in California.

1. *P. capitatus.* Western ninebark. Erect or spreading shrubs to 8 ft., bark thin, shredding; leaves simple, alternate, round-ovate, to 3 in. long and 1½ in. wide, 3-5-lobed and irregularly double-serrate, glabrous above and slightly to densely stellate pubescent beneath; flowers borne in dense, round clusters, petals about 1/6 in. long, white to pinkish. April-June.

Western ninebark is found along streams and in moist places from Santa Barbara Co. north to British Columbia, east to Montana and Utah. In the Sierra Nevada it is found from Tulare Co. north to Humboldt, Siskiyou and Trinity cos. It grows in many plant communities usually at elevations below 4500 ft.

P. capitatus is related to, and somewhat similar in appearance to *Spiraea.* An attractive deciduous shrub which tolerates ordinary garden care, but inland it should be given light shade or at least protection from the afternoon sun.

Regions: 1, 2, 3 (with shade).

PICKERINGIA. A genus of the Fabaceae with a single species restricted to California and adjacent Baja California.

1. *P. montana.* Chaparral-pea. Plate 12j. A very spiny, densely branched evergreen shrub to 6 ft., young stems short pubescent and ending in a stout spine; leaves alternate, small, 1-3-palmately divided, palmate nature of the leaves sometimes obscure, leaflets ¼-½ in. long, deep green on the upper surface, lighter beneath, pubescent on both surfaces; flowers large, pea-shaped, purple or rose-purple, ¾ in. long, solitary in the leaf-axils near the ends of the branches; fruit a flat, straight, linear, several-seeded pod to 2 in. long. May-June.

Chaparral-pea is native on dry, rocky, mountain slopes in the coast ranges from Mendocino Co. south to the Santa Monica and San Bernardino mts., and to north-

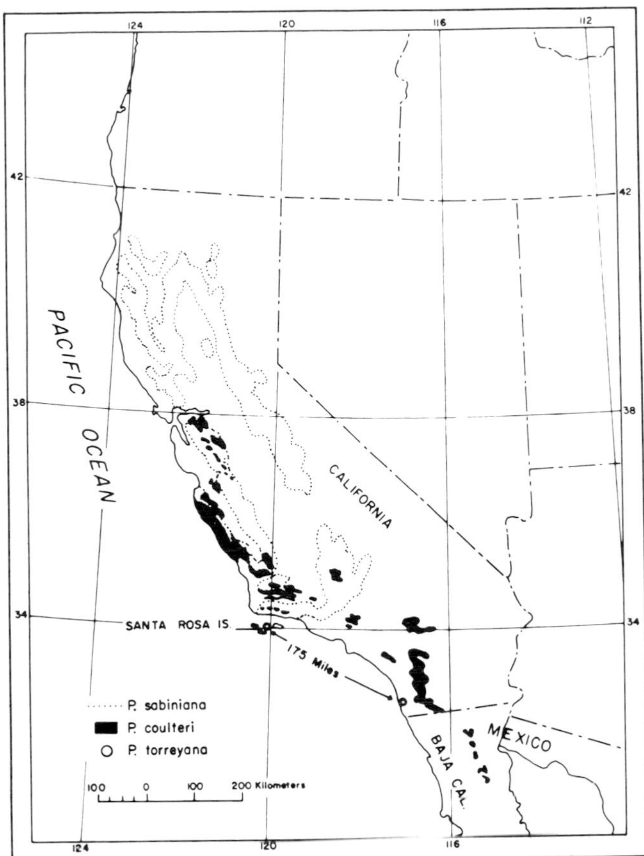

Fig. 55. The geographical distribution of *Pinus coulteri, P. sabiniana* and *P. torreyana.* (Mirov, 1967).

ern Baja California; also on Santa Cruz Island. It is rare in the Sierra Nevada where it is found from Butte Co. south to Mariposa Co. *P. montana* is found in the chaparral at elevations below 5000 ft.

According to Jepson, chaparral-pea is the most xerophytic of the shrubs found in the chaparral; it often flowers, but fruits very sparingly and mature pods are a rarity. The plants sprout from the root-crown after fires but not from the trunk-stem. Leafy shoots appear in abundance from roots wherever they are exposed, such as along road or trail cuts, or after landslides.

P. montana is a most attractive shrub when in flower and well worth the effort required in getting it into cultivation. Its rarity is perhaps due to the fact that the species seldom sets seed and stem cuttings have not been successful. At this garden, root cuttings are being tried. A deep red-flowered form has been reported.

Regions: 1, 2, 3, 6. Does well in the Central Valley.

PINUS. The pines form a large and diverse group of trees, or occasionally shrubs, with about 94 species native to the Northern Hemisphere; 55 are native to North America and there are 19 species in California, three of which are endemic to the state, with another eight that are basically Californian. Because many of the species are unsuited to our environmental conditions and others are susceptible to air pollution, only five species are recommended for our area. They include three species of the 'bigcone pines': *P. coulteri, P. sabiniana, P. torreyana*; and two pinyon pines, *P. monophylla* and *P. quadrifolia.*

The three species of bigcone pines have been placed in group *Macrocarpae* (Shaw, 1914) where they form a natural assemblage of low to medium elevation pines native to California and adjacent Baja California (Fig. 55). In addition to having a number of morphological characters in common they also are alike chemically in their turpentines, and all three contain alkaloids, a circumstance that is very unusual in pines (Mirov, 1967). At this botanic garden the three species have done well whether growing in heavy loam or in alluvial silt, sand and gravel, and while they are very drought tolerant, they respond well to additional watering so long as drainage is good. At Claremont trees growing in poor soil but with good drainage have been irrigated about three times during the summer months.

Digger pine, *P. sabiniana*; and to a lesser extent, Coulter pine, *P. coulteri*; are occasionally subject to a needle blight which may be controlled by spraying with a fungicide in the spring when the new candles emerge and again when the candles are full.

1. *P. coulteri.* Coulter or bigcone pine. Fig. 56a. A medium-sized tree usually with a single trunk, conical or more often spreading with long lower branches, bark dark brown with irregular netwood of ridges and fissures; needles in fascicles of 3, 5-12 in. long, stiff, dark bluish-green, with sharp points; cones very large, broadly ovoid, 9-12 in. long, 5-7½ in. broad, yellowish-brown, seeds nearly black, ½-⅔ in. long, wing about 1 in. long.

Coulter pine is restricted to California and is most abundant in the mountains of southern California, but it does extend north in the coast ranges to Contra Costa Co. and south into Baja California. Throughout its range it occupies dry, rocky slopes mostly at elevations of 4000-7000 ft. in the chaparral, foothill woodland and yellow pine forest communities. It has the most massive cone of all the pines.

Coulter pine forms a rather symmetrical, open tree, the lower branches persisting and often spreading widely. The growth rate is moderate to fast and the trees, which are not particular as to soil, are hardy and resistant to heat, drought and wind as well as to air pollution. One 28-year-old tree measured 68 ft. in height with a 36 ft. spread and a dbh of nearly 24 inches.

Regions: 1, 2, 3, 4, 6. Also does well in the Central Valley if protected from afternoon sun.

2. *P. sabiniana.* Digger pine. Fig. 56c. A medium-sized to large tree, trunk often branching into several to many secondary trunks which then form a broad, open crown; bark dark gray, roughly and irregularly fissured; needles in fascicles of 3, 6-12 in. long, gray-green, slender and usually drooping; cones broadly oblong-ovoid, 6-10 in. long, 5-7 in. broad, light chocolate-brown, seeds about ¾ in. long with very hard shell and a short wing.

Digger pine is a California species growing on the dry foothills and lower mountain slopes bordering the Central Valley, usually at elevations below 4500 ft. where it is a member of the foothill woodland community. It reaches southern California only in northern Ventura and Los Angeles cos. In many places its most common associate is the blue oak, *Quercus douglasii.* Digger pine is one of the most drought tolerant of California pines and withstands high summer temperatures as well as air pollution.

Digger pine is rather unusual in its very open crown and relatively sparse foliage, which Jepson described as "a thin gray cloud" which casts a very light shade. This species has not enjoyed the popularity it deserves, probably because of its gray-green foliage which some think looks dusty. One gardening manual has described the digger pine as very ornamental, another describes it as one of the most picturesque of all pines. At this garden one 28-year-old tree was 68 ft. tall with a 36 ft. spread and a dbh of nearly 24 inches.

Regions: 1, 2, 3, 4, 5, 6. Also does well in the Central Valley.

3. *P. torreyana.*★ Torrey pine. Fig. 56b. In its native habitat, a small to medium-sized tree with stout, spreading or ascending branches, bark irregularly fissured into broad flat ridges covered with thin compact red-brown scales; leaves in fascicles of 5, 7-11 in. long, stiff, dark green; cones broadly ovoid 4-6 in. long, about the same in diameter, seeds ¼-1 in. long with thick hard shell, wing ¼-½ in. long.

Torrey pine is one of the rarest of pines being known only from two small areas separated by about 175 miles; on the mainland it is found near Del Mar in San Diego Co. where it occurs on low, steep, eroded coastal bluffs and marine terraces on either side of Soledad Valley for a distance of about eight miles; it also occurs on Santa Rosa Island where it occupies a coastal strip about ½ mile long and ¼ mile wide. Nowhere is it found far from the ocean. It has been estimated that on the mainland there are about 6,000 trees and about 1,500 on Santa Rosa Island. Most of the mainland population is included within Torrey Pines State Reserve. Botanists have recognized that the two populations differ and some have felt that the island population should be accorded taxonomic recognition. Under cultivation the two populations are quite distinct; the island form tends to be shorter and to produce stout spreading and ascending branches rather than forming a single trunk as does the mainland form. Also the foliage of the island form is more dense and the leaf color dark green and glaucous, whereas that of the mainland form is gray-green in color. In our experience the foliage of the island form is longer than it is on the mainland form.

Professor Haller of the University of California, Santa Barbara is in the process (1980) of describing the insular form of the Torrey pine as a distinct subspecies. It will then be known as *Pinus torreyana* ssp. *insularis*★ and the mainland form as *Pinus torreyana* ssp. *torreyana.*★

Under cultivation Torrey pine is a fast-growing tree and at the botanic garden a specimen of the mainland form, 28 years old, measured 81 ft. in height with a 40 ft. spread and a dbh of 22 in. A specimen of the island form, 21 years old, measured 75 ft. in height with a 63 ft. spread and a dbh of almost 30 in. Although in nature Torrey pine is never found far from the ocean it has done very well in Claremont enduring both heat and drought; however, it does respond well to additional watering if the drainage is good. This species has never shown signs

Fig. 56. a, *Pinus coulteri,* 29 years old, 48 feet tall with a spread of 39 feet; b, *P. torreyana* (mainland form), 28 years old, 72 feet tall with a spread of 36 feet; c, *P. sabiniana,* 28 years old, 65 feet tall with a spread of 36 feet.

P of damage caused by air pollution. The island form has occasionally suffered broken branches from strong winds. Reported to be immune or highly resistant to oak root fungus.

Regions: 1, 2, 3, 6. Also does well in the Central Valley.

4. *P. monophylla.* Oneleaf pinyon. Fig. 58a, 59a. Most commonly a small round-headed tree with a short trunk, under cultivation usually with branches to the ground; needles borne singly, curving and sharp pointed, 1½-2 in. long, persisting 7-8 years; cones subglobose, chocolate-brown to yellowish, 2½-3½ in. in diameter, seeds dark brown, oblong, ¾ in. long, without wings.

Oneleaf pinyon is native to dry, rocky slopes at elevations of 3500-9000 ft. on the eastern slopes of the Sierra Nevada from Mono Co. south, and in the mountains in, and bordering, the Mojave Desert; at a few localities on the western slope of the Sierra Nevada and extending to Utah, Arizona and Baja California. It is found in the foothill woodland community and is an important component of the pinyon-juniper woodland community. According to Jepson, the oneleaf pinyon grows in the most arid situations of any California pine and because of its solitary leaves and what he called its "low apple-tree habit" it scarcely looks like a pine.

At this garden a 25-year-old specimen growing in poor soil and receiving only two irrigations during the summer months measured 18 ft. in height with a 16 ft. spread.

Regions: 1, 2, 3, 6 and portions of 4. Also does well in the Central Valley if protected from the afternoon sun.

5. *P. quadrifolia.* Parry or fourleaf pinyon. Fig. 58b, 59b. A small, much-branched, short-trunked tree of pyramidal form; in age with a wide crown and often twisted and gnarled branches; leaves mostly borne in fascicles of 4, ¾-1½ in. long, stiff somewhat incurved, bluish-green on the back, whitish on the inner surface; cones subglobose, 1¼-2 in. long, seeds about ⅝ in. long, thin-shelled, with a very narrow wing.

Parry pine is a rather rare pine confined to a strip some 220 miles long, the largest portion of which is in northern Baja California. The main Parry pinyon populations in California are in Riverside Co. where they are found in the Santa Rosa and San Jacinto mts. on the western edge of the Lower Colorado Valley. It is found on dry slopes at elevations of 3500-5500 ft. in the pinyon-juniper woodland community.

Parry pinyon is an extremely handsome, small, heavy-foliaged pine densely branched to the ground and of a very pleasing pyramidal shape, which in our experience it retains at least until it is about 40 years old. It has done well in both heavy loam and in alluvial sand and gravel. It is drought tolerant and has not been attacked by insect pests, disease or air pollution. Perhaps its greatest drawback, at least in certain situations, is its slow growth. At Claremont a 17-year-old specimen growing in poor soil and receiving only two irrigations during the summer measured 24 ft. in height with a 16 ft. spread.

Regions: 1, 2, 3, 6 and selected areas in 4. Also does well in the Central Valley if protected from the afternoon sun.

Hybrid Pines

Since 1944 the Rancho Santa Ana Botanic Garden has grown a number of hybrid pines produced at the Institute of Forest Genetics, Placerville. Of the hybrids, *P. ×attenuradiata* (*P. attenuata × radiata*) has been the most successful. This hybrid has been produced over a period of years at Placerville using different genetic stocks of the two parental species and the resulting hybrids have been variable in their response to the conditions at this botanic garden. All of them have been smog resistant and most fast growing. One tree 18 years old was 55 feet tall with a spread of 38 feet and a trunk diameter of about 18.5 inches.

One of the more recent introductions is shown in Plate 12k. This tree shows promise of being an exceptional ornamental and is being propagated asexually. The plant has a pleasing form, the branches are flexible and the needles are soft and of a lively shade of green. It would appear that it would be of use as a specimen or as a windbreak.

P. ×attenuradiata does well in the Central Valley if protected from the afternoon sun.

→

Plate 12. a, b, *Malosma laurina*; c, *Nolina wolfii*; d-f, *Ornithostaphylos oppositifolia*; g, h, *Osmaronia cerasiformis*; i, *Philadelphus lewisii*; j, *Pickeringia montana*; k, *Pinus ×attenuradiata*; l, *Pluchea sericea.*

Plate 12

P *PLATANUS.* A genus of the Platanaceae with about nine species of deciduous trees widely distributed in the Northern Hemisphere. One species in California.

1. *P. racemosa.* Sycamore or Western sycamore. Fig. 60. Trees often to 90 ft. with large wide-spreading branches, trunks often leaning and twisted or bent, bark smooth, exfoliating in reddish-brown plates exposing greenish or whitish areas; leaves to 10 in. wide, yellowish-green, parted into 3-5 broad spreading lobes, tomentose on both surfaces when young; flowers small, borne in 2-7 ball-like clusters distributed at intervals along a slender, pendulous stalk near the ends of the branches; fruiting balls about 1 in. in diameter, falling apart during the winter, releasing the small seed-like nutlets.

Sycamore is very common along streams and water courses from the Sierra San Pedro Mártir Mts. in Baja California north to central California usually at elevations below 4000 ft.

For many Californians the sycamore is one of the most picturesque and beloved of the native trees and it does have a place in today's landscaping. It can be used in large informal gardens, and with careful pruning the the plant may be made most attractive. It is probably at its best when used in parks or natural areas where its summer shade is much appreciated. The plants which require adequate moisture for good growth, tolerate both high temperatures and wind. Perhaps the greatest drawback to the species is the susceptibility of the plants to leaf blight or anthracnose which affects the leaves soon after they appear, causing them to look as though they had been frozen. The infected leaves soon fall and the second or even third flush of growth may also be attacked, leaving the trees nearly leafless well into late spring. The disease, caused by a fungus (*Gloeosporium platani*) can at times be controlled by spraying, but timing is critical and in actual practise many trees that have been sprayed will often still develop in disease.

A second drawback to the sycamore is the matter of leaves which are large and tend to fall throughout much of the year. In well-manicured gardens the raking of fallen leaves makes the sycamore a high labor-cost item.

Regions: 1, 2, 3, 6.

PLUCHEA. A genus of the Asteraceae with about 30 species of tall shrubs or herbs, usually with purplish flowers, mostly in warm temperate or tropical regions. Two species in California.

1. *P. sericea.* Arrowweed. Plate 121. A slender, leafy, willow-like shrub to 12 ft., plants silvery-silky; leaves alternate, entire, lanceolate, ¾-1½ in. long, tapering at both ends; flowers purplish, borne in heads, clustered in corymb-like terminal cymes, ray flowers lacking; fruit an achene. May-June.

Arrowweed is frequent in wet areas in cismontane southern California from Santa Barbara Co. south into Baja California and east to Texas. In California it is a component of the coastal sage scrub, creosote bush scrub

Fig. 57. Coiled root system. The result of late transplanting. This *Pinus radiata,* with a trunk diameter of about six inches, had the major portion of its root system confined to an area not much larger than the can from which the plant had originally been planted.

Fig. 58. a, *Pinus monophylla*; b, *P. quadrifolia*. (Sudworth, 1908). Reduced.

and desert oases communities. The species probably consists of distinct ecotypes.

P. sericea is recommended with reservations. It is extremely durable and it is one native about which the statement 'you can't kill it' applies. For that reason care must be exercised in its use. The plants spread underground and they can come up through black-topped roads. Because of its spreading habit, arrowweed can be used for soil stabilization and erosion control and it is tall enough to serve as a windbreak. Although common in moist areas, the plants when established are completely drought tolerant. When in flower the plants are quite attractive.

Regions: 1, 2, 3, 4, 5, 6.

POPULUS. A genus of the Salicaceae with about 30 species of fast-growing deciduous trees widely distributed in the Northern Hemisphere. Many ornamental and shade trees. Pistillate and staminate flowers borne in catkins separately and on different trees.

1. *P. fremontii* var. *fremontii.* Fremont cottonwood. Fig. 61a. Handsome trees often 90 ft. tall with trunks to 5 ft. in diameter, branches wide-spreading, forming massive crowns, bark white or whitish, roughly cracked, twigs stout; leaves triangular to roundish or deltoid, to 4 in. wide, broader than long, abruptly sharp-pointed at apex, coarsely and irregularly serrate-dentate, bright green, lustrous; staminate catkins 2-4 in. long, pistillate catkins 2 in. long becoming much longer in fruit, seeds copiously white-hairy.

Fremont cottonwood is common in moist situations in southern California and south to northern Baja California, east to Texas and Colorado. Also on Santa Catalina and San Nicholas islands. Usually found below 6500 ft.

P it is a component of numerous plant communities, notably the riparian woodland.

Regions: 1, 2, 3, 4, 5, 6.

2. *P. trichocarpa* var. *trichocarpa*. Black cottonwood. Fig. 61b. Tall, broad, open-crowned trees to as much as 100 ft. with trunks to 3 ft. in diameter, bark dark or light usually with a yellowish cast, deeply furrowed in age; leaves ovate, acute or tapering to a point at the apex, to 7 in. long, margins finely serrate, lustrous green above, rusty-brown beneath becoming white in age; staminate catkins 1-2 in. or eventually 5 in. long, pistillate catkins to 3 in. long at flowering and 10 in fruit.

Black cottonwood is widely distributed in western United States extending north to Alaska. According to Jepson, localities for black cottonwood are not, comparatively speaking, numerous and the growth in any one place is not extensive or abundant. In elevation the species extends to about 6000 ft. It is the tallest of all species of *Populus.*

Regions: 1, 2, 3, 4, 5, 6.

Cottonwoods are very useful especially in inland valleys with cold winters and hot dry summers where they may be used for shade or as windbreaks. They are best with regular deep watering, and drought tolerant if the roots grow deep enough to tap the water table. The roots are invasive so they should not be planted where they may reach water or sewer lines, septic tanks, etc. Too large for lawns or small gardens and should not be used for city streets.

At Claremont our specimens of Fremont cottonwood have been subject to trunk and branch canker, the exact cause of which has not been determined. The trees at the garden are grown under stressful conditions of drought and air pollution and it is possible that in such a situation the plants are more susceptible to disease than they would be under more ideal conditions. The wood of *Populus fremontii* is extremely brittle and when branches are weakened by canker they have little or no resistance to wind damage.

PROSOPIS. A genus of the Fabaceae with about 30-35 species of deciduous shrubs or trees found in temperate and tropical regions. Two species native in California.

1. *P. glandulosa* var. *torreyana*. Mesquite. Fig. 62a. A much-branched, deciduous shrub or tree as much as 37 ft. tall with a crown diameter of 40 ft., branches usually crooked and arching; thorns 1-2, axillary, 1/4-1 1/4 in. long; leaves bright green, 2-4 in. long, usually with 2 pinnae, each pinna with 9-18 pairs of linear, entire leaflets 3/8-1 in. long; petioles enlarged and glandular at the base; flowers in slender, cylindrical spikes, 2-3 1/2 in. long, greenish-yellow; fruit pods in drooping clusters of 1-6, linear, 3-8 in. long, 1/3-1/2 in. wide, curved, flat or becoming thickened, irregularly constricted between the seeds. April-June, but may bloom several times during the summer.

Mesquite is common in washes and low places in the Colorado and Mojave deserts, interior cismontane southern California, north to the upper San Joaquin Valley, south into Mexico. Mesquite is found in the creosote bush scrub and alkali sink communities.

The species is at its best in deep soil where the taproots will go great distances for water. According to Jepson, the plants are an almost infallible sign of water, either on the surface or at varying distances underground. The plants however will tolerate drought and alkaline conditions. In shallow, rocky soil they are more shrublike. According to Sunset's *New Western Garden Book* mesquite is one of the widespreading shade trees that help make outdoor living in the desert more comfortable. The trees cast a light airy kind of shade. Mesquite is also useful for windbreaks and because of its sprawling habit it serves to break wind-blown sand. It is an important browse plant for animals, particularly the seedpods which are also an important source of food for skunks, bobcats and coyotes.

Regions: 1, 2, 3, 4, 5, 6. Reported to do well in the Central Valley.

2. *P. pubescens*. Screwbean, screwbean mesquite, tornillo. Fig. 62b. A shrub or small tree to 30 ft. with a narrow crown and ascending branches, spines stout, 1/4-1/2 in. long; leaves puberulent, usually with 2 pinnae, each pinna with 5-11 pairs of oblong leaflets 1/8-1/2 in.

→

Plate 13. a, d, *Prunus andersonii*; b, e, *Prunus ilicifolia*; c, f, *Prunus subcordata*; g, h, *Prunus virginiana* var. *demissa*; i, black knot disease of *Prunus*; j, *Ptelea crenulata*; k, *Purshia glandulosa*; l, *Rhus ovata.*

Plate 13

long; flowers small, numerous, yellowish in slender cylindrical spikes 2-3 in. long; fruits 2-15 from each flower spike, each pod tightly coiled into a narrow, straight, cylindrical body 1-1½ in. long. May-July.

Screwbean is rather common along the Colorado River bottoms, extending north into Death Valley in washes and canyons, east to Utah and Texas and south into Mexico. It is usually found in the creosote bush scrub community at elevations below 2500 ft.

Cultural requirements and uses of *P. pubescens* are the same as for *P. glandulosa* var. *torreyana.*

Regions: 1, 2, 3, 4, 5, 6. Also does well in the Central Valley and in southern Arizona.

PRUNUS. A genus of the Rosaceae with perhaps 150 species of trees and shrubs native mostly to the Northern Hemisphere and containing many species of horticultural importance. Shrubs or trees with simple leaves and flowers borne in clusters, or somtimes singly, from lateral buds borne on wood of the previous season, appearing either before or with the leaves; fruit a drupe, usually 1-seeded. Eight species native to California.

1. *P. andersonii.* Desert peach. Plate 13a, d. A spreading, divaricately branched deciduous shrub to 6 ft.; leaves fascicled on short lateral, very thorny branchlets, oblong or oblong-lanceolate, ½-1 in. long, ⅛-⅜ in. wide, finely serrulate, glabrous; flowers solitary, about ½ in. across, appearing with the leaves, variable in color from pale pink to deep rose-colored; fruit globose-compressed, about ½ in. long, covered with a brown pubescence. March-April.

Desert peach is common on arid desert mountain slopes and mesas on the eastern side of the Sierra Nevada from Modoc to Lassen cos., southward to Inyo and Kern cos., at elevations of 3500-7500 ft. where it is a component of the sagebrush scrub and yellow pine forest communities. It also occurs in western Nevada.

P. andersonii is extremely variable in flower color and

Fig. 59. a, *Pinus monophylla*, 27 years old, 34 feet tall with a spread of 19 feet; b, *P. quadrifolia*, 19 years old, 21 feet tall with a spread of 16 feet.

for horticultural purposes only the deeply colored forms should be used. The plants prefer coarse well-drained soils and once established are very drought tolerant. Because of the thorny branches the plants are valuable for barrier plantings. May also be grown in heavy soils if care is taken in watering.

Regions: 2, 3, 4, 6, and selected areas in 7. Does well in the Central Valley.

2. *P. fasciculata.* Desert almond. An intricately and densely but softly branched deciduous shrub to 8 ft. with short, thorn-like branchlets; leaves fascicled on short bud-like suppressed branchlets, spatulate to linear-oblanceolate, ½-¾ in. long, ⅛-¼ in. wide, finely pubescent, entire or rarely with 1-2 teeth on each margin; flowers inconspicuous, white, solitary or 2-3 together, sessile; fruit sub-globose to ovoid, about ⅜ in. long, covered with light brown, short bristly hairs. March-May.

Desert almond is found on desert slopes and mesas at elevations of 2500-6000 ft. in the Mojave and Colorado deserts, east to Utah and Arizona. In California it is a member of the creosote bush scrub community and the pinyon-juniper woodland.

Cultural requirements are the same as in the preceding species. *P. fasciculata* is less thorny than *P. andersonii* or *P. fremontii* but useful for barrier plantings because of its very dense branching habit. It has rarely set fruit at this botanic garden.

Regions: 2, 3, 4, 5, 6.

3. *P. fremontii.* Desert apricot. An intricately branched shrub or small tree to 15 ft. with numerous spinescent branchlets; leaves borne on short lateral spurs, roundish to broadly ovate, serrate, ½-¾ in. long, to ¾ in. wide, glabrous on both surfaces; flowers solitary or in clusters of 2-10, petals white to creamy-pink; fruit elliptic-ovoid, ⅓-½ in. long, minutely puberulent. February-March.

Desert apricot is found on the lower arid canyon slopes and flats of mountains bordering the western edge of the Colorado Desert from Palm Springs south to Baja California, at elevations below 4000 ft. where it is a member of the creosote bush scrub and pinyon-juniper woodland communities.

Prefers coarse well-drained soils and is drought tolerant. Another good barrier plant.

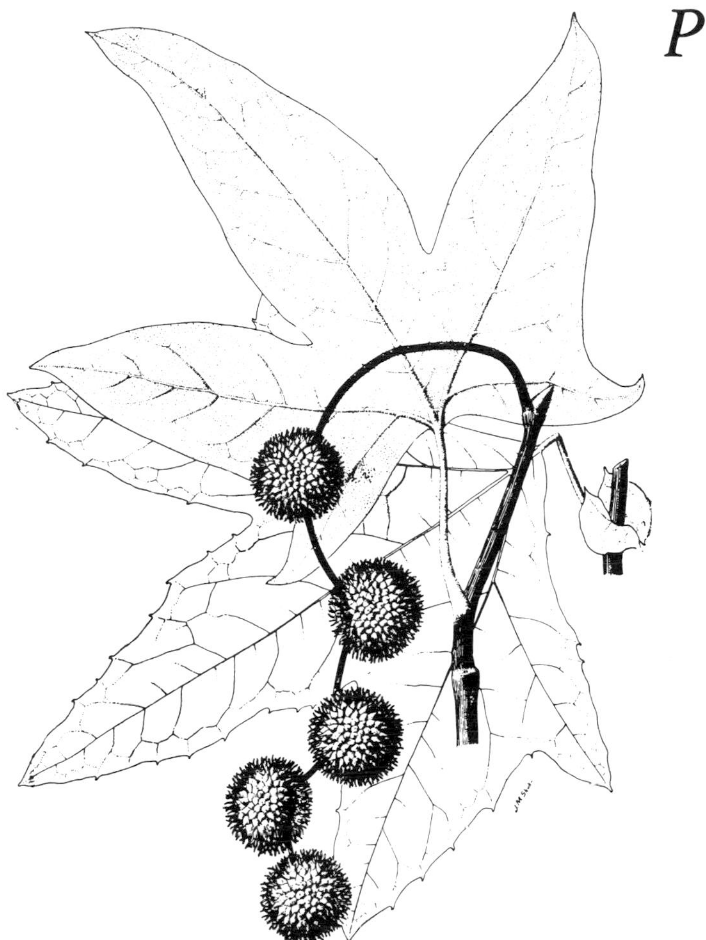

Fig. 60. *Platanus racemosa.* (Sudworth, 1908). Reduced.

Regions: 2, 3, 4, 5, 6. Does well in the Central Valley.

4. *P. ilicifolia.* Hollyleaf cherry, islay, evergreen cherry. Plate 13b, e. Densely branched evergreen shrub or tree to 30 ft.; leaves coriaceous, ovate to elliptic, acute or rounded at the apex, dark green and holly-like, ¾-2 in. long, ¾-1½ in. wide, glabrous and shiny above, paler and glabrous beneath; flowers white, in racemes 1-2 in. long; drupe dark red, purple or black, about ½ in. in diameter; pulp very thin, sweet when ripe but with a bitter aftertaste. May-June.

Hollyleaf cherry inhabits the dry foothills, lower mountain slopes and rich valleys from Napa Co. to the Santa Lucia and Tehachapi mts., and cismontane southern California south into Baja California. It is also found on San Clemente and Santa Catalina islands. It is found

Fig. 61. a, *Populus fremontii*; b, *P. trichocarpa.* (Sudworth, 1908). Reduced.

in a number of plant communities including the chaparral.

In writing of the hollyleaf cherry, Jepson (1936) noted that when the plants were growing in the chaparral they formed low bushes 3-5 ft. tall, assuming very well the spiny look and rigid stems of its xerophytic associates, whereas when growing on rich valley floors or along streams the plants became tree-like to 30 ft. with a crown almost as broad and a trunk as much as 16 in. in diameter.

P. ilicifolia has long been in cultivation and can be grown in many types of soil; however, the plants do best in coarse well-drained soil and once established they are very drought tolerant and fast growing. The general appearance of the plants is improved by deep but infrequent watering. In planting, great care should be taken to insure that the plants have not become rootbound in the can.

The hollyleaf cherry may be used in many ways, as a small tree, tall screen or as a formal trimmed hedge from 3-10 ft. tall. This and the following species should not be used as street trees or in patios because the fruits, when they fall, create a nuisance or even a hazard for those using the sidewalk. This species is reported to be unusually resistant to oak root fungus (*Armillariella*).

Regions: 1, 2, 3, 4, 5, 6. Reported to do well in the Central Valley if protected from afternoon sun.

5. *P. lyonii.* Catalina cherry. Fig. 63. A large evergreen shrub or small tree to 45 ft.; leaves coriaceous, 2-5 in. long, ovate to ovate-lanceolate, dark green, glabrous, usually entire; flowers white, borne in racemes 2-5 in. long. May-June.

Catalina cherry is known only from Santa Catalina, San Clemente, Santa Rosa and Santa Cruz islands where it is found in the chaparral.

P. lyonii has long been in cultivation and may be used in the same way as the hollyleaf cherry. In gardens the two species hybridize, and many plants offered in the trade are of hybrid origin. Catalina cherry is not particular as to soil, water or exposure and throughout the year the plants present a clean, bright green foliage. It has been reported that on Catalina Island old trees 45 ft. tall often resemble the California live oak, *Quercus agrifolia.* Reported to be highly resistant to oak root fungus.

Regions: 1, 2, 3, 4, 5, 6. Also does well in the Central Valley.

6. *P. subcordata.* Sierra plum. Plate 13c, f. Deciduous shrub or small tree to 20 ft. with stiff, crooked branches and thorn-like branchlets; leaves ovate, elliptic, obovate or nearly round, ¾-2 in. long, ½-1½ in. wide, finely serrulate; flowers in clusters of 2-4, white, fragrant; drupe oblong, pubescent or glabrous, bright red with a white bloom, or sometimes yellow, ¾-1 in. long. March-April.

Sierra plum is widespread in California from Tulare to Modoc cos. at elevations of 2500-4500 ft., and from Monterey to Siskiyou cos. at 500-3500 ft. Also in Oregon. In California it is a member of the yellow pine forest community.

This species is quite variable and numerous varieties have been described. In dry or gravelly soils the plants tend to be shrubby, but in more fertile soils they become larger and are often tree-like. In the north the fruits are larger, more juicy and quite edible, farther south they tend to be drier and bitter, and around the San Francisco Bay the plants seldom set seed. The plants, which spread by suckers, may be used for erosion control.

Regions: 2, 3, 6, and selected areas in 7.

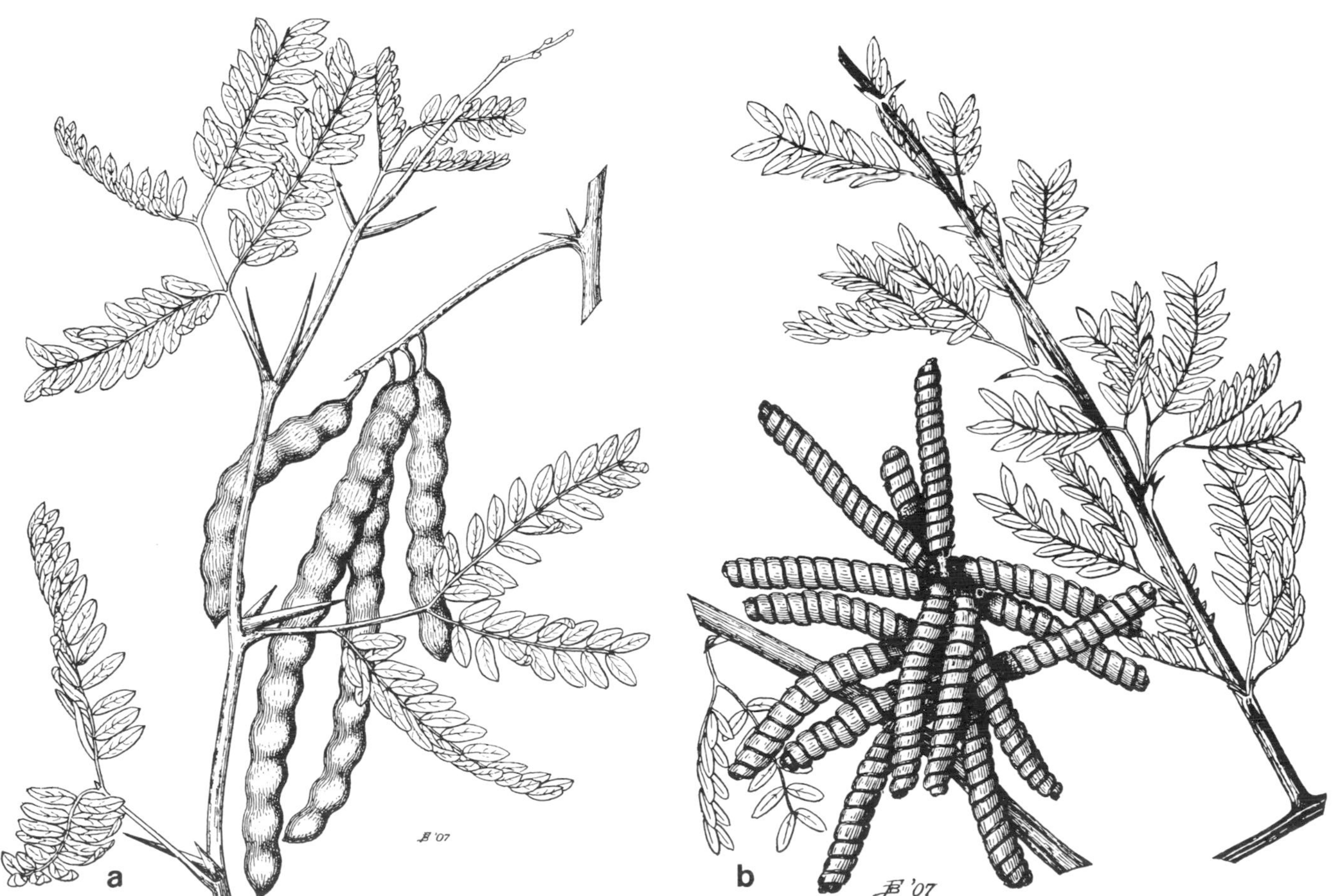

Fig. 62. a, *Prosopis glandulosa* var. *torreyana*; b, *P. pubescens.* (Sudworth, 1908). Reduced.

7. *P. virginiana* var. *demissa.* Western chokecherry. Plate 13g, h. An erect, deciduous shrub to 15 ft. but sometimes a small tree to 20 ft.; leaves ovate or broadly elliptic, sometimes broader above the middle, 1½-3½ in. long, ¾-3½ in. wide, glabrous or slightly pubescent, finely serrate; flowers numerous in elongated racemes 2-5 in. long terminating lateral leafy stems; fruit globose, ¼-½ in. in diameter, red to deep purple, edible and sometimes used for making jams and jellies. April-May.

Western chokecherry is widely distributed in California in moist places in woods, on brushy slopes and flats at elevations below 8200 ft. in the chaparral, yellow pine forest and foothill woodland communities.

The plants are drought and heat tolerant and in some areas give a good display of autumn foliage color. This species suckers freely and they may be used for erosion control. The berries are attractive to birds.

Regions: 1, 2, 3, 4, 6.

The deciduous species of *Prunus* are often attacked by black knot disease (*Dibotryon morbosum*) which causes twig and branch die-back (Plate 13i), thus making the plants unsightly, although the disease does not generally cause the death of the entire plant. Both *P. ilicifolia* and *lyonii* may at times be attacked by peach leaf curl virus.

Fig. 63. *Prunus lyonii* used as a hedge. (M. & M. Carothers).

PTELEA. A genus of deciduous shrubs or small trees belonging to the Rutaceae with about seven to ten species native to the United States and Mexico. One species in California.

1. *P. crenulata.* Western hop tree. Plate 13j. A shrub or small tree to 15 ft., young twigs glandular pubescent, bark smooth, brown, bitter; leaves alternate, pinnately-trifoliate, aromatic when crushed, leaflets ovate, ellipti-

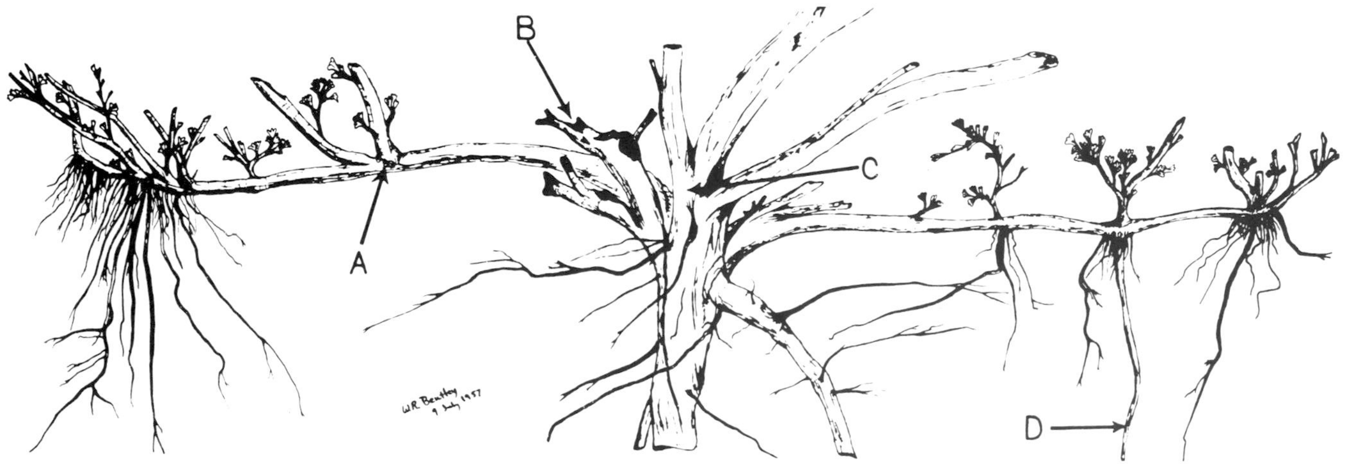

Fig. 64. Stem layering in bitterbrush. A, bud masses formed on the branches; B, C, fire scars on burned branches and crown; D, roots formed on the lateral branches. (Nord, 1959a).

Fig. 65. Road cut stabilized by bitterbrush. (Nord, 1959a).

cal, lanceolate, oblong-ovate or obovate, to 3½ in. long and about 1½ in. wide, margins entire or serrulate; flowers in axillary, paniculate cymes, small yellowish-green, pleasantly fragrant; fruit a 2-celled, 2-seeded samara, winged all the way around. March-May.

Western hop tree is found on moist flats and in canyons in the foothills of the coast ranges from Contra Costa Co. north to Shasta Co. and in the Sierra Nevada from Tehama Co. south to Calaveras Co., usually below 2000 ft. where it is a member of the foothill woodland and yellow pine forest communities.

P. crenulata is an attractive deciduous shrub, with glossy green foliage and fragrant yellowish-green flowers, which presents no cultural problems. At its best in partial shade and in moist situations it is, when established, quite drought tolerant even in poor soils. Easily propagated from cuttings, seeds and by division, it is free from diseases and insect pests.

Regions: 2, 3, 6. Also does well in the Central Valley if protected from the hot afternoon sun.

PURSHIA. A genus of the Rosaceae with two species native to western United States.

1. *P. glandulosa.* Waxy or desert bitterbrush. Plate 13k. An erect, widely branched, dark green shrub to 8 ft. with gray or brown bark, young branches pubescent; leaves ¼-⅜ in. long, divided into 3-5 linear lobes or sometimes merely toothed, slightly hairy above when young, glabrous in age, depressed-glandular above and on the margins, blades becoming revolute; flowers ½ in. or less in diameter, solitary and terminal on short lateral leafy branchlets, petals pale yellow or white. April-June.

Desert bitterbrush occurs in desert canyons and on mountain slopes bordering the western portion of the Colorado Desert, south into Baja California and northward in the mountains of the Mojave Desert to Inyo and Mono cos. and west as far as Mt. Pinos. Desert bitterbrush is a member of the Joshua tree and pinyon-juniper woodland communities.

Regions: 2, 3, 4, 5, 6.

2. *P. tridentata.* Antelope bitterbrush. An erect, widely branched silvery or gray shrub to 10 ft. with gray or brown bark, young branches pubescent; leaves wedge-shaped, ¼-½ in. long (sometimes 1 in. on new growth) with 3 oblong lobes at the apex, green and finely pubescent above, white-pubescent beneath, margins revolute; flowers about ⅔ in. across, usually solitary and terminating short lateral branchlets, petals pale yellow, ½-¾ in. long.

Antelope bitterbrush is found on dry slopes in the White Mts., the Sierra Nevada (especially on the eastern side) and on the western slopes from Tulare Co. north, then to British Columbia and east to Montana and New

Q

Fig. 66. *Quercus agrifolia.* (Greene, 1889). Reduced.

Mexico where it grows at elevations from 3000-10,000 ft. It is found in many plant communities from the sagebrush scrub to the subalpine forest.

Regions: 2, 3, 4, 6, 7.

Both species of bitterbrush grow on many types of soil but do best in coarse soils having a pH of 6.0-7.3 and with very good drainage.

Propagation is by seed, which requires stratification, stem layering and by cuttings. Because of the plant's ability to produce roots wherever the stems touch the ground (Fig. 64), bitterbrush is particularly valuable in stabilizing road cuts (Fig. 65), stream washes and 'blowouts.' Desert bitterbrush is reported to regenerate after fires, whereas antelope bitterbrush does so relatively infrequently. Under natural conditions bitterbrush plants may live to be at least 160 years old (Nord, 1959). Bitterbrush is one of the most widely distributed of all western shrubs and is considered in many areas to be the

and the seed is important in the diet of birds and rodents.
most important single browse plant for many animals,

QUERCUS. A large genus of about 300 species of evergreen or deciduous trees or shrubs native principally to the Northern Hemisphere, 16 species in California. The genus is divided into two sections: the 'white oak' group, which in California includes *Quercus chrysolepis, Q. douglasii, Q. engelmannii, Q. garryana, Q. lobata* and *Q. tomentella*; the 'black oak' group includes *Q. agrifolia, Q. kelloggii* and *Q. wislizenii.* Interspecific hybridization is common within, but not between, the sections, with the result that specific identification of individual trees is sometimes difficult. Many handsome species are in cultivation.

Propagation is usually by seed preferably sown as soon as ripe where the tree is wanted, but the plants can also be grown in rows in the field where the seedlings may grow 2-4 ft. the first season, or the seed may be planted in containers and the young plants later transplanted. According to some reports *Q. agrifolia* and *Q. tomentella* can be propagated by cuttings. With one exception (*Q. kelloggii*) oaks of almost any size may be successfully transplanted since cutting the tap root does not seriously impair the plants, due to the fact that they normally develop a strong fibrous root system. In transplanting larger trees the plants should be carefully balled or boxed before moving. Although most of the California oaks are tolerant of a wide variety of soil types, many of them do best and develop most rapidly given a deep loam and, in southern California, periodic deep watering during the summer. Although often considered slow growing, many of the oaks, given favorable conditions, grow rather rapidly.

1. *Q. agrifolia* var. *agrifolia.* Coast live oak, California live oak or encina. Figs. 66, 67a. Medium-sized to large evergreen trees, usually with a broad crown, trunk short, dividing into numerous massive branches, bark deep gray, smooth, or on old trunks broken into small plates; leaves thickened, coriaceous, oval to oblong, 1½-2 in. long, convex or cupped, margins with numerous small spines, sometimes spineless, dark green above, lighter beneath and slightly pubescent with stellate hairs; acorns maturing the first season, nuts 1-2 in. long,

Fig. 67. a, *Quercus agrifolia*, 20 years old, 39 feet tall with a 19 foot spread; b, *Q. tomentella,* 12 years old, 26 feet tall with a 19 foot spread; c, *Q. chrysolepis,* 26 years old, 36 feet tall with a 38 foot spread. All three growing under identical conditions of soil and moisture.

Fig. 68. *Quercus chrysolepis.* (Greene, 1889). Reduced.

slender, pointed, somewhat longitudinally striped, rich brown.

The coast live oak ranges from Mendocino Co. south to the Sierra San Pedro Mártir Mts. in Baja California. Common in valleys and on not too dry slopes below 3000 ft., mostly in the southern oak woodland and foothill woodland communities.

According to landscape architect Ralph Cornell (1938), the California live oak is "one of the most valuable of all our trees from a landscape viewpoint. Whether growing wild or in the garden, it is unsurpassed. Resistant to heat and cold alike, to drought or seasons of heavy rain, growing in gravelly soil or on heavy adobes, and repelling many of the insect pests that destroy other trees, the live oak displays greater hardiness and ability to survive than might be expected of a tree so lovely." It must be remembered that the coast live oak normally grows in situations where the ground becomes dry during the summer, and it should not be planted at the edge of lawns where it will receive regular and frequent shallow watering. Although tolerant of drought, under favorable situations it is one of the fastest growing of the native oaks. At Claremont a 20-year-old tree growing in good soil measured 39 ft. in height with a 42 ft. spread and a dbh of almost 18 inches. The California live oak has been described as having greedy roots, and because of its habit of shedding some leaves throughout the year, it is not a tree for the meticulously tidy gardener. Periodically it may be attacked by twig girdler that causes patches of brown foliage to occur over most of the tree. It may also be attacked by a mildew that causes a type of witches-broom. In pruning, care should be taken not to open up the trees to the extent that a great deal of light reaches the inside of the crown as that promotes the excessive production of small twigs from the larger branches. It is reported that this species can be sheared to form an attractive 10-12 ft. hedge.

Regions: 1, 2, 3, 6. Does well in the Central Valley if protected from the afternoon sun.

In interior Riverside and San Diego cos. there is a form with leaves densely pubescent beneath that has been described as var. *oxydenia.*

2. *Q. chrysolepis.* Canyon live oak or maul oak. Fig. 67c, 68. A small to medium-sized, round-headed or spreading evergreen tree, bark whitish and fissured into narrow, flat or more or less scaly ridges; leaves extremely

Fig. 69. *Quercus douglasii.* (Greene, 1889). Reduced.

variable in shape even on a single tree, 1-2 or even 4 in. long, ½-1¾ in. wide, ovate or oblong, thick, green above, yellowish tomentose beneath becoming grayish tomentose or glaucous in age, edges entire or irregularly spiny-toothed; acorns maturing the second year, cup resembling a golden turban, nuts 1-1½ in. long, ovoid-oblong, or cylindric, rounded at the apex or sharply pointed.

Canyon live oak is the most widely distributed oak in California extending from southern Oregon south to Baja California where it is found in canyons and on moist slopes usually below 6500 ft. and in many plant communities. It is also found occasionally at higher elevations in the Mojave Desert. *Q. chrysolepis* is the most variable of the western oaks both in leaf and in acorns, and the color of the underside of the leaf is probably more diagnostic than the size or shape of the leaf.

This species appears to be adaptable to a variety of

Fig. 70. *Quercus engelmannii*, 24 years old, 32 feet tall with a 24 foot spread.

Fig. 71. *Quercus engelmannii*. (Greene, 1889). Reduced.

soils and the growth rate is moderate, particularly if the trees receive some summer moisture. At this botanic garden, canyon oak has not been attacked by insect pests but at times has suffered from a twig blight of unknown cause.

Regions: 1, 2, 3, 6, and selected areas in 4, 7.

3. *Q. douglasii.* Blue oak. Fig. 69. A small to medium-sized deciduous tree with a rounded crown, bark whitish or light gray checked into thin plates; leaves 1-4 in. long, ½-2½ in. broad, oblong to ovate, entire or coarsely and often unequally few-toothed, or shallowly lobed, bluish-green above, pale beneath; acorns ripening the first season, nut variable in shape, commonly ovoid, ¾-1¼ in. long, often narrowed at the base.

Fig. 72. *Quercus garryana.* (Greene, 1889). Reduced.

Blue oak inhabits dry, rocky slopes usually below 3500 ft., mostly on the slopes of the hills bordering the interior valleys from northern Los Angeles Co. north to the headwaters of the Sacramento Valley. It is a member of the foothill woodland community and in places is the only woody species on the rolling grassy hills. The blue oak is remarkably uniform in aspect, color of leaves and the character of the bark but remarkably variable in the size and shape of the leaves.

Because of its drought tolerance this oak can be recommended for very hot and dry situations where it will be slow growing. Given a good soil and additional water the growth rate is moderate. At Claremont a 23-year-old tree growing in poor soil measured 18 ft. in height with a 14 ft. spread. It is important that young trees be carefully pruned in order for them to develop into well-shaped specimens. At Claremont this species has been free of insect pests and disease.

Regions: 1, 2, 3, 6. Also reported to do well in the Central Valley.

4. *Q. engelmannii.* Engelmann or mesa oak. Figs. 70, 71. A medium-sized evergreen tree with spreading branches forming a round-topped crown, bark with thin grayish scales; leaves plane, 1-3 in. long, ½-1 in. broad, entire to sinuate-dentate, blue-green in color; acorns maturing the first year, cup shallow to bowl-shaped, nuts oblong cylindric to 1 in. long.

Engelmann oak is native to the dry fans and foothills from eastern Los Angeles Co. south to eastern San Diego Co. and on into Baja California, always away from the coast. This species retains its leaves until the appearance of the new leaves in the spring, at which time the old leaves are dropped.

Mesa oak tolerates drought but growth is more rapid when it is grown in good soil with additional water. At this garden a 24-year-old tree growing in good soil was 32 ft. tall with a 24 ft. spread. This species has not been attacked by either insect pests or disease.

Regions: 1, 2, 3, 6. Also reported to do well in the Central Valley.

5. *Q. garryana.* Oregon oak or Oregon white oak. Fig. 72. A small to medium-sized deciduous tree, trunk dividing into widespreading branches which support a broad rounded crown, bark white, smooth or fissured into longitudinal bands that are transversely checked into small squarish scales; leaves leathery, obovate to oblong, deeply 5-9-lobed, the lobes rounded or unequally toothed, 3-6 in. long, 2-5 in. broad; acorns maturing the first year, cup shallow, nut ovoid or subcylindric, ¾-1 in. long, rounded at the apex.

Oregon oak is found from Vancouver Island south to Marin Co. and in the Cascades to eastern Shasta Co. It occupies wooded slopes mostly between 1000-5000 ft. in the northern oak woodland and the mixed evergreen forest communities.

Although widely distributed geographically, the Oregon oak is, after the valley and Kellogg oaks, the least variable of the western species. The most distinct variant has been called var. *semota,* or the Kaweah oak. It is a rounded-headed, deciduous shrub found on dry slopes in the chaparral and yellow pine forest from Plumas Co. south to Los Angeles Co.

The Oregon oak is reported to have a deep, non-aggressive root system. Although not native to our area, this species has done well at Claremont even in very poor soil, where one specimen 16 years old measured 12 ft. in height with a 10 ft. spread. Another specimen 20 years old had a height of 12 ft. with an 8 ft. spread and was multiple branched. This species has not been attacked by insect pests or disease.

Regions: 1, 2, 3, 6.

6. *Q. kelloggii.* California black oak. Figs. 73a, 74. A graceful medium-sized to large deciduous tree with mostly erect or ascending main branches, bark black or dark gray, deeply checked into small plates; leaves deeply and mostly sinuately parted with about 3 lobes on each side, ending in 1-3 or more coarse bristly-tipped teeth, bright green above, lighter beneath, 4-10 in. long, 1½-6 in. broad; acorns maturing the second season, cup large, nut oblong in outline, very rounded at the apex, 1-1¼ in. long and ¾ in. wide, covered at first by a fine fuzz and deeply set into the brown cup.

California black oak is found in the hills and mountains from Oregon south to San Diego Co. at elevations of 4000-8000 ft. where it is a member of the yellow pine forest community.

Quercus kelloggii is one of the most handsome and desirable of the California oaks, the unfolding leaves being pink, crimson or purple for a period of about two weeks, maturing to a bright shiny green, and in autumn turning yellow or orange before falling. Geographically this species is widespread, its altitudinal range is usually between 4000-8000 ft. and it occurs in a wide variety of soils; gravelly floor of the low valley, clay foothills, alluvial slopes and on rocky ridges mostly in areas with over 25 inches of precipitation per year. According to

Fig. 73. a, *Quercus kelloggii*, 28 years old, 32 feet tall with a 23 foot spread; b, *Q. lobata* (elm stage), 26 years old, 43 feet tall with a 36 foot spread.

Q

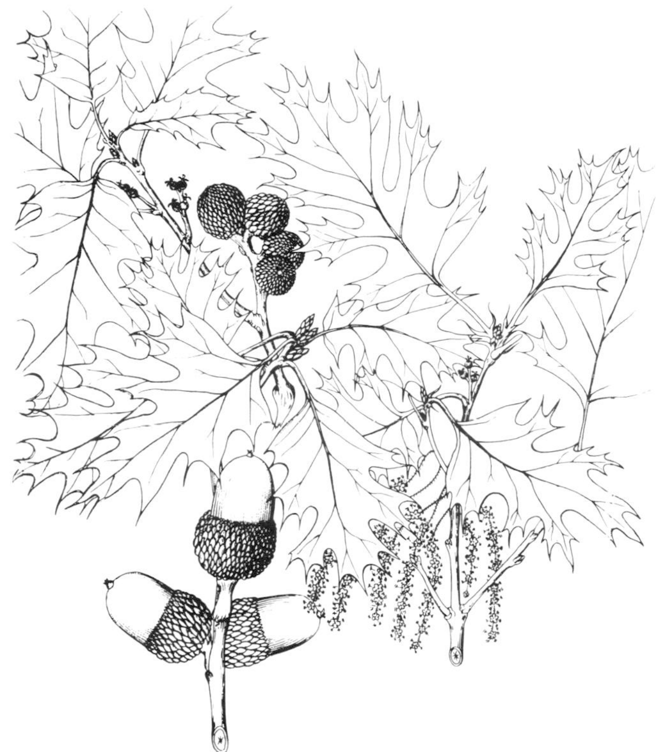

Fig. 74. *Quercus kelloggii.* (Greene, 1889). Reduced.

Jepson, it is the least variable of the California oaks. This species develops a strong taproot and this may explain the reason for failures in transplanting young trees. For that reason propagation should be by acorns planted where the trees are wanted, or they may be sown in containers and the young plants transplanted while still small. At this botanic garden a 25-year-old tree growing in good soil measured 26 ft. tall with a 21 ft. spread. The dbh was 7½ inches.

Regions: 2, 3, 6, 7.

7. *Q. lobata.* Valley oak, California white oak or roble. Figs. 73b, 75. A large, graceful, deciduous tree with a rounded crown, often broader than tall and at maturity often with slender pendulous branchlets, bark dark brown or ashen-gray, checked into plates 1-2 in. across; leaves 3-5-lobed, lobes usually broadened toward the ends, 3-4 or even 6 in. long, 2-3 in. broad, green above, paler beneath; acorns maturing the first year, cup deeply hemispherical and warty, ½-¾ in. deep, larger in diameter than the nut which is long and conical, 1½-2¼ in. long, maturing to a mahogany or chestnut-brown color.

Valley oak is native to the fertile parts of the Sacramento, San Joaquin and adjacent valleys and in the foothills of the Sierra Nevada and the middle and inner coastal ranges where it is a member of the foothill woodland and southern oak woodland communities. It also extends south to the San Fernando Valley and as far east as San Marino. It is usually found below 2000 ft. elevation and is reported to be the largest of the American oaks. The valley oak favors hot valleys away from the influence of the ocean, deep, rich soil with the water table 10-40 ft. below the surface. It is however remarkable in its ability to thrive under extremes of moisture and drought. According to Jepson, it exhibits four marked stages of growth: the pole stage, the elm stage (Fig. 73b), the weeping stage which is only developed on fertile loams, and second youth. Under ideal conditions the valley oak grows rapidly. A specimen at the botanic garden 25 years old growing in good soil was 38 ft. tall with a 36 ft. spread. This oak is one of the most handsome of California trees and is highly recommended. Because of its eventual large size, care should be given in its placement. At Claremont it has not been attacked by insect pests or disease.

Regions: 1, 2, 3, 6.

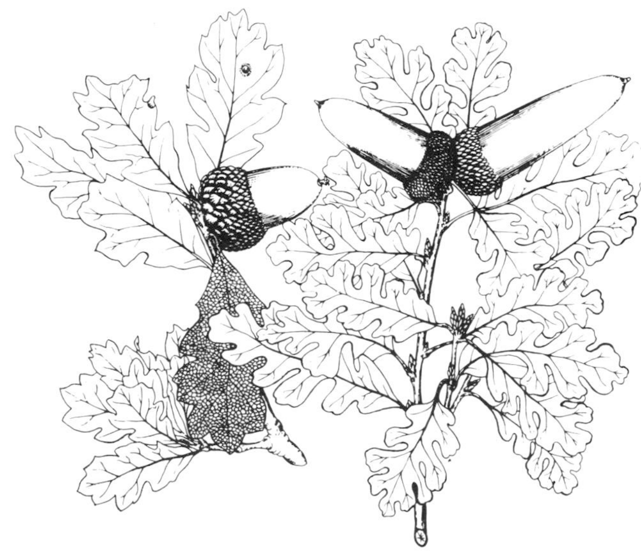

Fig. 75. *Quercus lobata.* (Greene, 1889). Reduced.

8. *Q. tomentella.*★ Island oak. Figs. 67b, 76. A very handsome, small to medium-sized evergreen tree of a narrow upright habit, bark grayish-brown, smoothish, or checked; leaves thick and leathery, elliptic to oblong, often revolute, light green above, paler beneath, densely tomentose when young, 2-3½ in. long, ¾-1¾ in. broad, nerves regular, parallel and very strong beneath, ending in teeth on the margin, or margin sometimes smooth; acorns maturing the second year, cup 1-1½ in. broad, ½-¾ in. deep, covered with dense tomentum, nuts subglobose with a short point, about 1 in. long.

Island oak is confined to Guadalupe and the California offshore islands and is the rarest of the California oaks. This species is a component of what Philbrick and Haller (1977) consider to be the island woodland community which in addition to the island oak contains such unique insular species as *Arctostaphylos insularis, Ceanothus arboreus, Quercus ×macdonaldii* and *Lyonothamnus floribundus.*

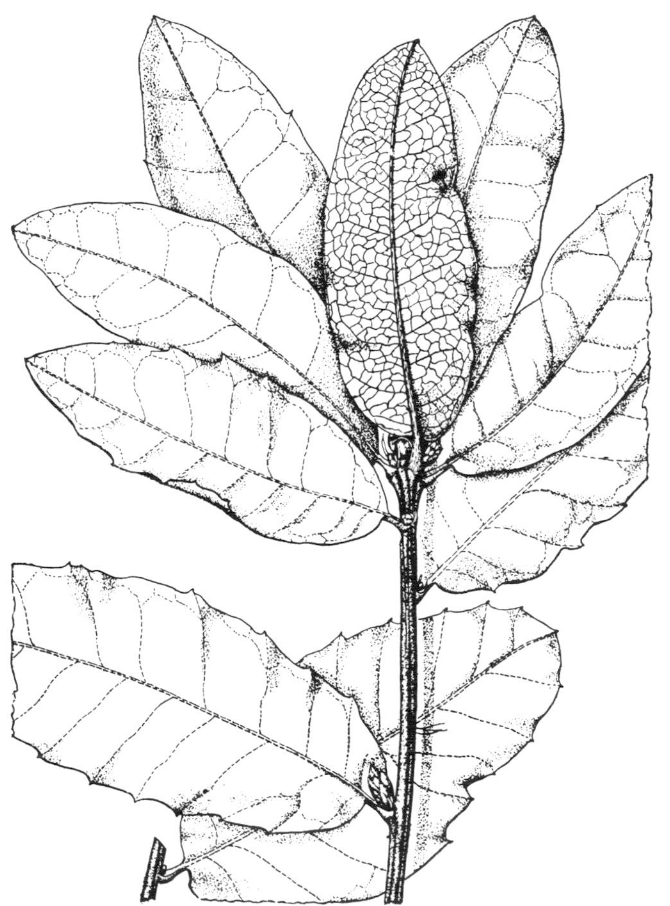

Fig. 76. *Quercus tomentella.* (Sudworth, 1908). Reduced.

Fig. 77. *Quercus wislizenii* var. *wislizenii.* (Greene, 1889). Reduced.

Although scarcely known to horticulturists, the island oak is an exceptionally handsome tree, and because of its rather narrow upright habit it is admirably suited for street planting as well as for home use. At the Rancho Santa Ana Botanic Garden a 20-year-old tree growing in good soil had a height of 23 ft. with a 15 ft. spread. Although it tolerates poor soil and drought, growth is faster when it receives summer irrigation. At Claremont it has not been troubled by either insect pests or disease. Although there is no record of when this species was introduced into horticulture, it was grown at this garden as early as 1930 when five seedlings were brought from Catalina Island by Carl Wolf. It is reported that this species can be propagated from cuttings.

Regions: 1, 2, 3, 6. Does well in the Central Valley if protected from the afternoon sun.

9. *Q. wislizenii* var. *wislizenii.* Interior live oak. Fig. 77. A medium-sized to large evergreen tree with stout spreading branches forming a round-topped crown, bark smooth or roughly fissured; leaves plane, stiff leathery, oblong, 1-4 in. long, ¾-1¾ in. broad, glabrous, dark green and glossy above, pale yellowish-green beneath, entire or spiny-toothed; acorns maturing the second year, cup cup-shaped or turbinate, covering ¼-½ of the nut, nut slender, oblong, 1-1½ in. long.

Interior live oak is native to the lower slopes of the Sierra Nevada and inner coast ranges from Ventura to Shasta and Siskiyou cos. usually at elevations below 5000 ft. where it is a member of the foothill woodland community.

This species is the most tolerant of shade of the California oaks. It is most closely related to *Q. agrifolia* with which it may be confused when young, but it may be readily distinguished from the coast live oak by the leaves which are always plane, whereas those of the coast live oak are cupped or twisted. At the botanic garden a 12-year-old multiple branched tree growing in poor soil measured 16 ft. in height with a 14 ft. spread.

Regions: 1, 2, 3, 6. Also reported to do well in the Central Valley.

A shrub form of *Q. wislizenii* native to the chaparral of southern California and extending south as far as the Sierra San Pedro Mártir Mts. has been described as var. *frutescens.*

Some species of California oak are subject to attack by the oak twig girdler (*Agrilus angelicus*). These include *Quercus engelmannii, Q. wislizenii* and *Q. agrifolia,* but it is the latter that suffers the greatest damage. The girdler causes patches of dead foliage to appear scattered throughout the entire canopy of the tree. The adult insect is a brownish-bronze beetle and the female, after mating, lays eggs on the young twigs. When the eggs hatch the young larvae bore directly into the twigs and bore in the direction of the older growth, later girdling the twig in a spiral fashion. After nearly two years in the twig the larvae pupates and later emerges as an adult, usually between May and September. Fortunately this pest can be controlled by spraying.

RHAMNUS. A genus of about 100 species belonging to the Rhamnaceae native to the Northern Hemisphere with a few species in South America. The shrubs or small trees may be either evergreen or deciduous and the flowers which are small and inconspicuous, and either 4- or 5-parted may or may not have petals and are either perfect or unisexual, if unisexual male and female flowers borne on separate plants. The fruit is a berry-like drupe with 2-4 nutlets. Five species in California.

Taxonomically the California species have been accorded different treatments by different authors. Here we follow Munz (1974).

The California species of *Rhamnus* constitute an important group of native shrubs that have not received the attention from horticulturists that they deserve. They are hardy and not particular as to soil, disease free and with no serious insect pests. Although the flowers are rather inconspicuous, fruiting plants are most attractive and the berries are an important food source for many native birds. The plants are drought tolerant but if grown in a coarse well-drained soil they benefit from a little summer irrigation. If grown in heavy soil they should have no summer watering. The plants may be grown from seed, which if not fresh, may require pretreatment to hasten germination. They may also be propagated by cuttings. The species of *Rhamnus* tolerate pruning and they may be used as formal hedges. Plantings of species with unisexual flowers must include both male and female plants to insure having berries.

The following species is extremely variable and distinct geographical races have been accorded subspecific status.

1. *R. californica* ssp. *californica.* Coffeeberry. Plate 14a. An evergreen shrub to 15 ft., bark of young twigs usually reddish; leaves oblong-elliptic, 1-3 in. long, ½-1 in. wide, 1-veined from the base, the lateral veins curving somewhat upward, shiny, dark green and glabrous above, paler and glabrous beneath, plane or more or less revolute, entire to serrate, flowers inconspicuous, perfect, usually 5-merous and borne in 6-50-flowered umbels on the current season's growth; berries ¼-⅜ in. in diameter, green, then red and finally black when ripe. April-June.

Coffeeberry is widely distributed along the coast from southern Oregon south to northern Baja California and inland to the San Bernardino and San Jacinto mts. Occurring usually below 3500 ft. it is found in many plant communities. Found in all types of soil, it prefers sandy and rocky places along the coast and hillsides and ravines farther inland.

R. californica ssp. *californica* 'Eve Chase' is a dense compact form to 8 ft. with an equal spread; another clone, 'Seaview,' can be kept below 2 ft. if upright growth is pruned out. Both of these clones, which must be propagated asexually, have foliage that is broader, flatter and greener than the typical form of ssp. *californica.*

Regions: 1, 2, 3, 6. Also the Central Valley if protected from the afternoon sun.

Ssp. *crassifolia.* Thickleaf coffeeberry. A shrub to 10 ft. with branchlets white-tomentose; leaves oval or elliptical, 1½-4 in. long, 1-2 in. wide, very thick and leathery, gray-green above, whiter beneath with dense, short, soft pubescence, margins nearly entire to serrulate.

Thickleaf coffeeberry is found in the inner north coast ranges from Napa and Lake cos. north to Trinity Co. where it grows on dry slopes and in ravines in the chaparral at elevations below 2500 ft. This variety is the best of the gray-leaved coffeeberries. If grown in partial shade the leaves will be more luxurious.

Regions: 2, 3, 6.

Ssp. *tomentella.* Chaparral coffeeberry. An erect shrub to 15 ft. with densely pubescent new growth, leaves oblong-elliptic to elliptical, 1½-3 in. long, tomentose to nearly glabrous above, white tomentose or silvery beneath, entire or obscurely serrulate, margins slightly revolute.

Chaparral coffeeberry is widely distributed in the interior chaparral areas throughout California, southward to San Luis Obispo and Fresno cos. and on the desert slopes of the Tehachapi, San Gabriel, San Bernardino, San Jacinto, Liebre and Providence mts. and south into northern Baja California, usually at elevations below 3000 ft.

Regions: 1, 2, 3, 6, and portions of 4.

Ssp. *ursina.* Rounded shrub of 15 ft.; leaves elliptical to oval, 1¼-2¼ in. long, ⅜-1⅛ in. wide, dull or bright green, glabrous above, whitish beneath with short dense hairs intermingled with long coarse hairs, margins sharply dentate, slightly revolute.

This subspecies occurs in creek bottoms and sheltered ravines in the Providence, New York and Clark mts. of eastern San Bernardino Co., eastward to Nevada, Arizona and New Mexico. In California it is found in the Joshua tree and pinyon-juniper woodland communities at elevations of 4000-7000 ft.

Regions: 2, 3, 4, 6.

2. *R. crocea.* Buckthorn, redberry. Plate 14b. A low, densely spreading evergreen shrub to 6 ft. with numerous short, rigid and spinose branchlets, rooting where touching the ground; leaves often clustered, rigidly coriaceous, obovate, elliptic, round or broadly ovate, ¼-½ in. long, nearly as wide, dark or pale green, glabrous and commonly brown or yellowish beneath, finely serrulate or glandular denticulate; flowers unisexual, small, few in axillary sessile umbels, usually 4-merous, petals absent; berry globose, scarlet, glabrous. February-April.

Redberry is found in dry washes and canyons in the coastal ranges from Lake Co. south to northern Baja California and in many plant communities. It usually occurs at elevations below 3000 ft.

In cultivation the form of the plant may be improved by careful pruning. Redberry is most attractive from August to October when the plants may be covered with small bright scarlet fruits. To insure fruiting, both staminate and pistillate plants must be included in the plantings. *R. crocea* is reported to be less drought tolerant than the following.

Regions: 1, 2, 3, 6. Does well in the Central Valley if protected from the afternoon sun.

3. *R. ilicifolia.* Hollyleaf coffeeberry. Evergreen shrub, often tree-like to 15 ft., or branched from the base; leaves abundant, holly-like, broadly ovate to nearly round, ¼-1¼ in. long, ½-1 in. wide, dark green and glabrous above, paler and often brownish beneath, spinulose-toothed, deeply serrate or entire; flowers small, in sessile umbels, petals usually absent; fruit oval, about ¼ in. in diameter, bright red. March-June.

Hollyleaf coffeeberry is found on dry slopes from northern Baja California to the Providence Mts. in eastern San Bernardino Co., north in the middle and inner coast ranges to northern California where it is common

R in the chaparral and yellow pine forest communities, usually below 5000 ft.

R. ilicifolia is drought tolerant and a good plant for dry banks and for informal screens or hedges in hot dry areas.

Regions: 1, 2, 3, 6, and selected areas in 4.

4. *R. pirifolia.* Island coffeeberry. Tree-like to 30 ft., evergreen; leaves abundant, elliptical to almost round, 1-2 in. long, mostly crenulate to entire. March-June.

Island coffeeberry is an insular species known only from Santa Catalina, San Clemente, Guadalupe and Santa Cruz islands where it grows in the coastal sage scrub and island chaparral communities, usually away from the immediate coast.

Regions: 1, 2, 3, 6.

RHUS. A genus of perhaps 120 species belonging to the Anacardiaceae. Shrubs or trees with resinous, acrid or milky sap; leaves simple or compound, alternate; flowers inconspicuous; fruit a small dry or semi-fleshy drupe. Widely distributed. Four species in California.

1. *R. integrifolia.* Lemonade berry. Fig. 78. A medium-sized aromatic evergreen shrub or sometimes tree-like to 30 ft., twigs stout; leaves variable in shape, rarely lobed or trifoliate, blades elliptical or ovate to nearly orbicular, 1-2½ in. long, ¾-1½ in. wide, rounded at both ends, stiff and leathery, edges smooth or with shallow lobing and sharp teeth, dark green and shiny above, paler and prominently veined beneath, usually flat; flowers in small, dense, terminal compound spikes, white to pinkish, subtended by roundish, hairy bracts; fruit a pubescent reddish somewhat flattened drupe, ¼-½ in. long, usually covered with a waxy secretion. February-March.

Lemonade berry is found on bluffs and mesas along the coast from Santa Barbara Co. south to northern Baja California and inland as far as Cahuenga Pass, Cucamonga Wash and the foothills east of San Diego. It is also found on some of the offshore islands. A member of the coastal sage scrub and chaparral communities it is found at elevations below 2500 ft.

R. integrifolia is a valuable garden plant in coastal areas and although it is drought tolerant, it responds well to good garden care and additional water during the

Fig. 78. *Rhus integrifolia.* This hedge was planted at the Santa Barbara Botanic Garden in the 1940's.

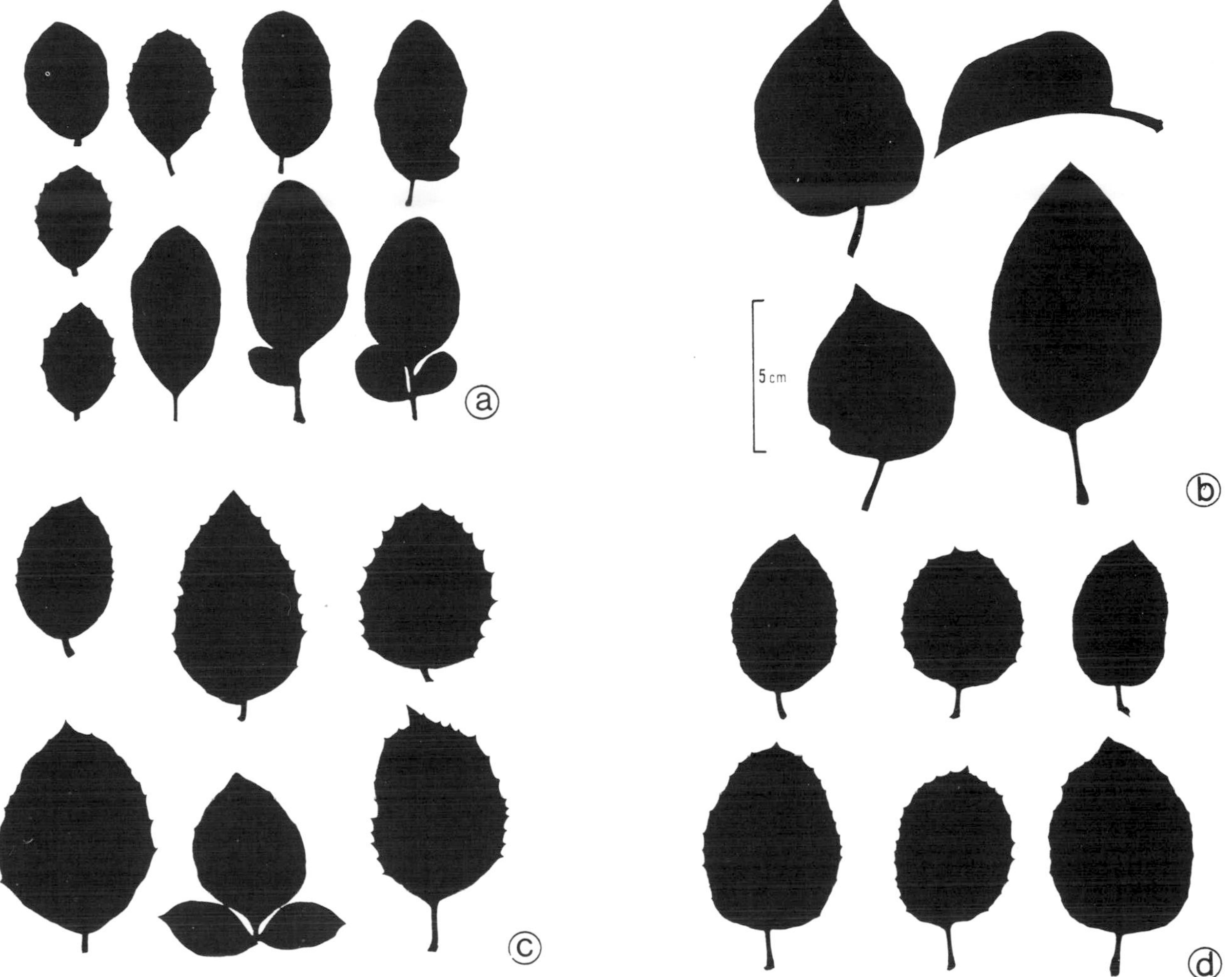

Fig. 79. Range of variability in leaf morphology in *Rhus*; a, *R. integrifolia*; b, *R. ovata*, note folded leaf; c, putative hybrid between *R. integrifolia* and *R. ovata*; d, artificial hybrids. (Young, 1974).

summer. It is reported to take ordinary garden watering if drainage is good. It is useful on banks and slopes and tolerates salt-laden winds. It can be used as a windbreak, for screening purposes as a clipped hedge or espaliered. It is reported to be very susceptible to verticillium wilt.

Regions: 1, 2, 3. Also the Central Valley if protected from afternoon sun.

2. *R. laurina.* Laurel sumac. See *Malosma laurina.*

3. *R. ovata.* Sugarbush. Plate 131. A large, rounded, densely branched aromatic shrub or small tree to 20 ft. with stout branches; leaves simple, leathery, folded along the midrib trough-like, deep green and shiny above, lighter beneath, broadly ovate with an acute tip, cuneate or obtuse at the base, 2-4 in. long, 1-3 in. wide; flowers borne in a rather dense, terminally compound spike, buds reddish, flowers small, creamy with tinges of red; fruit a pubescent drupe with red glandular hairs. March-May.

Sugarbush is abundant on dry bluffs and south-facing slopes in inland canyons mainly away from the coast from Santa Barbara Co. south well into Baja California at elevations of less than 4000 ft. It is also found on some of the offshore islands and in central Arizona. It is a member of the coastal sage scrub and chaparral communities.

Sugarbush remains an attractive green shrub throughout the year and the colorful red buds, which form early in the season, give added color to the plants for long periods of time before the creamy-white flowers open. Once established the plants are drought tolerant and they are not particular as to soil. Occasionally a twig blight affects the plants both in the wild as well as in the garden. Reported to be susceptible to oak root fungus.

R. ovata may be used much the same as the lemonade berry for which, in inland areas, it should be substituted.

Regions: 1, 2, 3, 4, 5, 6. Also does well in the Central Valley if protected from the afternoon sun and in southern Arizona.

Although generally occupying slightly different habitats, sugarbush and the lemonade berry are in some areas in southern California sympatric and it has been shown (Young, 1974) that the two species hybridize rather frequently, particularly in the Santa Monica Mts. and on Catalina Island. Young believes that much of the confusion that has existed in the past in delimiting *R. integrifolia* and *R. ovata* is due in large part to hybridization between the species and the backcrossing of the hybrids with the parental species, *i.e.,* introgressive hybridization (Fig. 79). This is of importance to the horticulturist since some of the hybrids are exceptionally handsome plants worthy of perpetuation. A particularly attractive one grown at this botanic garden has been designated *R.* 'Claremont' and is propagated from cuttings. If uniformity of plantings of *R. integrifolia* and *R. ovata* is to be achieved, seed must be collected where the two species do not grow together.

4. *R. trilobata* var. *trilobata.* Squawbush. A diffusely branched and spreading shrub to 8 ft. and equally broad, bark smooth, brown; leaves deciduous, trifoliate or 3-5-parted, usually 3 in. or less in length, terminal leaflet ¾-2 in. long tapering to the base, lobes variously toothed; flowers pale yellow, appearing before the leaves, or with the leaves; fruit a fleshy berry-like reddish drupe covered with a viscid secretion.

Var. *anisophylla.* A dwarf, stout, short-branched shrub with puberulent twigs becoming glabrous in age, leaves small, distinctly trifoliate; spikes of flowers short, almost head-like; fruit bright crimson.

This variety is found on dry slopes in the mountains of the Mojave and Colorado deserts at elevations between 3500-5000 ft. where it is a member of the creosote bush scrub and juniper-pinyon woodland communities. It extends eastward to Utah and Arizona. At Claremont var. *anisophylla* has not spread to form the large clumps characteristic of var. *quinata.*

Var. *pilosissima.* A diffusely branched, spreading, deciduous shrub to 5 ft., often as broad or much broader, branches often turning down and rooting at the tips, strongly scented when crushed, ultimate branchlets rather coarse, heavily pubescent; leaves trifoliate or apparently 3-5-parted, leaflets rhombic-ovate to cuneate-obovate, obtuse, crenate, terminal leaflet ⅜-1¼ in. long and about as broad, larger than the lateral leaflets; flowers small, in clustered spikes, yellowish, appearing before the leaves; fruit a fleshy, berry-like drupe, reddish, covered with a viscid secretion.

This variety is widely distributed in cismontane southern California to the Little San Bernardino Mts., north to Butte Co. It occurs in canyons and washes especially in the interior valleys at elevations below 3500 ft. It is a member of the coastal sage scrub, chaparral and southern oak woodland communities.

Var. *quinata.* Plate 14c. The ultimate twigs are slender and puberulent, the terminal leaflet is usually 3-lobed so that the entire leaf appears to be 5-lobed. This variety, which is uncommon, is found on dry slopes and in thickets from northern Baja California throughout cismontane California north to Oregon. In California it is found in the coastal sage scrub, chaparral and foothill woodland communities. The plants spread to form large clumps.

Squawbush has done well at Claremont both in heavy soil and in rocky and sandy alluvial deposits and has been free from both insect pests and disease. Because of their fibrous root systems and the fact that the plants spread by rooting of stem tips that touch the soil, as well as by suckering, squawbush is recommended for soil stabilization and erosion control.

→

Plate 14. a, *Rhamnus californica* ssp. *californica*; b, *Rhamnus crocea*; c, *Rhus trilobata* var. *quinata,* yardstick shown for scale; d, *Ribes aureum* var. *gracillimum*; e, *Ribes sanguineum* var. *sanguineum*; f, *Ribes speciosum*; g, *Ribes viburnifolium*; h, i, *Romneya coulteri*; j, *Rosa gymnocarpa*; k, *Rosa minutifolia*; l, *Salazaria mexicana.*

Plate 14

R. trilobata is easily propagated by cuttings or root sections and less easily by seed which, in addition to having impervious seedcoats also has dormant embryos which require pretreatment to hasten germination.

Regions: 2, 3, 6, and selected areas in 4, 5. Also does well in the Central Valley if protected from the hot afternoon sun. The variety *anisophylla* is reported to do well in Region 1.

RIBES. A genus of the Saxifragaceae containing about 120 species native to the North Temperate Zone and the Andes. The shrubs are either unarmed or with nodal spines and sometimes internodal bristles. Those with spines are known as gooseberries and are sometimes placed in a separate genus, *Grossularia.* The unarmed species are known as currants. Thirty-one species in California.

All members of the genus listed here make satisfactory garden plants and are drought tolerant as well as tolerant of many soil types. In southern California they are best given filtered shade or protection from the afternoon sun. They respond well to good garden care and given additional irrigation the deciduous forms will retain their leaves for a longer period of time. Indeed the major complaint about members of the genus is that (except for the one evergreen species) because they are deciduous, the plants are unattractive for part of the year.

Propagation is by seed, cuttings or division. In this area insect pests and diseases have not been problems.

1. *Ribes aureum* var. *aureum.* Golden currant. An intricately branched, unarmed shrub to 8 ft. with gray or brown bark; leaves deciduous, clustered on short lateral branchlets, obovate to round-reniform in outline, ½-2 in. broad, mostly 3-lobed, the lobes rounded, entire or toothed, glossy bright green; racemes 1-2½ in. long, 5-15-flowered, flowers yellow with a spicy odor; berry globular, about ¼ in. in diameter, mostly red or black. February-March.

Golden currant is common in moist places in many areas of California at elevations of 2500-7800 ft. The species extends to the Rocky Mts. and north to British Columbia.

Regions: 2, 3, 6.

Var. *gracillimum.* Plate 14d. Differs from the preceding by flowers lacking fragrance, by having a more slender perianth tube, and flowers soon turning red or reddish and with berries usually orange or yellowish. The variety is found in brushy often alluvial places in the southern oak woodland community from western Riverside Co. north to central California.

Regions: 2, 3, 6. Also reported to do well in the Central Valley if protected from the afternoon sun.

2. *R. indecorum.* White-flowered currant. An erect, open, deciduous shrub to 6 ft. with dark brown, shreddy bark on older growth and with tomentose and glandular young growth; leaves round in outline, 3-5-lobed, ¾-1½ in. in diameter, thick, stipitate-glandular, finely rugose, dark green above, whitish tomentose beneath, the lobes obtuse and crenulate; flowers white, about ¼ in. long in compact glandular-pubescent racemes 1-2 in. long; berry globose, about ¼ in. in diameter, viscid-pubescent and with stalked glands.

White-flowered currant occurs in interior washes and canyons from Santa Barbara Co. south to northern Baja California at elevations below 2000 ft. in the coastal sage scrub and chaparral communities.

R. indecorum is one of the earliest native shrubs to come into bloom and may be in flower as early as November.

Regions: 1, 2, 3, 6. Also the Central Valley if protected from the afternoon sun.

3. *R. malvaceum* var. *malvaceum.* Chaparral currant. An erect, deciduous shrub to 8 ft. with straight, stout stems from the base, young branches tomentose with glandular, bristly hairs, bark brown; leaves roundish in outline, rather thick, rugose, ¾-3 in. wide, dull green and rough above, with stalked glands, glandular and gray-pubescent beneath, 3-5-lobed, lobes obtuse and doubly toothed; racemes drooping, 2-4 in. long, 10-25-flowered, flowers light pink to bright rose-colored, or sometimes nearly white, about ½ in. long; berry globose, about ¼ in. in diameter, blue with a glaucous bloom and more or less glandular-hairy. February-April.

Chaparral currant occurs from Los Angeles Co. north to Tehama Co. and is also found on Santa Cruz Island. It is found on dry wooded slopes or on open hills in the chaparral, foothill woodland and closed-cone pine forest communities at elevations below 2500 ft.

Var. *viridifolium.* Differs from the preceding in having leaves greener beneath and more or less scabrous with a coarser more glandular pubescence. The variety is found in dry gullies and canyons from the Santa Monica Mts. to the San Jacinto and Santa Ana mts. where it is found in the chaparral at elevations below 5000 ft.

Regions: 1, 2, 3, 6. Also does well in the Central Valley if protected from the afternoon sun.

4. *R. sanguineum* var. *sanguineum.* Red-flowered currant. Plate 14e. An erect, spreading, deciduous shrub usually 4-6 ft. tall with brownish shreddy bark; leaves round-reniform, often wider than long, 1-2½ in. wide, 3-5-lobed, lobes obtuse, irregularly toothed, dark green and puberulent above, whitish-pubescent to tomentose beneath; racemes erect or ascending, 2-4 in. long, 10-20-flowered, flowers deep pink to deep red; berry globular, ¼-⅜ in. in diameter, blue-black, covered with a whitish bloom. January-March.

Red-flowered currant occurs from Lake Co. north to British Columbia in moist, shady places in the yellow pine and red fir forest communities at elevations between 2000-6000 ft.

Var. *glutinosum.* Differs from the preceding in having pendulous racemes and leaves that are more sparsely pubescent and greenish beneath. It occurs from Santa Barbara Co. north to Del Norte Co. in open places and among brush and trees in the chaparral, foothill woodland, close-cone pine and mixed evergreen forest communities at elevations below 2000 ft.

Red-flowered currant is perhaps the most beautiful of all the native species of *Ribes* and it has been common in cultivation for many years. Because of the great variation in flower color, numerous forms have been named and propagated. For southern California var. *glutinosum* is recommended. Although fairly drought tolerant, the plants do better if they receive moderate amounts of water and in southern California partial shade.

Regions: 1, 2, 3, 6.

5. *R. speciosum.* Fuchsia-flowered gooseberry. Plate 14f. A tall shrub to 10 ft. with numerous horizontal branches, stems with 3 very sharp spines at the nodes, and more or less densely bristled between the nodes; leaves round, oblong or obovate, thick and leathery, dark green and shiny above, paler beneath, ¾-1½ in. long, slightly 3-5-lobed, or toothed at the apex, entire along the lower half of the leaf, glabrous or slightly glandular-hairy above; flowers deep crimson, 1-4 on drooping peduncles from short lateral branchlets; stamens much exserted; berry ovoid, about ⅜-½ in. in diameter, densely glandular-bristly, becoming dry. January-May.

Fuchsia-flowered gooseberry occurs near the coast from Santa Clara Co. south to northern Baja California where it is found in the coastal sage scrub and chaparral communities at elevations below 1500 ft.

In full flower, *R. speciosum* is certainly one of California's most spectacular native shrubs and the common name is very appropriate, the long red flowers hang thickly along the undersides of the long horizontal branches very much like those of *Fuchsia magellanica.* If possible the plants should be sited so that the view is directed upward into the plants rather than down onto them. Hummingbirds are fond of the fuchsia-flowered gooseberry and are commonly seen feeding from the flowers.

Drought tolerant and free from insect pests and disease, perhaps the greatest drawback of this species is its spines which are both sharp and produced in large numbers.

Regions: 1, 2, 3, 6. Also does well in the Central Valley if protected from the afternoon sun.

6. *R. viburnifolium.* Catalina currant, viburnumleaf currant. Plate 14g. A straggling, unarmed, evergreen shrub to 6 ft. or more, stems tending to spread horizontally, young stems and branches resinous-glandular; leaves leathery, fragrant, ovate, obovate to nearly round, dark green and glabrous above, lighter and resinous-dotted beneath, not lobed, ¾-1½ in. in diameter; racemes 2-3 from each node, few-flowered, flowers small, rose-colored; berry globular, about ¼ in. in diameter, glabrous, red. February-April.

Viburnumleaf currant is known only from Santa Catalina Island and on the mainland at All Saints Bay in Baja California. It grows among shrubs in canyons in the chaparral community.

R. viburnifolium is a very un-currant-like currant which is often mistaken for a member of several other related and unrelated genera. Viburnumleaf currant makes an admirable ground cover for shady areas inland and sunny to lightly shaded areas along the coast. It is particularly good under oaks where excessive watering is undesirable. It is also useful for erosion control. The plants are disease-free but they may at times be attacked by spider mites, especially in coastal areas. The small flowers are produced in such abundance as to give the entire plant a rosy color.

Regions: 1, 2, 3, 6. Also reported to do well in the Central Valley if protected from the hot afternoon sun.

ROMNEYA. A genus of the Papaveraceae with two species of coarse, woody-based perennials native to southern California and northern Baja California.

1. *R. coulteri.* Matilija poppy. Plate 14h, i. Large, bushy, somewhat glaucous perennials to 8 ft., spreading widely by underground rootstocks, stem leafy, branched; leaves alternate, firm, gray-green, round-ovate, to 8 in. long, pinnately divided into 3-5 main lanceolate to ovate divisions, these in turn dentate, to 2-3-cleft; sepals and summit of peduncle glabrous, peduncle not conspicuously leafy near the summit; flowers large, to 9 in. in diameter, petals 5-6, white, crepe-like in texture, fragrant; fruit a dry capsule, seeds numerous. May-July.

Matilija poppy is found in dry washes and canyons away from the immediate coast from San Diego Co. north to the Santa Ana Mts. at elevations below 4000 ft. where it is a member of the coastal sage scrub and chaparral communities.

2. *R. trichocalyx.* Differs from the preceding in having sepals and summit of the peduncles setose and peduncles that are leafy to the top. May-July.

R. trichocalyx extends from northern Baja California north to Ventura Co. at elevations below 3600 ft. A member of the coastal sage scrub and chaparral communities.

Matilija poppy is a very attractive and unusual plant which finds many uses in both ornamental and environmental horticulture. Because of its size and the invasive and spreading nature of the plants care should be taken where it is planted. Probably best used for stabilizing hillsides, along roadways, etc. Plants should have full sun and they will tolerate a variety of soil types and amounts of water. The appearance of the plants is improved if in late summer the old stems are cut back nearly to the ground. New attractive growth will appear following the first autumn rains.

If cut in the bud stage the flowers will last for several days. Propagation may be by seed or by root cuttings taken during the autumn. Hybrids between the species are known. One in the horticultural trade is 'White Cloud.'

Regions: (both species) 2, 3, 5, 6. *Romneya trichocalyx* does well in Region 1.

ROSA. A genus of the Rosaceae with an unknown number of species (over 100), of erect, spreading or climbing, deciduous or evergreen shrubs, usually with prickly stems; widely distributed in temperate and subtropical regions. Leaves pinnately-compound, flowers large, solitary, in corymbs or panicles, in ours mostly rose-pink, petals usually 5 but as many as 8; fruit a fleshy, berry-like structure called a hip. Nine species in California.

Horticulturally the California species are much alike, requiring the same treatment, and all are useful for environmental landscaping where they may be used for barrier plantings, for soil stabilization and for supplying wildlife cover and food. In addition the plants are very attractive both in flower and fruit. Because of their wide distributions it is reasonable to expect that the species may consist of ecotypes, and propagating material should be obtained from populations adapted to the environment into which the plants are being placed.

1. *R. californica.* California rose. Stout, much-branched shrubs to 6 ft., branchlets glabrous or pubescent, nonglandular, or with gland-tipped hairs, prickles few to many, mostly stout and recurved; leaflets 5-7, ovate to oblong, to 1½ in. long, glabrous or pubescent above, paler beneath and finely pubescent to distinctly glandular, margins singly- or doubly-serrate; flowers few to many in much-branched clusters, petals about ¾ in.

→

Plate 15. a, *Salvia apiana*; b, *Salvia clevelandii*; c, *Salvia leucophylla*; d, *Salvia pachyphylla*; e, *Sambucus mexicana;* f-i, *Simmondsia chinensis*, f, plant habit, g, male flowers, h, female flowers, i, fruit; j, *Spiraea douglasii*; k, *Styrax officinalis* var. *fulvescens*; l, *Tetracoccus dioicus*.

Plate 15

R

long; fruits globose to ovoid, with persistent foliaceous appendages. This species is extremely variable. May-August.

California rose is common everywhere throughout California on moist valley flats, along streams, rivers and around springs, often forming sizeable thickets. The species also extends south to Baja California and north to southern Oregon. Many plant communities.

Regions: 1, 2, 3, 6, 7. Also the Central Valley if protected from the afternoon sun.

2. *R. gymnocarpa.* Wood rose. Plate 14j. Slender, deciduous shrubs to 3 ft., stems with long, straight, slender prickles and numerous bristles; leaflets 5-7, elliptic-ovate or roundish, to 1 in. long and ½ in. wide, usually glabrous on both surfaces, dark green and shiny above, paler beneath, margins doubly-serrate; flowers solitary or in clusters of 2-4, pink, petals about ½ in. long; fruit globose or elliptical, without foliaceous appendages. May-July.

Wood rose is found in shady woods from the Palomar and San Gabriel mts. north to British Columbia and east to Montana. In California it is usually at elevations below 5000 ft.

Regions: 1, 2, 3, 6.

3. *R. minutifolia.* Small-leaf rose. Plate 14k. Diffusely-branched shrub to 4 ft.; prickles on the branches numerous, straight, stout, bristles present; leaflets 5-7, minute, about 3/16 in. long with 5-9 shallow lobes and a few gland-tipped hairs, pubescent on the lower surface and slightly pubescent on the upper surface; flowers about 1¼ in. in diameter, petals deep rose-purple; fruit densely spiny. December-May.

Small-leaf rose is found in arroyos and on mesas near the coast in northwestern Baja California where it is a member of the maritime desert scrub community.

R. minutifolia is unknown in horticulture. The species was first grown at this botanic garden in the early 1950's from cutting grown material. Since the plants do not set seed it is reasonable to assume that all the cuttings were taken from a single plant. Small-leaf rose is a very attractive plant that has continued to spread so that at the present time it forms a large thicket. Grown on heavy loam, the plants do not receive any summer irrigation and we have had no losses. Besides being an unusual rose it is ideally suited for erosion control and soil stabilization in areas along the coast. Difficult to propagate from cuttings, seed should be collected from native populations.

Regions: 1, 2, 3.

RUBUS. A large genus with an undetermined number of species native to many parts of the world. Many are important horticultural plants (blackberry, loganberry, boysenberry, etc.). Taxonomically a difficult group.

1. *R. vitifolius.* California blackberry. An evergreen mound-builder, trailer or partial climber with long tip-rooting canes as much as 20 ft. long with straight, bristle-prickles; leaves pinnately 3-5-foliate, rarely simple and lobed, 3-6 in. long, leaflets oblong-ovate to triangular, 2-3 in. long, sharply and doubly serrate, midribs and lateral veins very prickly; flowers white to pinkish, ¾-1½ in. across, functionally unisexual; fruit more or less oblong, drupelets adhering and falling with the fleshy receptacle. March-May.

California blackberry is widespread throughout much of California, usually in woods and somewhat damp places below 4000 ft. in many plant communities.

R. vitifolius is a rampant growing, coarse, very prickly plant perhaps best used as a living barrier either growing on and over fences or spreading on the ground. When ripe, the black, juicy fruits are sweet and are often gathered for jams and jellies.

Propagation by cuttings is easy and the plants grow rapidly.

Regions: 2, 3, 6.

2. *R. ursinus.* California blackberry. Differs from the preceding mainly in having dull or gray-green leaves which are felted-tomentose beneath.

R. ursinus is widely distributed in California, usually at elevations below 3000 ft. and in many plant communities.

Horticulturally the two species are interchangeable. The boysenberry, loganberry and youngberry are supposed to have been developed from the species.

Regions: 1, 2, 3, 6.

SALAZARIA. A genus of the Lamiaceae with a single species of shrubs native to southwestern United States, north to Utah and south to Sonora and Baja California.

1. *S. mexicana.* Bladdersage. Plate 14l. An intricately-branched soft-woody shrub to 3½ ft. with spinescent branchlets; leaves oblong-ovate to broadly lanceolate, to ¾ in. long and ¼ in. wide, entire or irregularly toothed, glabrous; flowers borne in loose, spike-like clusters, to 4 in. long, corolla bi-lipped, deep purple, calyx in fruit becoming enlarged and bladder-like. April-May.

Bladdersage occurs in dry washes and rocky canyons in desert areas from Inyo Co. to Riverside Co., north to Utah, east to Texas and south to northern Mexico. In California it is usually found below 5000 ft. where it is a member of the creosote bush scrub and Joshua tree woodland communities.

S. mexicana is an attractive and unusual shrub suitable for desert landscaping where it can be used for bedding, as accent plants or as a medium-height ground cover.

The plants require good drainage and benefit from occasional watering during the summer.

Regions: 4, 5. Also southern Arizona and northern Sonora.

SALVIA. A very large genus of the Lamiaceae with perhaps 500 species of mostly aromatic herbs or shrubs; widely distributed. Some used as ornamentals or for flavoring. Flowers 2-lipped and borne usually in whorls in more or less interrupted spikes, panicles or racemes, ovary deeply 4-lobed, developing into 4 nutlets. The species treated here belong to section *Audibertia,* all members of which are native to the arid portions of the southwest.

Our species very frequently hybridize and the identification of individual plants at times is difficult. Horticulturally a valuable group of plants.

1. *S. apiana.* White sage. Plate 15a. An erect, whitish shrub to 8 ft., growth of the current year a few long, straight, almost herbaceous stems from a woody base; leaves opposite, crowded at the base, oblong-lanceolate to oval, to 4 in. long and 1½ in. wide, whitened on both surfaces with minute appressed hairs, finely wrinkled above, margins crenulate; flowers few in sessile loose clusters arranged in branching spike-like panicles to 4 ft. long, corolla white, often spotted with lavender, lower lip much enlarged and 3-lobed, middle lobe cupped, rounded and fringed, upper lip very short, stamens long-exserted. May-September.

White sage is very common and widespread on mesas and in canyons at lower elevations from Santa Barbara Co. south to northern Baja California, extending inland to the western edge of the Colorado Desert and Cajon Pass, usually below 5000 ft. it is a member of the coastal sage scrub, chaparral and yellow pine forest communities.

S. apiana is easily recognized by its white leaves and flowers borne on long willow-like stems. An important browse plant it also is important as a bee-plant (as are the other species of *Salvia*). This species hybridizes with nearly all the species of *Salvia* with which it comes in contact.

Regions: 1, 2, 3, 6.

2. *S. clevelandii.* Cleveland sage. Plate 15b. A low, compact, rounded shrub to 4 ft. with a very strong fragrance, young stems reddish, covered with short hairs, leaves opposite, elliptical to obovate, to 2 in. long and ⅓ in. wide, rugose, grayish-green above, tomentulose beneath, margins crenulate to entire; flower heads many-flowered, solitary to 2-3, separated from one another by 1-2 inches, corolla rich bluish-purple or violet-blue, lower lip with 2 lateral lobes, upper lip reflexed, stamens exserted. May-August.

Cleveland sage occurs on dry slopes in the mountains of middle and western San Diego Co., south into Baja California, usually below 3000 ft. in the coastal sage scrub and chaparral communities.

S. clevelandii is one of the finest of the California sages and is highly recommended. Easily propagated from seed, the plants prefer a well-drained soil and a sunny situation. They should be kept dry during the summer. Hard-pruning following blooming will improve the plant's overall appearance and increase the number of flowers produced the following year.

One of the most desirable features about Cleveland sage is the delightful fragrance which is quite different from that of the other native salvias. Unfortunately *S. clevelandii* hybridizes with several other species and some

Plate 16

of the hybrids are less attractive than the species itself. To insure having pure *S. clevelandii* the plants should be propagated from cuttings which root readily.

Regions: 1, 2, 3, 6. Also the Central Valley if protected from afternoon sun.

Salvia 'Allen Chickering' is a very attractive hybrid between *S. clevelandii* and *S. leucophylla* which appeared spontaneously in the botanic garden in Orange Co. It is doubtful if the true F_1 hybrid is still in existence and plants now bearing that name have all been grown from seed and therefore are segregates of the original hybrid, some of which may approach in appearance the original 'Allen Chickering.'

3. *S. leucophylla.* Purple sage. Plate 15c. An upright, much-branched shrub to 6 ft. with grayish-white tomentose stems; leaves opposite, ovate to oblong-lanceolate, to 3 in. long and ¾ in. wide, finely wrinkled, grayish above, white-tomentose beneath, margins crenulate; flowers numerous in 3-5 compact, whorled clusters, corolla light purple. May-June.

Purple sage is found on dry foothills of the coastal ranges from San Luis Obispo Co. south to Orange Co. Also in Kern Co. It is usually found in the coastal sage scrub community below 2000 ft. elevation.

Regions: 1, 2, 3, 6.

4. *S. mellifera.* Black sage. An erect, very aromatic shrub to 5 ft. with a woody base and puberulent, more or less glandular herbaceous branches; leaves oblong to oblong-elliptic, to 3½ in. long and ¾ in. wide, green and wrinkled above, paler and tomentose beneath, margins crenulate; flowers numerous, in compact whorled clusters arranged in an interrupted spicate inflorescence, flowers pale blue to lilac, occasionally white. April-May.

Black sage is common on foothills and dry lower mountain slopes from Contra Costa Co. southward to northern Baja California. It extends inland as far as the San Jacinto Mts. and Cajon Pass. Also found on Santa Catalina and Santa Cruz islands. It is a member of the coastal sage scrub and chaparral communities.

Regions: 1, 2, 3, 6.

5. *S. pachyphylla.* Mountain desert sage. Plate 15d. A compact, rounded shrub to 2 ft. with spreading and ascending branches and scurfy-pubescent branchlets; leaves mostly obovate, to 1¼ in. long, rounded at the apex, margins entire, densely white-pubescent on both surfaces; flowers in dense clusters arranged in rather tight spikes to 5 in. long, flowers subtended by showy, rosy or purplish bracts to ¾ in. long, these more or less ciliate on the margins, corolla dark violet-blue or sometimes rose-colored. July-September.

Mountain desert sage is found on dry slopes and flats in the Panamint, Kingston, Clark, New York, San Bernardino and Santa Rosa mts. south into Baja California usually at elevations of 5000-10,000 ft., where it is a member of the pinyon-juniper woodland and yellow pine forest communities.

S. pachyphylla is an extremely handsome species well worth growing at the higher elevations. In nature it usually is found on well-drained slopes growing in decomposed granite. At lower elevations the species tends to be short-lived.

Regions: 4 (higher elevations), 6, 7.

All the salvias are useful for covering dry banks and for controlling soil erosion. Their cultural requirements are much the same; well-drained soil and no summer watering. The plants will tolerate water and may even look better than those kept dry, but the plant's life will be shortened. All are easily propagated by seed or by cuttings.

SAMBUCUS. A genus of the Caprifoliaceae with about 20 species of deciduous herbs (rarely), shrubs or trees, widely distributed in temperate and subtropical regions. Taxonomically a difficult and confusing group with many names being used. Five species in California.

1. *S. mexicana.* Mexican elderberry. Plate 15e. A deciduous shrub or small tree to as much as 30 ft. with very pithy branches; leaves opposite, odd-pinnately compound with 3-5 leaflets, leaflets ovate or oblong-lance-

←

Plate 16. a, b, *Trichostema lanatum*; c, *Vitis californica*; d, e, *Umbellularia californica*; f, *Washingtonia filifera*; g, *Xylococcus bicolor*; h, *Yucca baccata* var. *baccata*; i, *Yucca schidigera*; j, *Yucca whipplei* ssp. *parishii*; k, *Yucca brevifolia* var. *brevifolia* showing clump formation from underground rootstocks; l, *Yucca brevifolia* var. *jaegeriana*. No clumping. The two varieties were grown under identical conditions.

S

olate, terminal one often much larger than the lateral ones, yellowish-green, glabrous, margins finely serrate; flowers numerous, small, cream-colored, borne in flat-topped branched clusters; fruit a berry-like drupe about ¼ in. in diameter, bluish with a white bloom. March-September.

Mexican elderberry occurs on open flats and in cismontane valleys and canyons from central California south to northern Baja California, and in the desert mountains east to Arizona. In California it occurs in the coastal sage scrub, chaparral and southern oak woodland communities below 4500 ft.

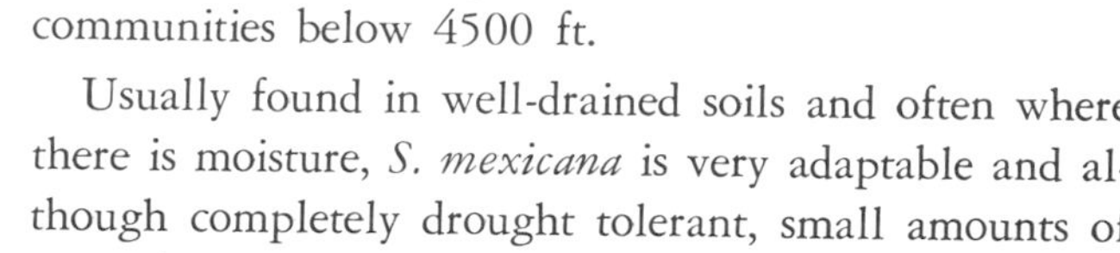

Usually found in well-drained soils and often where there is moisture, *S. mexicana* is very adaptable and although completely drought tolerant, small amounts of water help hold foliage on the plants which otherwise drop their leaves by about September.

Mexican elderberry is a fast-growing plant whose ap-

Fig. 80. a, *Sequoiadendron giganteum,* juvenile form; b, more mature form; c, *Sequoia sempervirens.*

pearance is enhanced by hard-pruning during the dormant season. It may be used for screening, as a specimen or for background plantings. The fruits, although not very flavorful, are edible. The plants are used as browse by livestock and the fruits are a valuable source of food for birds.

Regions: 1, 2, 3, 4, 5, 6. Also does well in southern Arizona.

S. caerulea, a species morphologically very similar to *S. mexicana* differing from the latter mainly in having 5-9 leaflets rather than 3-5, is found in the yellow pine forest in southern California. From southern California the species extends north to British Columbia, Alberta and Idaho, east to Arizona and south to Sonora and Chihuahua, Mexico. *S. caerulea* has not been successful in Claremont and may represent in southern California, a pine-belt ecotype of a wide-ranging, polymorphic species.

Region: 7.

SEQUOIA. A genus of the Taxodiaceae with a single species.

1. *S. sempervirens.* Coast redwood. Fig. 80c. Tall evergreen trees with narrow crowns and columnar trunks, branching horizontally or branches sweeping downward, bark thick, red, fibrous; leaves linear, mostly ½-¾ in. long, spreading right and left forming flat sprays, sharp-pointed, dark green and shiny above or on adult trees leaves sometimes short, linear or awl-shaped strikingly like those of *Sequoiadendron giganteum*; cones oval, reddish-brown, to 1¼ in. long, borne in clusters or at the ends of branchlets, mostly toward the top of the tree, seed maturing the first autumn.

Coast redwood is confined to a narrow strip 450 miles long and averaging 20 miles in width from Curry Co., Oregon south to Monterey Co. at elevations to 3000 ft. It is the dominant species in the redwood forest community.

The coast redwood is one of the best known of California trees and it has been planted in many parts of the world. Growing outside of its native habitat, in southern California the trees may be subject to stress caused by lack of sufficient water, poor quality of the water, high temperatures, low humidity, drying winds and perhaps air pollution. The plants may respond by showing yellowing of the foliage, especially new growth, due to lack of iron, slow stunted growth from insufficient water and the burning of the leaves caused by the high salt content of the water.

Morphologically, *S. sempervirens* is a variable species, especially in plant habit and the texture and color of the foliage. In recent years some of the more striking variants have been propagated asexually and are on the market as named varieties.

Best in coastal areas. Sunset's *New Western Garden Book* details many ways this species may be used in ornamental horticulture.

Regions: 1, 2, 3.

SEQUOIADENDRON. A genus of the Taxodiaceae with a single species.

1. *S. giganteum.* Giant sequoia, bigtree. Fig. 80a, b. Massive evergreen trees with thick trunks even from an early age, in nature often 80-225 ft. to the first branch but under cultivation often branched to the ground, at least for a number of years, bark thick, red, deeply furrowed; leaves scale-like or awl-like, appressed with only the tips free, blue-green; cones ovoid, reddish-brown, to 3¾ in. long, seed maturing the second autumn.

Giant sequoia occurs on the western slopes of the Sierra Nevada from Placer Co. south to Tulare Co., a range of 250 miles. The trees are mostly restricted to locally favorable or protected spots where the soil is rich, deep and moist. These isolated populations are referred to as groves.

S. giganteum does best when grown in deep soil and watered infrequently but deeply. When stressed from drought, high temperatures and perhaps air pollution, the trees may be attacked by *Botryosphaeria ribis* which causes branch dieback, and if the growing tip is infected the fungus will eventually cause the death of the entire tree. This species is of limited value in southern California.

Regions: 2, 7.

SHEPHERDIA. A genus of the Elaeagnaceae containing three species, one of which is found in California.

1. *S. argentea.* Silver buffaloberry. An erect, spiny,

S deciduous shrub to 20 ft., or sometimes tree-like, bark gray, shaggy, twigs brown or silvery-scurfy terminating in a sharp thorn; leaves simple, usually opposite, mostly oblong, ¾-1¾ in. long, ¼-⅝ in. wide, silvery-scurfy on both sides, paler beneath, entire; flowers small, inconspicuous, in clusters at the nodes of the twigs, pistillate, sometimes solitary; fruit drupe-like, fleshy, red to orange, about ¼ in. in diameter, sour but edible. Attractive to birds. April-May.

Silver buffaloberry is found along streams in a few localities in Ventura, Kern, Santa Barbara and San Bernardino cos., north to Oregon and east to the Rocky Mts. In California it is usually found between 3500-6500 ft. in the sagebrush scrub, pinyon-juniper and northern juniper woodlands.

S. argentea is easily grown in almost any type of soil and it tolerates considerable alkali and any amount of cold and wind. Once established the plants are quite drought tolerant. The silvery-gray foliage is quite unlike that of any other native shrub. In order to insure fruiting both male and female plants must be included in the plantings. At this botanic garden silver buffaloberry seldom flowers and it has never fruited, the reasons being unknown as the plants have grown very well. The plants may require more chilling than they receive at Claremont.

The plants spread by suckering and they may be used for erosion control and as medium-height barrier plantings.

Regions: 1, 2, 3, 4, 6.

SIMMONDSIA. A genus of the Buxaceae with a single species native to arid regions of southwestern United States and northern Mexico.

1. *S. chinensis.* Jojoba, goatnut. Plate 15f-i. A rigidly-branched evergreen shrub to 10 ft. and equally broad, young stems yellowish-green, pubescent, branching dichotomously; leaves simple, opposite, held in an upright position, oblong-ovate, rounded at both ends, to 2 in. long and 1 in. wide, thick and leathery, pale green or more often greenish-yellow, minutely pubescent on both surfaces, margins entire; flowers unisexual and without petals, borne on separate plants, staminate flowers small, greenish, inconspicuous, in sessile or short-peduncled head-like clusters, pistillate flowers about ½ in. long, solitary on short peduncles; fruit a smooth cylindrical capsule somewhat resembling an acorn, about ¾ in. long, containing 1-3 brown to black seeds.

Jojoba is found on arid slopes in San Diego, Imperial and Riverside cos., east to Arizona and southward to Sonora and Baja California, usually at elevations below 5000 ft. in the creosote bush scrub, Joshua tree woodland and chaparral communities. It is also found on San Clemente and Cedrus islands.

S. chinensis presents no cultural problems when grown in hot inland areas and after establishment the plants can survive with only natural precipitation, although growth is more rapid, plants look better and seed production is increased if they do receive some additional water. At this garden we have had no losses among the plantings of this species which is free from insect pests and disease.

Jojoba is one of the few desert shrubs adapted to formal landscaping; the plants with their dense, mounding growth tolerate clipping and shaping and they may be used for formal hedges, accent plants, barriers or windbreaks. Also good along freeways. Highly recommended.

In recent years considerable interest has been shown in jojoba as a potentially valuable cash crop for arid lands. Information on the economic aspects of the species may be obtained from the Office of Arid Lands Studies, University of Arizona, Tucson. The natural history and cultural aspects of *Simmondsia chinensis* have been considered by Gentry (1958); the agronomic aspects of jojoba in California have been dealt with by Yermanos (1974).

Regions: 1, 2, 3, 4, 5, 6. Also good in southern Arizona and in some areas in the Central Valley.

SPIRAEA. A genus of the Rosaceae with perhaps 70 species distributed throughout the Northern Hemisphere. Many species of horticultural merit. Two species in California.

1. *S. douglasii.* Douglas spiraea. Plate 15j. Erect, deciduous shrubs to 4 ft., branches usually rather straight, ascending, bark on older stems brown, pale brown and pubescent on young branches; leaves simple, alternate, oblong ovate to elliptic, rounded at both ends, to 3 in.

long and 1½ in. wide, 1-veined from the base, glabrous above, white-tomentose beneath, margins serrate above the middle, or sometimes only at the apex; flowers small, rose-colored or pink, in rather narrow, elongated panicles 2-5 in. long; fruit a small, dry pod about ⅛ in. long. June-August.

Douglas spiraea occurs on valley flats, meadows, swales and along streams or near seepages in the mountains of northern California north to British Columbia, usually below 6000 ft., and in many plant communities from the redwood forest to the red fir forest.

S. douglasii has long been in cultivation and is one parent of numerous garden hybrids. The species presents no cultural problems, the plants require adequate moisture during the summer and in hot inland areas should be given protection from the afternoon sun. Easily propagated from stem cuttings or by division. The plants spread by suckering and soon form large clumps. Useful for stabilizing moist banks.

The rose color of the flowers is quite unlike that of almost any other California shrub.

Regions: 1, 2, should have partial shade in 3, 6.

STAPHYLEA. A genus of the Staphyleaceae containing 11 species of shrubs or small trees native to North America, Europe and temperate Asia. Several species in cultivation. One species in California.

1. *S. bolanderi.* Sierra bladdernut. Shrubs to 10 ft. or occasionally a small tree to 20 ft.; leaves opposite, pinnately trifoliate, leaflets round-ovate to orbicular, to 2½ in. long, and 1½ in. wide, glabrous, crenulate or serrulate; flowers small, white, in drooping terminal clusters, petals about ½ in. long; fruit a 3-celled bladdery capsule to 2 in. long, each cell with 1-4 globose seeds. April-May.

Sierra bladdernut is rather rare, occurring on canyon slopes in the Sierra Nevada from Tulare Co. north to Shasta and Siskiyou cos. at elevations of 1000-4500 ft. in the chaparral, foothill woodland and yellow pine forest communities.

S. bolanderi grows rapidly in many types of soil and should be given some additional water during the summer. Reported to prefer full sun. It is easily propagated by cuttings, layering or by taking root-shoots which the plants produce in abundance. An attractive deciduous shrub with unusual seedpods.

Regions: 2, 3, 6.

STYRAX. A genus of the Styracaceae with perhaps 100 species of deciduous or evergreen shrubs or trees distributed in the warmer areas of Eurasia, Malaysia and the Americas. One species in California.

1. *S. officinalis* var. *fulvescens.* California styrax, snowdrop bush. Plate 15k. An erect, deciduous shrub to 12 ft. with grayish twigs; leaves alternate, round-ovate to obovate, entire, 1-3½ in. long and ¾-3 in. broad, leathery, dull dark green, somewhat pubescent above, densely stellate-tomentose beneath; flowers numerous, in few-flowered clusters at the ends of the branchlets, white and somewhat resembling orange blossoms, petals 4-8, ½-1 in. long; fruit globose, about ½ in. in diameter, usually splitting into 3 valves exposing a single, large bony seed. April-May.

California styrax occurs on slopes and in canyons from San Luis Obispo to San Bernardino and San Diego cos. at elevations below 5000 ft. where it is a member of the southern oak woodland and chaparral communities.

When in flower California styrax is a most attractive shrub and is highly recommended for ornamental plantings. When established it is drought tolerant and withstands heat and rocky soils but inland it is best if protected from the afternoon sun.

According to some taxonomists the California plants are conspecific with those from Asia Minor.

Regions: 1, 2, 3, 6. Also the Central Valley if protected from the hot afternoon sun.

SYMPHORICARPOS. A genus of the Caprifoliaceae with about 17 species of low to medium height deciduous shrubs native to North America except for one species in China. Seven species in California.

1. *S. mollis.* Snowberry, creeping snowberry, spreading snowberry. A low, diffusely branched, decumbent, straggling shrub usually less than 18 in. tall but stems sometimes longer; leaves broadly oval or elliptical to nearly round, ½-¾ in. long, larger on sterile shoots, soft-pubescent on both surfaces, or nearly glabrous; flowers small, bell-shaped, pinkish, in pairs in the leaf-axils,

or in small clusters; fruit a porcelain-white berry-like drupe about ¼ in. in diameter. April-June, but off-season bloom is common.

Creeping snowberry is common on shaded slopes in the coast ranges, Sierra Nevada and the mountains of southern California and northern Baja California. Also on some offshore islands. Usually below 5000 ft. it is found in many plant communities.

S. mollis presents no cultural problems and it can be propagated by seed or division. The plants are drought tolerant but their appearance is better if they are given some summer water. They are at their best in light shade or on north-facing slopes. Creeping snowberry spreads by suckering and the plants soon form low, dense thickets which makes them ideally suited as bank or ground covers and for erosion control. They are also important in providing food and cover for wildlife. Widely adaptable.

Regions: 1, 2, 3, 6. A related species, *S. rivularis,* does well in the Central Valley.

TETRACOCCUS. A genus of the Euphorbiaceae with three species of small evergreen shrubs native to southern California, Arizona and northern Mexico.

1. *T. dioicus.* Tetracoccus. Plate 151. An erect, spreading shrub to 5 ft. with rather slender, reddish to grayish, glabrous branches; leaves alternate, opposite, ternate, or fascicled on short spur branches, coriaceous, lanceolate, oblanceolate to nearly linear, ½-1¼ in. long, glabrous, dull green above, brownish beneath, entire or weakly serrulate, sometimes revolute; flowers small, reddish, apetalous, staminate and pistillate flowers borne on separate plants, male flowers solitary or in few-flowered racemes, female flowers solitary, axillary; fruit a 4-lobed, 4-celled dry capsule about ¼ in. in diameter. April-May and often during the summer following rains.

Tetracoccus occurs on dry, stony slopes from southern Orange Co. to northern Baja California, to eastern San Diego Co., usually at elevations below 2500 ft. where it is a member of the chaparral community.

In cultivation tetracoccus has given little trouble and does well in coarse well-drained soils. The plant's appearance is improved however if they are given some summer irrigation. They may be pruned or sheared and should be given full sun to prevent them from becoming leggy. Although the flowers are small and without petals they are produced in such abundance that the entire plant takes on a reddish tinge.

Regions: 1, 2, 3, 6.

Tetracoccus hallii which is very abundant around Cottonwood Springs in the Eagle Mts. of Riverside Co. has been grown by this garden and the species should be again brought into cultivation. Its appearance is quite different from that of *T. dioicus.*

TRICHOSTEMA. A genus of the Lamiaceae with about 16 species of strong scented herbs or shrubs. Five species native to California.

1. *T. lanatum.* Woolly blue-curls, romero. Plate 16a, b. Rounded subshrub to 3½ ft. with many brittle stems from a woody base, often sprawling in age; leaves simple, opposite, lance-linear, to 2½ in. long, about ¼ in. wide, revolute, bright green above, glabrous, lanate-pubescent beneath; flowers borne in dense, separated clusters along the upper portions of the stems, flower parts and upper stem covered with copious pale lavender to deep purple wool, flowers bilaterally symmetrical, deep bluish-purple, stamens and pistils long exserted, curved. May-August, but some flowers nearly all year.

Woolly blue-curls is common on dry slopes in the coast ranges from Monterey Co. south to the ocean-facing slopes of San Diego Co. usually at elevations below 4500 ft. where it is a component of the chaparral. It is often associated with *Adenostema fasciculatum.*

T. lanatum has long been in cultivation in the south and is one of southern California's best known and admired natives. One reason for its popularity is that it is in flower nearly every day of the year. The plants are not particular as to soil so long as drainage is good and in poorly drained soils summer watering should be avoided. The plants in age tend to become floppy and unattractive but their appearance can be improved by cutting back the old growth almost to the base of the plant. There is great variation in the color of the wool which covers the stems and exterior parts of the flowers, and for garden use only those plants with deeply colored wool should be propagated. Propagation may be by seed or by cuttings.

Fig. 81. *Umbellularia californica.* Branched from the base, shrub typical of plants growing on poor soil and under dry conditions.

Regions: 1, 2, 3, 6. Also the Central Valley if protected from afternoon sun.

Another species, *Trichostema parishii,* tends to be neater in growth habit than *T. lanatum,* and although the species is widely distributed on dry slopes of the interior mountains of southern California from Los Angeles Co. to northern Baja California, it has never done well at this garden, perhaps because of some coastal influence affecting it at Claremont.

The flowers of *Trichostema* with their long-exserted curved stamens and pistils are admirably suited for hummingbird pollination, the pollen being deposited on the top of the bird's head.

By cutting back to strong side-growth as the flowers fade and before seed sets it is possible to maintain a continuous series of flower spikes over most of the year. This species make good cut-flowers for the home.

UMBELLULARIA. A member of the Lauraceae, *Umbellularia* consists of a single species native to California and southwestern Oregon.

1. *Umbellularia californica* var. *californica.* California bay, California laurel, pepperwood, Oregon myrtle. Plate 16d, e; Fig. 81, 82. A large, densely foliaged evergreen tree, often with a broad round-topped crown, sometimes shrubby, or on ocean-fronting bluffs even forming prostrate mats 1-2 ft. high; bark of large trunks thin, very dark reddish-brown, scaly, on young stems smooth grayish-brown; leaves alternate, greenish-yellow, glabrous, thick and leathery, oblong-lanceolate, 3-5 in. long, ¾-1½ in. broad, very pungent when crushed; flowers in small clusters, yellowish, pleasantly scented, appearing between December and March; fruit a 1-seeded drupe resembling an olive, yellowish-green becoming purplish but variable in color, about 1 in. long.

California bay is very widely distributed in California usually below 5000 ft. and is found in the chaparral, yellow pine, mixed evergreen and redwood forest communities. Its southernmost limit is on the eastern slopes of the Laguna Mts. in San Diego Co. According to Jepson, no broad-leaved tree in California has adapted itself to so great a variety of situations and the species may be suspected of consisting of a number of ecotypes.

Propagation of California bay is by seed gathered in October and November. After removing the thin flesh, the seeds should be planted immediately. The seedlings, which appear in 1-2 months, may grow to 1½ ft. the first season when they should be set into their permanent positions. Although highly tolerant of many soil types and of drought, growth will be more rapid in rich garden soil with summer irrigation and most rapid when given cool, moist growing conditions. Under ideal conditions the tree usually develops a single trunk from a large burl (Fig. 43), in poorer soils the tree usually becomes

Fig. 82. *Umbellularia californica* var. *californica* (right); a particularly fine form, 'Claremont' of *U. californica* var. *fresnensis* (left).

more shrub-like with multiple stems (Fig. 81). The plants are not tolerant of poorly drained soil where water may accumulate during the winter. At Claremont the plants have not been attacked by either insect pests or disease (except for root rot brought on by poor drainage).

In addition to propagation by seed, plants may also be grown from cuttings. *Umbellularia californica* has for many years been a favorite tree for California landscaping where it has been used for street planting, parks and for home use. It is tolerant of pruning and may be used as a hedge where its size may be kept under control for a number of years. It may also be grown in large containers for use in patios and malls.

Regions: 1, 2, 3, 6.

A form from Fresno Co. with leaves finely pubescent on the underside has been called var. *fresnensis. Umbellularia californica* var. *fresnensis* 'Claremont' is a clone of the variety which has very broad leaves and in our experience is to be preferred for horticultural purposes (Fig. 82).

The dried leaves of *Umbellularia californica* are sold as a substitute for those of sweet bay (*Laurus nobilis*).

VITIS. A genus of the Vitaceae with about 50 species of woody plants often climbing by tendrils. Chiefly tropical and subtropical in the Northern Hemisphere, two species in California.

1. *V. californica.* California wild grape. Plate 16c. Woody vines to 50 ft., sprawling and bush-like when not finding support, or often climbing to the tops of supporting trees; leaves deciduous, roundish in outline, to 5 in. in diameter, cordate at the base, thinnish and only thinly cobwebby-tomentose beneath, sometimes slightly 3-5-lobed, margins finely serrate, tendrils opposite the leaves, at least once-branched; flowers small, greenish, fragrant, numerous and borne in compound clusters; berry globose, to ½ in. in diameter, purplish with a white bloom. May-July.

California wild grape is found along streams and in canyons of the coast ranges from San Luis Obispo Co. north to Siskiyou Co., in the Great Valley and in the foothills of the Sierra Nevada from Kern Co. north to Shasta Co. It is found in many plant communities.

Regions: 1, 2, 3, 6.

2. *V. girdiana.* Desert wild grape. Differs from the preceding in having leaves that are thicker, more coarsely toothed and with more permanently cobwebby-tomentose beneath as well as berries that are smaller and with little or no bloom. May-June.

Desert wild grape is found from Santa Barbara Co. south to northern Baja California and occasionally on the desert edge from Inyo Co. south. Also on Santa Catalina.

Regions: 1, 2, 3, 4, 5, 6.

The wild grapes present no cultural problems, growing well in a wide variety of soils, but they do require some summer irrigation as they normally occur in places where water is available during the summer. Propagation is by seed or cuttings. Plants grow rapidly and may be kept within bounds by severe pruning. Ideal for covering chain-link fences, for screening, etc. The species used should be chosen on the basis of where the plants are to be grown; horticulturally the species are interchangeable.

WASHINGTONIA. A genus of the Arecaceae (Palmae in Munz and Keck, 1959) with two species native to southwestern United States and northwestern Mexico. One species in California.

1. *W. filifera.* California fan palm. Plate 16f; fig. 83. A tree with fibrous roots and massive columnar unbranched trunks to 75 ft., covered with leaf-scars or the bases of the leaf-stalks, bearing at the summit a large tuft of leaves, sometimes clothed to the ground with dead, persistent, recurved leaves; leaves gray-green, fan-shaped with much-folded blades, 3-6 ft. long with 40-60 folds, torn nearly to the middle, divisions copiously fibrous, petioles 2-5 ft. long, very stout, bearing on the margins irregular, sharp, straight or hooked teeth; flowers minute, perfect, borne in a large, branched inflorescence 8-10 ft. long, enclosed at the base by a spathe; fruit berry-like, spherical or elongated, with a pleasant taste but with thin pulp surrounding the seed, about ⅜ in. in diameter, black when ripe. August-September.

The California fan palm occurs in widely scattered groves in alkaline places in the Colorado Desert, east to western Arizona and south into Baja California, always in areas where the roots can reach a permanent source

Fig. 83. The two species of *Washingtonia*. The two front ones are *W. filifera*, the two in the rear are *W. robusta*. The two in the center are Canary Island date palms, *Phoenix canariensis*. (Bailey, *Gentes Herb.* 4: 82).

of water. Many stands are found along seepage areas of the San Andreas Fault. A member of the desert oasis community, it is found at elevations below 3500 ft. Naturalized in Kern Co.

W. filifera is widely cultivated throughout the subtropical areas of the world. Its cultivation presents no problems. Seed germinates readily and the young plants can be planted bareroot. Larger specimens are often moved but must be securely tied until the plants form a root system sufficient to support the plant. The fibrous roots penetrate deeply and spread great distances, the plants are best located where the roots will not disturb sidewalks and streets. With adequate water, the California fan palm grows rapidly and soon becomes too large for use in most home landscaping. They are ideal for use as a roadside tree. In parks or where there is sufficient space they could be used to form attractive naturalistic plantings using random spacing and trees of different ages. At this garden trees grown from seed planted in 1953, first flowered in 1977. After attaining blooming size the plants flower annually. Birds distribute the seeds and young plants may be found at great distances from the parent plant.

California fan palm is probably most attractive when allowed to form a natural 'skirt' of old dead leaves clothing the trunk. Unfortunately vandals often set fire to the skirts and, although the trees are usually not killed, their appearance is drastically altered.

W. filifera should not be planted near the coast as the young leaves are subject to rot caused by a fungus. Leaves may be damaged by temperatures in the low 20's.

Regions: 2, 3, 4 (warmer areas), 5. Also does well in southern Arizona and in the Central Valley.

The Mexican fan palm (*W. robusta*) is often planted in southern California and in general it is taller and has a more slender trunk than *W. filifera.* After carefully studying the genus, Bailey (1936) came to the conclusion that there was considerable variation in both *W. filifera* and *W. robusta* and at times it was difficult to determine the species to which an individual plant belonged. However he was able to present a series of technical characters by which the two could be distinguished. For the horticulturist the problem remains. Bailey did not rule out the possibility of hybridization between the species, saying only that it was not known whether the species did hybridize.

XYLOCOCCUS. A genus of the Ericaceae with a single species native to southern California. Allied to *Arctostaphylos* and separated from it primarily in having revolute leaves and bark that exfoliates more tardily than in *Arctostaphylos.*

1. *X. bicolor.* Mission manzanita. Plate 16g. Densely branched evergreen shrubs to 8 ft. with grayish-brown, usually somewhat persistent bark, branchlets tomentulose; leaves simple, alternate, thick and leathery, brittle, ovate, oval, or oblong-elliptic, ¾-2½ in. long, ½-1 in. wide, margins entire, revolute deep dark green above, white-tomentose beneath; flowers urn-shaped, white to pinkish, about ¼ in. long, borne in dense, few-flowered terminal panicles; fruit drupe-like, dry, about ¼ in. in diameter, becoming black when mature, nutlets fused into a solid stone. December-March.

Mission manzanita is found on dry slopes in scattered localities from near the coast in Los Angeles Co. south to northern Baja California, usually at elevations below 2000 ft. where it is a member of the chaparral community. It is also found on Santa Catalina Island.

X. bicolor is distinctive in appearance and with its dark green revolute leaves, which are held in an upright position, it is somewhat reminiscent of summer holly, *Comarostaphylos diversifolia* var. *diversifolia.* The flower buds for the following season's bloom are formed as early as July and remain on the plants in small, drooping clusters throughout late summer and fall, opening after the beginning of the winter rains.

Mission manzanita presents no cultural problems, growing well in gravelly and sandy soils or in heavy soil. In heavy soil, care must be taken in watering during the summer. Drought tolerant but looks better if it receives some additional water. Stump-sprouts.

Regions: 1, 2, 3, 6.

YUCCA. A genus of the Liliaceae with about 40 species of evergreen trees or shrubs native largely to arid areas of North America. Many of the species are variable and some have been separated into often poorly defined subspecific units. Four species in California. All have numerous, usually large white to cream-colored flowers borne in large clusters often on tall stalks.

1. *Y. baccata* var. *baccata.* Spanish bayonet. Plate 16h. Plants usually single, sometimes forming small open clumps with 2-6 short, procumbent stems; leaves pale bluish-green, 1-3 ft. long, in a rosette on the ground or from a short unbranched stem, leaf margins with coarse recurved threads, leaves ending in a spine; flowers borne on a short scape 2-3½ ft. long, the inflorescences often at least partially enclosed within the leaf rosette, flowers campanulate, reddish-brown on the outside, creamy-white within, segments to 4 in. long; fruit a fleshy capsule, ellipsoidal, often 6 in. long. March-June.

Spanish bayonet is uncommon on dry slopes in the desert ranges of eastern San Bernardino Co., north to Utah and east to Texas. In California it is usually found

between 3000-4000 ft. in the Joshua tree woodland community.

Regions: 1, 2, 3, 4, 6.

Var. *vespertina.* Differs from the preceding in that the plants are usually clustered into large colonies with many heads of leaves. This is the common form in California, being found in the mountains of eastern San Bernardino Co. at elevations of 4000-5000 ft. It also extends north to Utah and east to Arizona.

This species prefers coarse well-drained soils and when established, the plants are completely drought tolerant.

Regions: 2, 3, 4, 6.

2. *Y. brevifolia* var. *brevifolia.* Joshua tree. Plate 16k. Commonly a tree to 30 ft. with an open, round crown of arm-like branches, young trees unbranched until first flowering and covered nearly to the ground with spreading or reflexed leaves, trunks to 3½ ft. in diameter, bark dark brown, checkered into small squarish plates; leaves 6-9 in. long, to ¾ in. wide, more or less 3-sided, forming tufts and spreading in all directions from the ends of the branches, margins denticulate, without fibers; flowers greenish-white, congested in a heavy, often nodding, panicle to 14 in. long, perianth segments thick and leathery, to 1½ in. long; capsule ovoid, 2-4 in. long, dry, spongy and indehiscent. April-May.

Joshua trees are common on dry mesas and slopes in the Mojave Desert, north to Owens Valley and Utah, east to western Arizona, usually at elevations between 2000-6000 ft. in the Joshua tree woodland community.

Var. *herbertii.* Plants with many stems arising from scaly underground rootstock, forming clumps to 30 ft. in diameter. Western portion of Mojave Desert, Los Angeles Co., extending to Walker Pass in Kern Co.

Var. *jaegeriana.* Plate 16l. Plants usually 9-12 ft. tall, seldom over 18 ft., branching from about 3 ft.; leaves usually about 4 in. long. Clark to New York mts., eastern Mojave Desert, north to southwestern Utah and eastward to Arizona.

In cultivation, Joshua trees prefer coarse well-drained soils and at this garden, following seasons of high rainfall, losses have been noted among plantings growing in alluvial sand and gravel but in or along natural drainage channels.

Joshua trees continue to grow unbranched until after their first flowering. The inflorescence develops from the terminal growing tip, and two or more axillary buds below the inflorescence develop into new vegetative branches which continue growth until they bloom, when the branching process is repeated. In cultivation, *Y. brevifolia* var. *brevifolia* tends to produce underground rootstocks similar to those found in var. *herbertii* and this may be in response to the extra water the plants receive.

In 1951, 76 seedlings of *Y. brevifolia* var. *brevifolia* grown from seed collected in the Antelope Valley were planted in the plant community area of the Rancho Santa Ana Botanic Garden; by 1968, 68 plants remained, several of which were sending out suckers; by 1969, 4 plants had bloomed; and by 1976, at least 14 had flowered.

Eighty plants of *Y. brevifolia* var. *herbertii* were set out in 1957, and by 1966 many of the plants had produced numerous suckers, and by 1976 several had bloomed and were branching.

In 1953, 30 seedlings of *Y. brevifolia* var. *jaegeriana* were set out in the plant community area, and by 1973, 26 plants remained. None had bloomed and none had produced suckers (Plate 16l).

Propagation of the Joshua tree by seed is easy and young plants can be set out bare-root. It has been reported that large plants can be successfully transplanted, but our attempts have resulted in failures.

Regions: 3, 4.

3. *Y. schidigera.* Mojave yucca. Spanish dagger. Plate 16i. Trunk simple or shortly-branched to 15 ft., or sometimes with almost no stem, often clothed almost to the ground with living leaves; leaves light yellowish-green, concave, to 3½ ft. long and 1½ in. wide, with a sharp-pointed tip, margins with long, pale filaments; flowers in a dense, sessile or short-stalked panicle, creamy-white, outer surfaces often stained purple, perianth segments to 1¾ in. long; capsule cylindric, fleshy to 4 in. long and 1½ in. thick. April-June.

Mojave yucca is found on dry, rocky slopes and mesas

throughout the Mojave Desert, San Bernardino Valley, the San Jacinto and Santa Rosa mts. and west almost to the coast in San Diego Co. It also occurs in Baja California and in Arizona and southern Nevada. It is found in the chaparral and creosote bush scrub communities.

Mojave yucca prefers coarse well-drained soils and when established is completely drought tolerant. The plants will tolerate some summer watering. At this garden losses have been experienced following winters of exceptionally high rainfall. In Orange Co. some plants were lost from a heart-rot believed brought on by the use of overhead sprinklers.

Plants of this species have bloomed when eight years old.

Regions: 2, 3, 4, 6. Does well in the Central Valley.

4. *Yucca whipplei* ssp. *whipplei.* Chaparral yucca, quixote plant. Plants simple, without obvious stems, with a usually large, many-leafed basal rosette; leaves gray-green, to 2 ft. long and ¾ in. wide, linear-subulate, attenuate to a sharp terminal spine, margins finely serrulate; flowers creamy-white, open campanulate, pendulous, segments to 2 in. long. May-June.

The chaparral yucca is found in dry, often stony slopes from the Santa Ana and San Jacinto mts. to northern Baja California, usually at elevations of 1000-4000 ft., a member of the chaparral community but also found in the coastal sage scrub and creosote bush scrub. After blooming the entire plant dies.

Ssp. *caespitosa.* Differs from the preceding in having stems branching above the ground to form crowded, caespitose clumps with several branches blooming the same season. Found on the edge of the Mojave Desert from the San Bernardino Mts. to Walker Pass and the western base of the Sierra Nevada. Usually at 2000-4000 ft., this species occurs in the chaparral, Joshua tree and pinyon-juniper woodland communities.

Ssp. *intermedia.* Stems branching by axillary buds but only after the first flower stalk is formed. Flower stalks averaging about 9 ft. in length. Santa Monica and Santa Susanna mts. where it occurs at elevations below 2000 ft. in the coastal sage scrub and chaparral communities.

Ssp. *parishii.* Plate 16j. This subspecies has a single basal rosette and flower stalks often very tall, averaging 12 ft. Some forms with flowers stained purple on the outside. On slopes and alluvial fans along the southern slopes of the San Gabriel and San Bernardino mts. at elevations of 1000-8000 ft. where it is a member of the coastal sage scrub, chaparral and yellow pine forest communities. The most spectacular of the subspecies of *Y. whipplei* the entire plant dies following flowering.

Ssp. *percursa.* In this subspecies the plants spread by underground stems to form large colonies, flowering stalks averaging 9 ft. Some plants have flowers stained purple on the outside and a fragrancc similar to orange blossoms.

This subspecies is found from the Santa Ynez Mts. in Santa Barbara Co. north to Monterey Co. at elevations below 2000 ft., where it is a member of the coastal sage scrub and chaparral communities.

In cultivation *Y. whipplei* presents no problems. Usually found in well-drained situations it has grown at this garden on a north-facing slope in heavy loam where the plants have received a considerable amount of watering during the summer. The species occurs in numerous plant communities from coastal sage scrub to the Joshua tree woodland, yellow pine forest and creosote bush scrub. On the chance that *Y. whipplei* consists of ecotypes the subspecies chosen should perhaps be one native to the plant community that most nearly approximates that into which the plants will be grown.

Some plants of *Y. whipplei* are monocarpic; *i.e.*, the entire plant dies following blooming, in other instances the plants produce offsets either (1) by branching near the base, but above ground, and (a) either before the central leaf rosette blooms, or (b) only after the central rosette has formed a flower stalk, or (2) plant spreading by means of underground stems to form large colonies.

Although these characteristics have been used in attempting to define subspecific units, there appears to be considerable variation in the expression of these characteristics, at least among plants in cultivation.

Y. whipplei is among the best known and most often photographed of California plants and when in flower in late May and June *Y. whipplei* ssp. *parishii* is one of the state's most beautiful species. J. O. Hickox of the

Carnegie Institution of Washington Mt. Wilson Observatory lived on Mt. Wilson for 40 years and during that time made a careful long-term study of *Y. whipplei.* In 1905 he transplanted a group of small plants to an area in front of the laboratory. During the following 61 years only 14 plants bloomed, the remaining were large and healthy but unbloomed.

It has been known that when the flower stalk appears it grows very rapidly, and Hickox measured plants, recording their daily growth, which he found to be as much as 14 inches per day, slowing on days that were foggy and cool. On one plant he counted 696 capsules which contained on an average of 180 seeds each, thus the seed production for the one plant was over 125,000 seeds, about 50% of which had been eaten or damaged by the pronuba moth larvae. In the early days at this garden, *Y. whipplei* failed to set fruit as the plants are self-incompatible, and unless pollinated by the pronuba moth (or hand pollinated) with pollen from another plant, seed is not set. In recent years some plants at the garden have fruited, indicating that the pronuba moth is now established in the garden, probably originally migrating from surrounding hillsides where *Y. whipplei* is native.

Regions: 1, 2, 3, 4, 6.

ZIZYPHUS (= *Condaliopsis*). A genus of the Rhamnaceae with 7 species native to the warmer parts of North and South America. Two species in California. Shrubs or small trees with spinose branchlets; leaves small, alternate, early deciduous; flowers small, inconspicuous; fruit a 1-seeded drupe.

1. *Z. obtusifolia* var. *canescens* (= *C. lycioides* var. *canescens*). Desert condalia. A very rigid, spinose shrub to 6 ft. with grayish-green puberulent branchlets; leaves narrow-elliptical or oblong-ovate, ⅜-⅝ in. long, 1/16-¼ in. wide, grayish-green, pubescent on both surfaces, tending to be cupped; flower clusters short-peduncled, few-flowered, flowers minute; fruit globose, about ⅓ in. in diameter, dull black. January-February.

Desert condalia is uncommon in California occurring in sandy places in a few localities along the northern borders of the Colorado Desert. It extends east to New Mexico and south into Mexico. In California it is a member of the creosote bush scrub.

Regions: 3, 4, 5.

2. *Z. parryi* var. *parryi* (= *C. parryi*). Large rounded shrubs to 12 ft., with ultimate branchlets glabrous, spinose and flexuous; leaves coriaceous, mostly fascicled, obovate to oblong-elliptical, to ¾ in. long and ½ in. wide, light green, glabrous on both sides; flowers single or borne in sessile clusters of 2-6; drupe ellipsoid-ovoid, to ¾ in. long, usually distinctly beaked. February-April.

Parry condalia is local on dry slopes and in canyons along the western edge of the Colorado Desert from Morongo Pass south into Baja California at elevations of 1200-3500 ft. It is found in the Joshua tree woodland and pinyon-juniper woodland as well as the upper creosote bush scrub.

Regions: 2, 3, 4, 5.

The condalias require good drainage and when established the plants should receive no summer water. Free from insect pests and disease. Recommended for difficult areas in the desert where it may be used for barrier plantings.

Chapter 5

SURVIVAL OF PLANTS WITHOUT CARE

AT PRESENT great interest is being shown in native trees and shrubs which, after they are established, can be expected to grow with little or no care and no more water than what they receive naturally. Until now there has been little or no specific information available.

The Rancho Santa Ana Botanic Garden was established in 1927 and until 1951 occupied a site in Santa Ana Canyon in interior Orange Co. (For a history of the botanic garden see Lenz, 1977.) During the 24 years that the garden was located in Orange Co. a large proportion of the native trees and shrubs of the state were planted on the grounds. No exact record is available of the number of species that were grown, but a good estimate would be that most of the native trees and perhaps 75% of the shrubs were at one time or another under cultivation, often in very large numbers. When the garden moved to Claremont in Los Angeles Co. in 1951 the original site was abandoned and title to the property reverted to Rancho Santa Ana. Between 1951 and 1977 the plants received no care and the only water was what they received from natural rainfall. In 1977 a survey was made of the site to determine what species might still be growing (Tilforth, 1977). It was thought that this information might be of value to those wishing to know what plants can survive, at least in Orange Co., without care or irrigation. On 28 October 1980 the remaining plants at the Orange Co. site were destroyed by a fire that swept portions of San Bernardino, Riverside and Orange cos.

The original garden site consisted of about 200 acres of slopes, steep hills and deeply-cut canyons with south, southeast and southwest-facing slopes. Elevations ranged from about 450-1100 ft. and the average rainfall was 17.43 inches per year. According to the Storie and Weir system the soils would be classified as *Cnm,* terrace lands with moderately dense subsoils; *Esc*, rolling to steep upland with residual soils of moderate depth; and *En*, rolling hilly to steep upland with residual soils of medium to shallow depth, all with low to intermediate amounts of water.

The following is a list of the plants that grew at the old garden site 26 years after being abandoned.

Acacia greggii (Creosote bush scrub, pinyon-juniper woodland). Many plants at the base of a south-facing slope, 6 ft. tall, in good condition. 30 years old.

Acalypha californica (Chaparral). Healthy young plants, apparently naturalized.

Adenostema sparsifolium (Chaparral). About 52 plants in good to poor condition, largest ones about 12 ft. tall. 43 years old.

Adolphia californica (Chaparral). Many plants in good condition. They have grown together to form a tangled mat. 46 years old.

Aesculus californica (Foothill woodland). About 45 trees growing on a steep, west-facing slope in excellent condition. "The finest specimens of this species I have ever seen." 46 years old.

Agave shawii (Coastal sage scrub). Descendents of cuttings brought in in 1928. 49 years.

* *Arctostaphylos glauca* (Chaparral). Two plants 8 ft. tall in good condition. Could be from lot planted in 1946. Thirty-two hundred plants representing 24 species of *Arctostaphylos* were planted in 23 years. The two

* Native to the Santa Ana Mts. (Lathrop & Thorne, 1978).

plants reported here are the *only* manzanitas discovered during this survey.

Atriplex lentiformis ssp. *breweri* (Coastal sage scrub). Excellent plants on steep bank, possibly from 1931 plantings, or descendents of the original plants.

Baccharis pilularis ssp. *consanguinea* (Coastal sage scrub). Common in all parts of the garden site. Two original introductions made.

Carpenteria californica (Foothill woodland). One plant perhaps of a group of 35; in 1946, eight plants were alive. 39 years old.

* *Ceanothus crassifolius* (Coastal sage scrub). Many plants on east-facing slope, 6 ft. tall, in very good condition. 31 years old.

Ceanothus cuneatus (Chaparral, pinyon-juniper woodland). Many plants in good condition. 32 years old. Two other plantings, 31 and 34 years old.

* *Ceanothus megacarpus* (Chaparral). Several plants on east-facing slope, to 12 ft. with large multiple trunks. 48 years old. Other plantings 31 years old.

Ceanothus ramulosus (Chaparral). Few plants in fair condition. 32 years old.

* *Ceanothus spinosus* (Chaparral, coastal sage scrub). The largest specimens of *Ceanothus* found during the survey. At least 20 ft. tall with a large basal burl, fair condition but showing age. 47 years old. Two other plantings grown from different seed accessions. One plant in poor condition, 22 ft. tall, the other planting in fair condition. Both 35 years old.

Ceanothus verrucosus (Chaparral). Many plants in fair to good condition. 40 years old. Other plantings 31 and 30 years old.

Cercidium floridum (Creosote bush scrub). Many plants in good condition. 35 years old.

Cercocarpus ledifolius (Pinyon-juniper woodland, sagebrush scrub). Many plants in good condition. 30 years old.

Chilopsis linearis (Creosote bush scrub, Joshua-tree woodland). One plant 8 ft. tall and much broader. Could be 31 years old.

Cneoridium dumosum (Chaparral, coastal sage scrub). A large number of plants solidly covering a west-facing hillside. Excellent condition, to 6 ft. 48 years old.

* *Comarostaphylis diversifolia* var. *planifolia* (Chaparral). Several small trees to 20 ft., excellent condition. 43 and 41 years old.

Condaliopsis lycioides var. *canescens.* (Creosote bush scrub). One plant 46 years old. A second planting in very good condition having grown together to form a solid mass 3-5 ft. tall. 43 years old.

Crossosoma californicum (Chaparral). About 80 plants on south-facing slope. Fruiting heavily but no evidence of seedlings. 43 years old.

* *Cupressus forbesii* (Chaparral). Nine different plantings were found. The oldest is 48 years old and the trees 30 ft. tall with very large spreads. In good condition. Two plantings 41 years old, four plantings 40 years old, one planting 39 years old and one planting 37 years old. Groups differ from poor to good condition. Many dead ones on the ground.

Cupressus nevadensis (Foothill woodland). Four trees to 15 ft. 31 and 32 years old.

Cupressus sargentii (Chaparral, yellow pine forest). Only one tree still bore a label. In 1948 there were 10 plants in good condition. 40 years old. Another tree 41 years old. Three others 35 years old.

Dendromecon rigida ssp. *rhamnoides* (Chaparral). Two plants found in different areas with the same propagation number. One in fair condition, one in poor condition. 29 years old.

Fallugia paradoxa (Joshua tree woodland and pinyon-juniper woodland). One plant in good condition. 34 years old.

* *Fraxinus dipetala* (Chaparral, foothill woodland). Found in many areas, obviously has seeded itself and is doing well. Original planting 40 years old.

Fremontodendron californicum (Chaparral, yellow pine forest). Several plants seen but only two were identifiable by number. 31 years old.

Fremontodendron mexicanum (Chaparral). Two plantings found, all plants in good condition. One planting 39 years old, the other 33 years old.

Isomeris arborea var. *angustata* (Creosote bush scrub to coastal sage scrub). Many plants, both old and young indicating natural regeneration.

* *Juglans californica* (Southern oak woodland). Trees

found in many areas of the old garden, some in good condition, some poor. Fifty pounds of seed was sown in 1934 yielding 1970 seedlings; 140 seedlings planted in 1936 and 1937. 40 years old.

Justicia californica (Creosote bush scrub). Several plants appear to be very old. No evidence of naturalizing as they do in Claremont perhaps because they get supplemental watering or the young plants may be eaten by rodents.

Lycium andersonii var. *deserticola* (Pinyon-juniper woodland to coastal sage scrub). About 20 plants flowering and in good condition. 42 years old.

Lyonothamnus floribundus ssp. *floribundus* (Chaparral). Several trees observed to be in good condition. 43 years old.

Mahonia haematocarpa (Pinyon-juniper woodland). Many plants to 6 ft. and in good condition. 36 years old.

Mahonia nevinii (Coastal sage scrub, chaparral). Three propagations of this species; the youngest consisting of many plants in good to fair condition. 30 years old. The second group contains many plants in good condition. 40 years old. The third group, half of which burned in 1943 but grew back after the fire, is in good condition. 43 years old.

Olynea tesota (Creosote bush scrub). Two trees in excellent condition. Original planting of 64 plants declined to three plants in three years. Remaining trees 34 years old.

* *Pinus coulteri* (Yellow pine forest, foothill woodland). Five small trees on steep canyon side. An early note says that the area is too dry for the young seedlings. 34 years old.

Pinus quadrifolia (Pinyon-juniper woodland). Six trees in good condition and now about 20 ft. tall. Forty-seven plants were put out in 1932. Ten dug and moved to Claremont in 1951. 45 years old.

Pinus sabiniana (Foothill woodland). Three separate plantings were found, all trees in good condition. 31, 33 and 45 years old.

Pinus torreyana (Dry slopes near ocean). One grove of 71 trees of a uniform height of 40 ft. 47 years old.

* *Platanus racemosa* (Many communities). Many sycamores in washes and canyons. Native to the area.

Prosopis glandulosa var. *torreyana* (Creosote bush scrub). Two plants were found, both in good condition, remnant of a planting mostly destroyed by a slide. 39 years old.

Prunus fremontii (Creosote bush scrub to pinyon-juniper woodland). Many plants in good condition, to 6 ft. tall. 46 years old.

* *Prunus ilicifolia* (Chaparral). Several plants have survived on a south-facing slope, mostly in fair condition, a few in poor condition. 43 years old.

Ptelea crenulata (Foothill woodland, yellow pine forest). Two plantings have grown together on a steep west-facing slope, plants in excellent condition. 44 and 41 years old.

* *Quercus agrifolia* (Southern oak woodland). Many trees in scattered locations. 42 years old.

Quercus lobata (Foothill woodland). Fifty-eight trees on a gentle, southwest-facing slope and another group on a south slope, in fair condition and about 14 ft. tall. 40 years old. Another scattered group about 8 ft. tall. 32 years old. According to Tilforth, "The appearance of these valley oaks gives the impression that only their inherent toughness is keeping them alive and that they would be much happier somewhere else."

* *Rhamnus crocea* (Coastal sage scrub, chaparral). Many plants in good to fair condition, about 4 ft. tall. 35 years old.

* *Rhus ovata* (Coastal sage scrub, chaparral). This species is very much at home in the original garden site and some may be native plants. One group 35 years old.

Ribes viburnifolium (Island chaparral). Plants growing on a northfacing, shady slope. Could be 37 years old.

Shepherdia argentea (Sagebrush scrub, pinyon-juniper woodland). Small thicket on canyon floor. In 1952 all the plants were dug and taken to Claremont. This species suckers from the roots and the present thicket most likely developed from roots left remaining in the ground.

Simmondsia chinensis (Creosote bush scrub). Many plants, some obviously seedlings, all in good condition. Original planting now 48 years old.

Tetracoccus dioicus (Chaparral). Three plantings were found, all in good condition. 43, 35, 30 years old. There is evidence of some regeneration.

Table 2

Plant Community	Range of Precipitation Within the Community	Surviving Species
Northern coastal sage scrub	25-75	0
Coastal sage scrub	10-20	10
Closed-cone pine forest	20-60	0
Redwood forest	35-100	0
Mixed evergreen forest	25-65	1
Foothill woodland	15-40	8
Southern oak woodland	15-25	3
Chaparral	14-25	20
Yellow pine forest	25-80	4
Red fir forest	35-65	0
Lodgepole pine forest	30-60	0
Subalpine forest	30-50	0
Northern juniper woodland	10-30	0
Pinyon-juniper woodland	12-20	8
Joshua tree woodland	6-15	2
Sagebrush scrub	8-15	2
Shadscale scrub	3-7	0
Creosote bush scrub	2-8	10
Alkali sink	1½-7	0

Distribution of surviving species by plant community. Precipitation given in inches. (Plant communities are those recognized by Munz and Keck, 1949.)

Torreya californica (Mixed evergreen forest). Twelve small trees to about 12 ft. tall, good to fair condition. Seventy plants were originally planted. According to Tilforth, "It was a surprise to find these, but the relatively cool, north-facing slope of a canyon must have given the trees just enough protection from the heat for their survival." 47 years old.

* *Umbellularia californica* (Chaparral, foothill woodland). One group with a large number of plants, some of them probably seedlings. 35 years old. Another stand, 34 years old.

Viguiera deltoides var. *parishii* (Creosote bush scrub). Plants could be descendants of a planting made in 1930. In 1946 it was noted that volunteer plants were appearing.

Washingtonia filifera (Desert oasis). A row of palms was found in the bottom of a small wash. Some were dug and taken to Claremont in 1951. 33 years old.

From the findings of Tilforth's survey certain facts stand out: (1) of the 58 surviving species, 26% are native to the Santa Ana Mts.; (2) of the surviving species, 52% are native to the two plant communities found at the Santa Ana garden site, *i.e.*, coastal sage scrub and chaparral. Except for the four species found in the yellow pine forest community and the one species from the mixed evergreen forest community, all surviving species are native to plant communities where the precipitation is less than, or brackets that received at the Orange Co. site.

Of the four species from the yellow pine forest, none is confined to that community; *Fremontodendron californicum* is much more common in the chaparral than it is in the yellow pine forest and *Ptelea crenulata* and *Pinus coulteri* both occur in the foothill woodland as well as in the yellow pine forest. The survival of *Torreya californica* from the mixed evergreen forest was a surprise and the only explanation is that made by Tilforth in his original report.

These facts suggest that if plants, after they are established, are to be grown without irrigation or supplemental watering they should be *species native to plant communities whose normal range of precipitation is less than, or within the range of the amount received where the plants are to be grown.* Many plants will tolerate more water than they receive in their native habitats but often only if there is excellent drainage.

Chapter 6

USE OF NATIVE AND NON-NATIVE PLANTS

IN AN EARLIER CHAPTER it was pointed out that there are six areas of the world possessing mediterranean climate. The floras of these six areas have for the most part evolved from entirely different groups of plants responding to a similar set of environmental factors, *convergent evolution.* In addition to requiring much the same treatment under cultivation, plants, whether they come from the chaparral of California, the *maqui* of the Mediterranean, the *matorral* of Chile or the *fynbos* of South Africa, all present a similar physiognomy, or gross

Quercus agrifolia

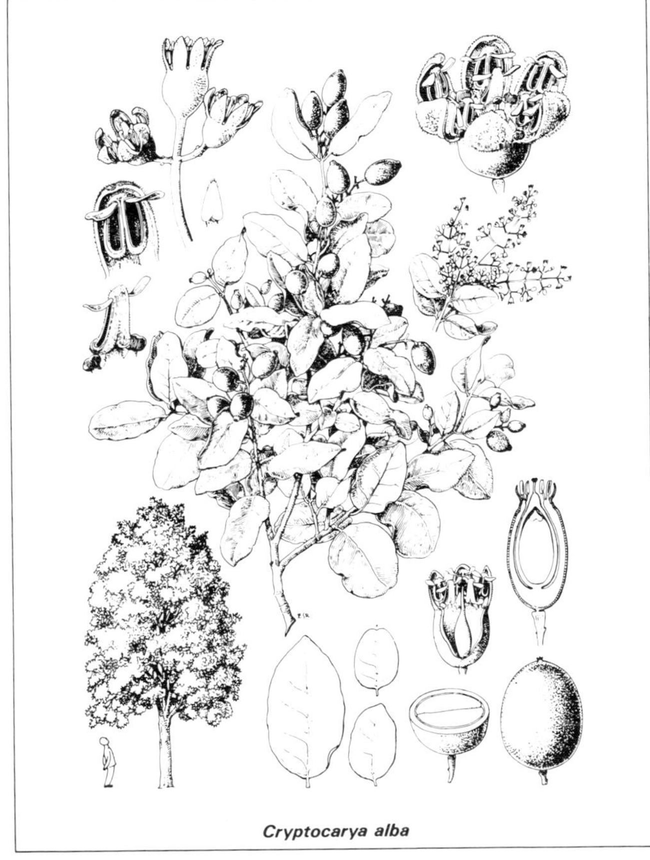

Cryptocarya alba

Fig. 85.1. Analog species. See text for explanation. (Thrower and Bradbury, 1977).

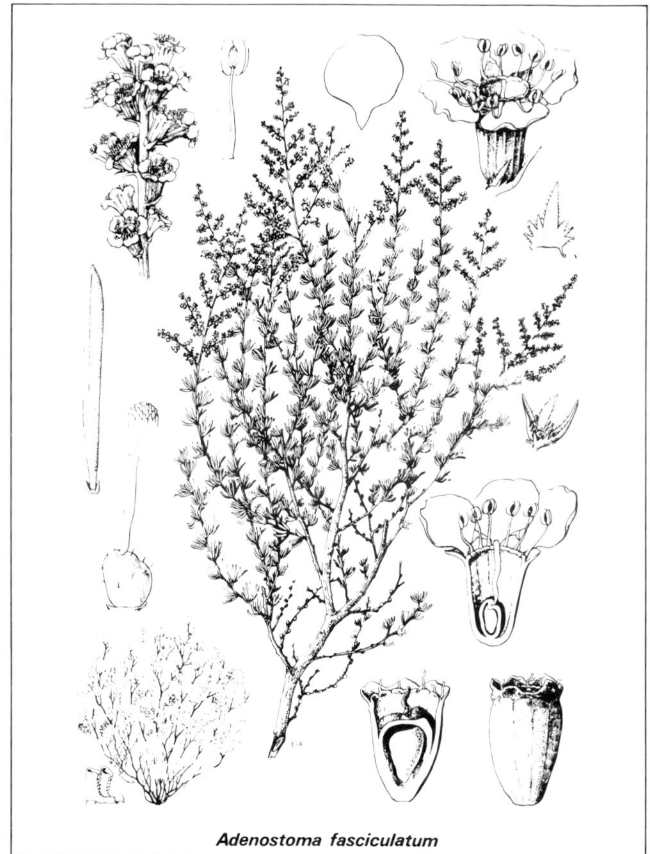

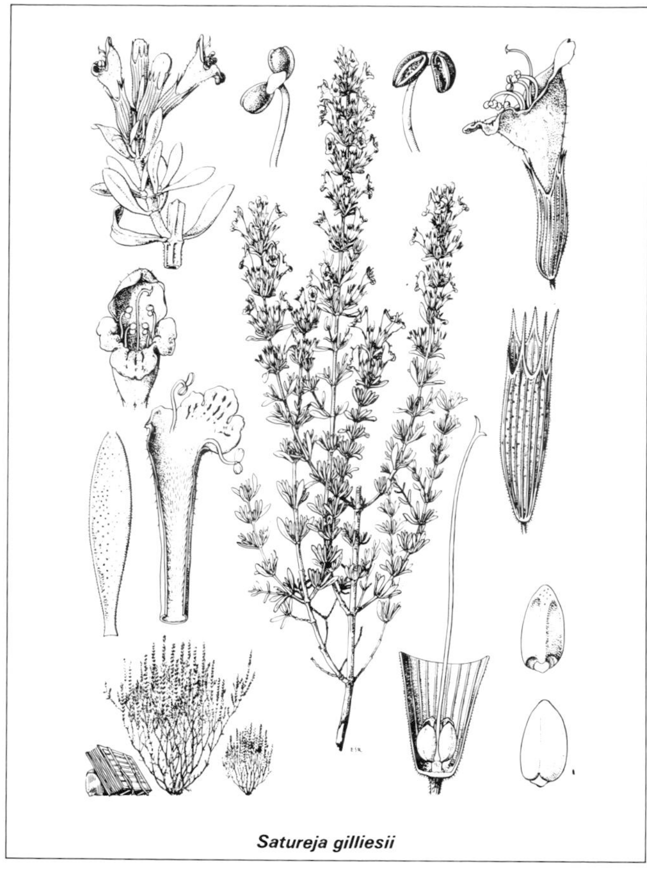

Fig. 85.2. Analog species. See text for explanation. (Thrower and Bradbury, 1977).

appearance, and species from all these areas may be combined into aesthetically harmonious groupings. Stebbins (1974) has commented upon the superficial similarity between the flora of lowland Chile and that of California saying, "Most of the Chilean shrubs and trees look like Californians from a distance, but close by, many of them turn out to be members of completely different families." Rafols (1977) has published a series of illustrations of 'look alike' species which he refers to an analog species. Each pair is made up of one species from California and one from Chile (Figs. 34, 35).

We believe that horticulturists would do well to seek out new plants from other areas with mediterranean climate for inclusion in plantings of California plants thus enriching the diversity and increasing the number of species with which the landscape architect has to work.

Chapter 7

DEVELOPMENT OF SUPERIOR CULTIVARS

THE DEVELOPMENT through selection and hybridization of superior cultivars of native trees and shrubs is a matter that should be of concern to all horticulturists using California plant materials but to date it has received little serious attention. Most selection has been for improved flower color rather than for improved plant habit or for physiologically superior types capable of tolerating garden conditions. Except for the work of Nobs (1963) on *Ceanothus,* few controlled hybridizations have been made in any of the genera of major importance to horticulturists. The named hybrids which have appeared in the literature and on the market in recent years have almost without exception resulted from chance hybridizations either in the wild or in gardens and proof of the hybrid nature of the cultivar or its parentage is mostly lacking.

Selection. Selection within native populations or within large populations grown under cultivation is the simplest method of obtaining improved forms. The bush anemone, *Carpenteria californica,* is a narrow endemic native to a few areas in Fresno Co. and it has been considered to be morphologically a rather uniform species and, since it has no near relatives, interspecific hybridizations are not possible. In 1925 the Royal Horticultural Society (England) awarded an A.M. to *Carpenteria californica* 'Ladhams' which was described as a very vigorous sport of the species with flowers 3¼ in. in diameter. Roderick (1977) has reported that he examined all the plants he could of the species growing in its native habitat and that he found one plant which was so different (small flowers produced in large clusters) that it stood out from all the rest. Cuttings were obtained and it is hoped that this selection may soon be introduced into horticulture. This same author also reports that he has made selections of the silktassel bush, *Garrya elliptica,* one clone of which 'Evie' has been awarded an A.M. by the California Horticultural Society. Many other examples of selection for superior forms could be listed.

Hybridization. Many groups of California trees and shrubs are noted for the amount of interspecific hybridization that occurs under natural (although sometimes disturbed) conditions. Genera in which natural hybridization has been shown to be of frequent occurrence includes the manzanitas, *Arctostaphylos* spp.; California lilacs, *Ceanothus* spp.; oaks, *Quercus* spp.; pines, *Pinus* spp.; barberries, *Mahonia* spp.; sticky monkeyflowers, *Diplacus* spp.; *Rhus* spp.; *Rhamnus* spp.; *Prunus* spp.; and *Salvia* spp., all of which are of importance to the horticulturist. Except for the pines and to a very limited extent the California lilacs, few scientifically controlled hybrids have been produced. Garden hybrids bred under controlled conditions have been made in some groups of subshrubs such as the sticky monkeyflowers, *Diplacus* spp., but more often they have been among the herbaceous perennials, sometimes in conjunction with scientific studies. Notable among the controlled hybrids produced between species of large woody shrubs are those between species of *Fremontodendron* (Dourley, 1977) and those in *Ceanothus* sect. *Cerastes* by Nobs (1963). In arborescent species the work on *Pinus* conducted at the Forest Genetics Laboratory at Placerville is outstanding.

Few areas offer greater promise of reward than that of scientifically controlled breeding programs using some of California's more interesting and useful trees and shrubs.

The Concept of the Ecotype. Species that occupy a series of contrasting environments develop genetically and physiologically distinct ecological races, *ecotypes,* which are suited to those environments (Clausen, Keck and Heisey, 1945). Many such species are also geographically widespread. In some instances ecotypes differ morphologically and can be recognized taxonomically as varieties, subspecies or even species. In other instances there are no morphological differences but different populations show distinct habitat preferences in different parts of their geographical range (Daubenmire, 1968). With the realization that species may contain large amounts of physiological variation, agronomists and foresters have searched for the most suitable ecotypes for cultivation in particular areas and under particular conditions (Davis and Heywood, 1973). To date, growers of California native plants have given little or no consideration to the sources of the material they are growing with the result that a species may be considered unsuited, or marginally suited, for a particular area when in reality the horticulturist may have only selected an ecotype that was unsuitable for that area.

Madrone, *Arbutus menziesii,* is most widely distributed in northern California but does extend south as far as northern Baja California. It is found in the redwood forest, mixed evergreen forest, Douglas fir forest, foothill woodland, northern oak woodland and southern oak woodland communities. With such an extensive area of distribution and as a component of so many plant communities it appears likely that the species consists of a number of distinct ecotypes. The horticulturist in southern California might do well to select propagating material from populations in the southern oak woodland of southern California rather than from the redwood forest or Douglas fir forest of northern California.

Mountain dogwood, *Cornus nuttallii,* is rare in the coastal ranges south of Marin Co. and in the Sierra Nevada its southernmost site is in northern Kern Co. In southern California the species is scattered in the San Gabriel, San Bernardino, Cuyamaca and Palomar mts. at elevations as low as 5800 ft. where it is found in the yellow pine forest. A difficult species in southern California, attempts should be made to collect material from southern localities to determine whether the plants would react more favorably to cultivation than do those from northern populations.

Toyon, *Heteromeles arbutifolia,* is widely distributed in California and is found in many plant communities. Roderick (1977), however, reports that homeowners living along the coast experience trouble with the species since the berries fail to color and later rot. He suggested that for those living in such areas propagating material be taken from well-berried forms that are native to coastal areas; ecotypes adapted to cool, cloudy areas rather than ecotypes adapted to the hot, dry chaparral of interior southern California.

Many other California natives with widespread distributions have populations in southern California and it is from those populations that horticulturists should select material for growing in southern California.

In contrast to those species of wide distribution which have developed distinct ecotypes in response to differing habitats is section *Cerastes* of the genus *Ceanothus. Ceanothus* is divided into two sections; one of them, *Cerastes,* includes those species with opposite leaves and usually persistent stipules and fruits usually with horns. According to Nobs (1963), the majority of the members of section *Cerastes* are sufficiently distinct not to be questioned as species on morphological grounds but biologically they are comparable to ecotypes or ecological races. As in ecotypes of other groups, the species in section *Cerastes* show specific habitat requirements that determine their geographical distribution. However many of the species within the section are found in two or more plant communities and it has not been shown whether they represent ecotypes with broad habitat tolerances or whether the species may possess subecotypes adapting them to specific plant communities.

Fig. 86. Well grown nursery stock: a, *Dendromecon rigida* ssp. *harfordii*; b, *Ceanothus* 'Skylark'; c, *Ceanothus gloriosus* 'Emily Brown.' (Courtesy of Skylark Nurseries, Santa Rosa, California).

Chapter 8

PLANTING LISTS

Species	Suitable for Region: 1	2	3	4	5	6	7
Acacia greggii	•	•	•	•	•	•	
Acalypha californica	•	•	•		•		
Acer macrophyllum	•	•	•			•	
Acer negundo ssp. *californicum*	•	•	•			•	
Adenostoma fasciculatum	•	•	•			•	
sparsifolium		•	•			•	
Adolphia californica	•	•	•				
Aesculus californica	•	•	•			•	
Agave deserti ssp. *deserti*	•	•	•	•	•	•	
shawii ssp. *shawii*	•	•	•			•	
Alnus oregona	•	•	•			•	
rhombifolia	•	•	•			•	•
Amorpha californica	•	•	•			•	•
fruticosa var. *occidentalis*	•	•	•	•		•	
Arbutus menziesii	•	•				•	
Arctostaphylos andersonii var. *andersonii*	•	•	•			•	
var. *imbricata*	•	•	•			•	
var. *pallida*	•	•	•			•	
auriculata	•	•	•			•	
catalinae	•	•	•			•	
crustacea var. *crustacea*	•	•	•			•	
var. *rosei*	•	•	•			•	
densiflora	•	•	•				
edmundsii	•	•	•				
elegans		•	•				
glandulosa ssp. *glandulosa*	•	•	•			•	•
ssp. *mollis*	•	•	•			•	
ssp. *crassifolia*	•	•	•			•	
glauca	•	•	•			•	
insularis	•	•	•				
hookeri ssp. *hookeri*	•	•	•				
ssp. *franciscana*	•	•	•				

Species	Suitable for Region: 1	2	3	4	5	6	7
manzanita ssp. *manzanita*	●	●	●			●	
obispoensis	●	●	●			●	
otayensis	●	●	●			●	●
pajaroensis	●	●	●			●	
parryana		●	●			●	●
pechoensis var. *pechoensis*	●	●	●				
var. *viridissima*	●	●	●				
pilosula	●	●	●				
pringlei var. *drupacea*		●	●			●	●
pumila	●	●	●				
pungens var. *pungens*		●	●			●	●
refugioensis	●	●	●			●	
stanfordiana ssp. *stanfordiana*	●	●	●				
ssp. *bakeri*	●	●	●				
subcordata	●	●	●			●	
tomentosa var. *tomentosa*	●	●	●				
var. *tomentosiformis*	●	●	●				
viridissima	●	●	●				
Aristolochia californica	●	●	●			●	
Artemisia tridentata ssp. *tridentata*	●	●	●	●		●	●
ssp. *parishii*			●	●		●	●
Atriplex canescens	●	●	●	●	●	●	
confertifolia	●	●	●	●	●	●	
hymenelytra				●	●		
lentiformis ssp. *lentiformis*	●	●	●	●	●	●	
ssp. *breweri*	●	●	●	●	●	●	
polycarpa			●	●	●	●	
Baccharis glutinosa	●	●	●	●			
pilularis ssp. *pilularis*	●	●	●	●		●	
ssp. *consanguinea*	●	●	●				
sarothroides			●	●	●		
sergiloides			●	●	●	●	
Betula fontinalis	●	●	●			●	●
Brickellia californica			●	●	●	●	
incana				●	●	●	
Calliandra eriophylla	●	●	●	●	●	●	
Calocedrus decurrens	●	●	●				●
Calycanthus occidentalis	●	●	●			●	
Carpenteria californica	●	●	●			●	
Cassia armata				●	●		
Castela emoryi			●	●	●	●	
Ceanothus arboreus	●	●	●			●	
cordulatus						●	●
crassifolius var. *crassifolius*	●	●	●			●	
cuneatus	●	●	●			●	●
cyaneus	●	●	●				

Species	Suitable for Region: 1	2	3	4	5	6	7
gloriosus var. *gloriosus*	●	●	●				
var. *exaltatus*	●	●					
var. *porrectus*	●	●	●				
greggii var. *vestitus*		●	●	●		●	●
var. *perplexans*		●	●	●		●	●
griseus var. *griseus*	●	●	●				
var. *horizontalis*	●	●	●				
impressus var. *impressus*	●	●	●				
var. *nipomensis*	●	●	●				
incanus		●	●			●	
integerrimus	●	●					
leucodermis	●	●	●			●	●
maritimus	●	●	●				
megacarpus ssp. *megacarpus*	●	●	●			●	
ssp. *insularis*	●	●	●			●	
oliganthus	●	●	●			●	
papillosus var. *papillosus*	●	●	●				
var. *roweanus*	●	●	●				
ramulosus var. *ramulosus*	●	●	●			●	
var. *fasciculatus*	●	●	●			●	
rigidus	●	●	●				
sorediatus	●	●	●			●	
spinosus	●	●	●				
thyrsiflorus var. *thyrsiflorus*	●	●	●				
var. *repens*	●	●	●				
tomentosus var. *olivaceus*	●	●	●			●	
verrucosus	●	●	●			●	
Celtis reticulata	●	●	●	●	●	●	●
Cephalanthus occidentalis var. *californicus*	●	●	●			●	
Cercidium floridum	●	●	●	●	●	●	
microphyllum		●	●	●	●	●	
Cercis occidentalis		●	●	●		●	
Cercocarpus betuloides	●	●	●			●	
intricatus		●	●	●		●	
ledifolius		●	●	●		●	●
minutiflorus	●	●	●			●	
traskiae	●	●	●			●	
Chamaebatiaria millefolium		●	●	●		●	●
Chilopsis linearis		●	●	●	●	●	
Chrysothamnus nauseosus ssp. *bernardinus*				●		●	●
ssp. *mohavensis*			●	●		●	
Clematis lasiantha	●	●	●			●	
Cneoridium dumosum	●	●	●	●	●	●	
Comarostaphylis diversifolia var. *diversifolia*	●	●	●			●	
var. *planifolia*	●	●	●			●	
Cornus glabrata	●	●	●			●	●

Species	Suitable for Region: 1	2	3	4	5	6	7
nuttallii		●				●	
stolonifera	●	●	●			●	●
Cowania mexicana var. *stansburiana*	●	●	●	●	●	●	●
Crataegus douglasii	●	●					
Crossosoma californicum	●	●	●				
Dalea spinosa					●		
Dendromecon rigida var. *rigida*	●	●	●	●		●	
var. *harfordii*	●	●	●	●		●	
var. *rhamnoides*	●	●	●	●		●	
Diplacus aridus	●	●	●			●	
aurantiacus	●	●	●			●	
clevelandii		●	●			●	
longiflorus	●	●	●			●	
parviflorus	●	●	●			●	
puniceus	●	●	●			●	
Encelia californica	●	●	●				
farinosa		●	●	●	●		
Ephedra californica	●	●	●	●	●	●	
nevadensis	●	●	●	●	●	●	
viridis			●	●	●	●	
Epilobium canum ssp. *mexicanum*		●	●			●	
Eriodictyon californicum	●	●	●			●	
crassifolium		●	●	●		●	
tomentosum		●	●			●	
traskiae ssp. *traskiae*	●	●	●			●	
trichocalyx ssp. *trichocalyx*		●	●	●		●	●
Eriogonum arborescens	●	●	●			●	
cinereum	●	●	●			●	
crocatum	●	●	●			●	
fasciculatum ssp. *fasciculatum*	●	●	●			●	
ssp. *foliolosum*	●	●	●			●	
ssp. *polifolium*		●	●	●		●	
giganteum ssp. *giganteum*	●	●	●			●	
ssp. *compactum*	●	●	●			●	
ssp. *formosum*	●	●	●			●	
Fallugia paradoxa	●	●	●	●	●	●	
Forestiera neo-mexicana	●	●	●	●	●	●	
Fouquieria splendens			●	●	●		
Fraxinus anomala		●	●	●		●	
dipetala	●	●	●			●	
latifolia	●	●	●			●	
velutina	●	●	●	●		●	
Fremontodendron californicum ssp. *californicum*	●	●	●	●		●	
ssp. *decumbens*	●	●	●	●		●	
ssp. *napense*	●	●	●	●		●	
mexicanum	●	●	●	●	●	●	

Species	Suitable for Region:						
	1	2	3	4	5	6	7
Galvezia speciosa	●	●	●			●	
Garrya congdoni		●	●			●	
elliptica	●	●	●				
flavescens var. *pallida*	●	●	●	●		●	●
fremontii	●	●	●			●	●
veatchii	●	●	●			●	
Gutierrezia microcephala			●	●	●	●	
sarothrae	●	●	●	●	●	●	
Haplopappus arborescens		●	●			●	
berberidis	●	●	●				
canus	●	●	●				
detonsus	●	●	●				
ericoides	●	●	●				
parishii		●	●			●	●
squarrosus ssp. *grindelioides*	●	●	●			●	
venetus ssp. *oxyphyllus*	●	●	●				
ssp. *vernonioides*	●	●	●				
Heteromeles arbutifolia var. *arbutifolia*	●	●	●			●	
var. *cerina*	●	●	●			●	
Holodiscus discolor var. *discolor*	●	●	●			●	
Hyptis emoryi	●	●	●	●	●		
Isomeris arborea var. *arborea*	●	●	●	●	●	●	
var. *angustata*	●	●	●	●	●	●	
var. *globosa*	●	●	●	●	●	●	
var. *insularis*	●	●	●	●	●	●	
Iva hayesiana	●	●	●			●	
Juglans californica	●	●	●			●	
Juniperus californica	●	●	●	●		●	
occidentalis ssp. *australis*		●	●			●	●
osteosperma	●	●	●	●		●	●
Justicia californica	●	●	●		●		
Keckiella antirrhinoides ssp. *antirrhinoides*	●	●	●			●	
cordifolia	●	●	●			●	
ternata ssp. *ternata*		●	●			●	
Larrea tridentata			●	●	●		
Lavatera assurgentiflora	●	●	●				
Leptodactylon californicum ssp. *californicum*	●	●	●			●	
Lithocarpus densiflora	●	●	●				
Lonicera involucrata var. *ledebourii*	●	●	●			●	
Lupinus albifrons var. *albifrons*	●	●	●			●	
Lycium brevipes var. *brevipes*	●	●	●		●		
cooperi			●	●	●		
Lyonothamnus floribundus ssp. *asplenifolius*	●	●	●				
ssp. *floribundus*	●	●	●				
Mahonia amplectens	●	●	●	●		●	
aquifolium	●	●	●			●	●

Species	Suitable for Region: 1	2	3	4	5	6	7
haematocarpa		●	●	●	●	●	
higginsae	●	●	●	●		●	
nevinii	●	●	●	●	●	●	
pinnata	●	●	●			●	
piperiana	●	●	●			●	
Malosma laurina	●	●	●				
Myrica californica	●	●	●				
Nolina bigelovii	●	●	●	●	●	●	
wolfii	●	●	●	●	●	●	
Olynea tesota					●		
Ornithostaphylos oppositifolia	●	●	●			●	
Osmaronia cerasiformis	●	●	●			●	●
Parkinsonia aculeata	●	●	●	●	●	●	
Philadelphus lewisii	●	●	●				
Physocarpus capitatus	●	●	●				
Pickeringia montana	●	●	●			●	
Pinus coulteri	●	●	●	●		●	
sabiniana	●	●	●	●	●	●	
torreyana	●	●	●	●			
monophylla	●	●	●	●		●	
quadrifolia	●	●	●	●		●	
Platanus racemosa	●	●	●			●	
Pluchea sericea	●	●	●	●	●	●	
Populus fremontii var. *fremontii*	●	●	●	●	●	●	
trichocarpa var. *trichocarpa*	●	●	●	●	●	●	
Prosopis glandulosa var. *torreyana*	●	●	●	●	●	●	
pubescens	●	●	●	●	●	●	
Prunus andersonii		●	●	●		●	●
fasciculata		●	●	●	●	●	
fremontii		●	●	●	●	●	
ilicifolia	●	●	●	●	●	●	●
lyonii	●	●	●	●	●	●	
subcordata		●	●			●	●
virginiana var. *demissa*	●	●	●	●		●	
Ptelea crenulata		●	●			●	
Purshia glandulosa		●	●	●	●	●	
tridentata		●	●	●	●	●	
Quercus agrifolia	●	●	●			●	
chrysolepis	●	●	●	●		●	●
douglasii	●	●	●	●		●	
engelmannii	●	●	●			●	
garryana	●	●	●			●	
kelloggii		●	●			●	●
lobata	●	●	●			●	
tomentella	●	●	●			●	
wislizenii	●	●	●			●	

Species	Suitable for Region:						
	1	2	3	4	5	6	7
Rhamnus californica ssp. *californica*	●	●	●			●	
ssp. *crassifolia*		●	●			●	
ssp. *tomentella*	●	●	●	●		●	
ssp. *ursina*		●	●	●		●	
crocea	●	●	●			●	
ilicifolia	●	●	●	●		●	
pirifolia	●	●	●			●	
Rhus integrifolia	●	●					
ovata	●	●	●	●	●	●	
trilobata var. *anisophylla*	●	●	●	●	●	●	
var. *pilosissima*		●	●	●	●	●	
var. *quinata*		●	●	●	●	●	
Ribes aureum var. *aureum*		●	●			●	
var. *gracillimum*		●	●			●	
indecorum	●	●	●			●	
malvaceum var. *malvaceum*	●	●	●			●	
var. *viridifolium*	●	●	●			●	
sanguineum var. *sanguineum*	●	●	●			●	
var. *glutinosum*	●	●	●			●	
speciosum	●	●	●			●	
viburnifolium	●	●	●			●	
Romneya coulteri		●	●			●	
trichocalyx	●	●	●		●	●	
Rosa californica	●	●	●			●	●
gymnocarpa	●	●	●			●	
minutifolia	●	●	●				
Rubus vitifolius		●	●			●	
ursinus	●	●	●			●	
Salazaria mexicana				●	●		
Salvia apiana	●	●	●			●	
clevelandii	●	●	●			●	
leucophylla	●	●	●			●	
mellifera	●	●	●			●	
pachyphylla				●		●	●
Sambucus mexicana	●	●	●	●	●	●	●
coerulea							●
Sequoia sempervirens	●	●	●				
Sequoiadendron giganteum		●					●
Shepherdia argentea	●	●	●	●		●	
Simmondsia chinensis	●	●	●	●	●	●	
Spiraea douglasii	●	●	●			●	
Staphylea bolanderi		●	●			●	
Styrax officinalis var. *fulvescens*	●	●	●			●	
Symphoricarpos mollis	●	●	●			●	
Tetracoccus dioicus	●	●	●			●	
Trichostema lanatum	●	●	●			●	

Species	Suitable for Region: 1	2	3	4	5	6	7
Umbellularia californica var. *californica*	●	●	●			●	
var. *fresnensis*	●	●	●			●	
Vitis californica	●	●	●			●	
girdiana	●	●	●	●	●	●	
Washingtonia filifera		●	●	●	●		
Xylococcus bicolor	●	●	●			●	
Yucca baccata var. *baccata*	●	●	●	●		●	
var. *vespertina*		●	●	●		●	
brevifolia var. *brevifolia*			●	●			
var. *herbertii*			●	●			
var. *jaegeriana*			●	●			
schidigera	●	●	●	●		●	
whipplei ssp. *whipplei*	●	●	●	●		●	
ssp. *caespitosa*	●	●	●	●		●	
ssp. *intermedia*	●	●	●	●		●	
ssp. *parishii*	●	●	●	●		●	
Zizyphus obtusifolia var. *canescens*			●	●	●		
var. *parryi*		●	●	●	●		

The following lists are not definitive but are plants to which we would like to call attention.

SOIL STABILIZATION & EROSION CONTROL

Adenostoma fasciculatum, A. sparsifolium, Adolphia californica, Armopha californica, A. fruticosa, Artemisia tridentata, Atriplex confertifolia, A. polycarpa, Baccharis pilularis ssp. *pilularis, B. p.* ssp. *consanguinea, B. sarothroides, B. sergiloides, B. glutinosa, Brickellia californica, B. incana, Carpenteria californica, Celtis reticulata, Cercidium floridum, C. microphyllum, Cercocarpus betuloides, C. ledifolius, C. minutiflorus, Chamaebatiaria millefolium, Chilopsis linearis, Chrysothamnus nauseosus, Cornus glabrata, C. stolonifera, Eriodictyon* spp., *Eriogonum* spp. *Fallugia paradoxa, Forestiera neo-mexicana, Gutierrezia microcephala, Haplopappus ericoides, H. parishii, H. venetus, Iva hayesiana, Malosma laurina, Purshia glandulosa, P. tridentata, Rhus trilobata, Ribes viburnifolium, Rosa* spp. *Salvia* spp., *Shepherdia argentea, Spiraea douglasii, Symphoricarpus mollis, Tetracoccus dioicus, Xylococcus bicolor.*

ATTRACTIVE TO BIRDS & OTHER WILDLIFE

Adenostoma sparsifolium, Atriplex canescens, A. confertifolia, Betula fontinalis, Chilopsis linearis, Comarostaphylos diversifolia, Epilobium canum, Heteromeles arbutifolia, Ornithostaphylos oppositifolia, Osmaronia cerasiformis, Prosopis spp., *Prunus* spp., *Rhamnus* spp., *Rhus integrifolia, R. ovata, Ribes aureum, Rosa* spp., *Sambucus* spp., *Symphoricarpus mollis.*

LIVING BARRIERS

Acacia greggii, Adolphia californica, Agave spp., *Atriplex confertifolia, Castela emoryi, Lycium* spp., *Mahonia* spp., *Prosopis spp. Prunus andersonii, P. fasciculata, P. fremontii, Rosa* spp. *Rubus* spp., *Shepherdia argentea, Simmondsia chinensis, Yucca* spp., *Zizyphus obtusifolia, Z parryi.*

TOLERATES SHEARING

Arctostaphylos edmundsii, Atriplex canescens, Baccharis pilularis ssp. *pilularis, Lavatera assurgentiflora, Malosma laurina, Myrica californica, Prunus ilicifolia, P. lyonii, Rhamnus* spp., *Rhus integrifolia, R. ovata, Tetracoccus dioicus, Umbellularia californica.*

TOLERATES ALKALINE CONDITIONS

Alnus oregona, Atriplex canescens, A. lentiformis, Baccharis sarothroides, Forestiera neo-mexicana, Iva hayesiana, Parkinsonia aculeata, Prosopis glandulosa, var. *torreyana.*

INVASIVE ROOT SYSTEMS

Acer negundo ssp. *californica, Eriodictyon* spp., *Fraxinus* spp., *Pluchea sericea, Populus* spp., *Romneya* spp.

Glossary

Abrupt. Terminating suddenly; not tapering.
Acaulescent. Stemless or essentially so.
Achene. A small, dry, hard, indehiscent, 1-seeded fruit.
Accessory. Additional to the usual number of organs.
Acuminate. Gradually tapering to a short point.
Acute. Sharp-pointed but less tapering than acuminate.
Adaxial. Located on the side nearest the axis.
Adnate. Grown together with an unlike part.
Alternate. Any arrangement of parts along the axis other than opposite or whorled.
Anther. The pollen-bearing part of the stamen.
Apical. Situated at the tip.
Appendage. Any attached supplementary or secondary part.
Appressed. Pressed flat against another organ.
Arborescent. Tree-like in tendency.
Ascending. Rising obliquely or curving upward.
Attenuate. Slenderly tapering or prolonged; more gradual than acuminate.
Auricle. An ear-shaped appendage.
Auriculate. Bearing auricles.
Axil. Upper angle formed by a leaf or branch with a stem.
Axillary. In an axil.
Basal. Relating to, or situated at, the base.
Berry. A pulpy indehiscent fruit with no true stone.
Bi- or *Bis-.* Latin prefex signifying two, twice, or doubly.
Bifid. Two-cleft to about the middle.
Bilabiate. Two-lipped.
Bipinnate. Doubly or twice pinnate.
Bipinnatifid. Twice pinnately cleft.
Bladdery. Thin and inflated.
Blade. The expanded part of a leaf or petal.
Bract. A reduced leaf subtending a flower, usually associated with an inflorescence.
Bristly. Bearing stiff hairs.
Calyx. The external, usually green, whorl of a flower, contrasted with the inner showy corolla.
Calyx-lobe. In a gamosepalous calyx, the free projecting parts.
Campanulate. Bell-shaped.
Canescent. Covered with grayish-white or hoary fine hairs.
Capsule. A dry, dehiscent fruit composed of more than one carpel.
Carpel. A simple pistil, or one of the modified leaves forming a compound pistil.
Catkin. A scaly deciduous spike.
Caespitose. In little tufts or dense clumps; said of low plants of turfy habit.
Channeled. Deeply grooved longitudinally, like a gutter.
Chartaceous. With the texture of writing paper.
Ciliate. Fringed with hairs on the margin.
Cinerous. Ash-colored; light gray.
Circumpolar. Occurring around the pole.
Cismontane. This side of the mountains, or west of the main Sierra Nevada crest, as opposed to the deserts.
Clasping. Leaf partly or wholly surrounding the stem.
Claw. The narrow petiole-like base of some petals and sepals.
Coalescent. Said of organs of one kind that have grown together.
Compound. Having two or more similar parts in one organ. *Compound leaf*: one with two or more separate leaflets.
Compressed. Flattened laterally.
Concave. Hollow.
Congested. Crowded together.
Conic. Cone-shaped, with the point of attachment at the broad base.
Constricted. Tightened or grown together.
Contracted. Narrowed in a particular place; the opposite of open or spreading.
Convex. Rounded on the surface.

Convolute. Rolled up longitudinally.

Cordate. Heart-shaped with the notch at the base and ovate in general outline.

Coriaceous. Leathery in texture; tough.

Corolla. The inner perianth of a flower, composed of colored petals, which may be almost wholly united.

Corymb. A flat-topped or convex racemose flower-cluster, outer flowers opening first.

Corymbose. In corymbs.

Crenate. Having the margin cut with rounded teeth, scalloped.

Crenulate. The diminutive of crenate.

Crisped. Irregularly curled.

Crown. The top of a tree.

Crustaceous. Of brittle texture.

Cuneate. Wedge-shaped; triangular, with the narrow part at point of attachment.

Cupulate. Cup-shaped, the cup of an acorn.

Cuspidate. Tipped with a cusp, or sharp, short, rigid point.

Cyme. A flat-topped or convex paniculate flower-cluster with central flowers opening first.

Cymose. Arranged in cymes.

Deciduous. Falling off, as petals fall, or leaves of nonevergreen trees.

Decumbent. Lying down but with the tip ascending.

Decurrent. Extending down the stem below the insertion; said of leaves or ligules.

Dehiscent. Opening spontaneously when ripe to discharge contents.

Deltoid. Equilaterally triangular.

Dentate. Having the margin cut with sharp teeth not directed forward.

Denticulate. Slightly and finely toothed.

Depressed. Low, as if flattened from above.

Dichotomous. Repeatedly forking in pairs.

Diffuse. Scattered; widely spread.

Dilated. Flattened and broadened.

Dioecious. Having staminate and pistillate flowers on different plants.

Disarticulating. Separating joint by joint at maturity.

Disk flower. In Asteraceae, the tubular flowers of the head as distinct from the ray flowers.

Dissected. Deeply divided into numerous fine segments.

Distichous. In two vertical rows or ranks.

Divaricate. Widely divergent.

Divided. Separated to the base.

Drooping. Erectish at base but bending downward above.

Drupe. A fleshy one-seeded indehiscent fruit containing a stone.

Ecotype. Those individuals that are fitted to survive in only one kind of environment occupied by the species.

Exfoliating. Coming away as scales or flakes.

Edaphic. Pertaining to, or influenced, by soil conditions.

Ellipsoid. An elliptic solid.

Elliptic. In the form of a flattened circle more than twice as long as broad.

Entire. Undivided; the margin continuous, not incised or toothed.

Erect. Upright in relation to the ground, or sometimes perpendicular to the surface of attachment.

Evergreen. Remaining green throughout the winter.

Exserted. Protruding, as stamens projecting beyond the corolla.

Falcate. Sickle-shaped.

Fascicle. A close cluster or bundle of flowers, leaves, stems or roots.

Fasciculate. Clustered, parallel, erect branches.

Filiform. Thread-like.

Flaccid. Weak, limp.

Fleshy. Thick and juicy; succulent.

Flexuous. Zigzag.

Floret. The individual flower of the Asteraceae.

Floriferous. Bearing or producing flowers.

Foliaceous. Leaflike; in texture or appearance resembling leaves.

Foliolate. Having leaflets.

Follicle. A dry, monocarpellary fruit, opening only on the ventral suture.

Free. Not joined to other organs.

Fruit. The ripened pistil with all its accessory parts.

Funnelform. Gradually widening upwards, like a funnel.

Glabrate. Almost glabrous.

Glabrous. Without hairs.

Glabrescent. Becoming glabrous.

Gland. A depression, protuberance, or appendage on the surface of an organ which secretes a usually sticky fluid.

Glandular. Bearing glands or glandlike.

Glaucous. Covered or whitened with a bloom.

Globose. Spherical or rounded.

Globular. Somewhat or nearly globose.

Glutinous. With a gluey exudation.

Habit. General appearance of a plant.

Habitat. The normal situation in which a plant lives.

Hastate. The shape of an arrowhead but with the basal lobes turned outward.

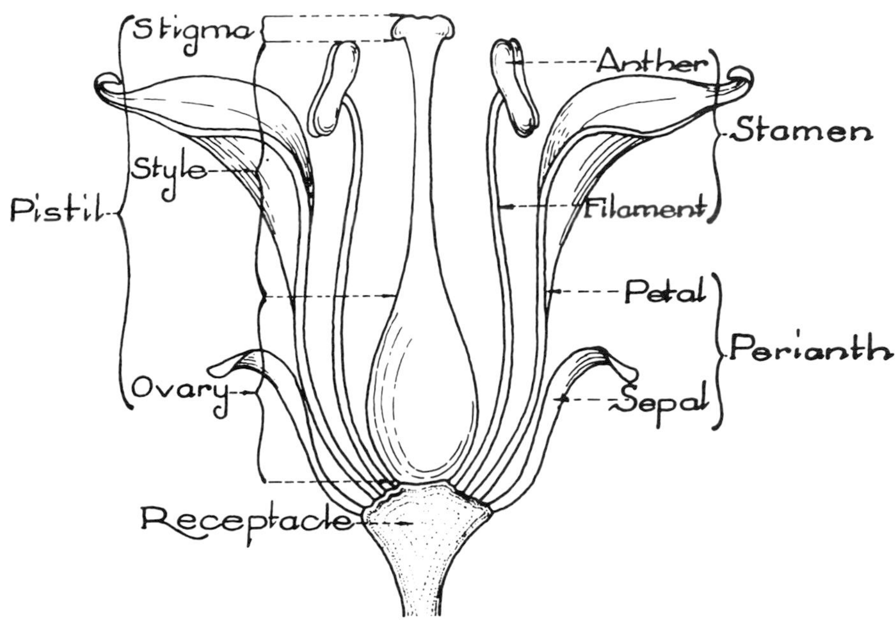

Fig. 87. Structure of a complete flower. (McMinn & Maino, *Pacific Coast Trees,* 1937).

Head. A dense globular cluster of sessile or subsessile flowers arising essentially from the same point on the peduncle.
Herbage. Collectively, the green parts of a plant.
Heteromorphic. Of more than one kind of form.
Hirsute. Rough with coarse or shaggy hairs.
Hispid. Rough with stiff or bristly hairs.
Hoary. Covered with white down.
Imbricate. Overlapping as shingles on a roof.
Imparipinnate. Odd-pinnate, having a terminal leaflet.
Indehiscent. Not splitting open.
Inflated. Blown up; bladdery.
Inflorescence. The flower-cluster of a plant.
Involucre. A whorl of bracts subtending a flower-cluster, as in the Asteraceae.
Irregular. Showing a lack of uniformity; asymmetric, a zygomorphic flower.
Laciniate. Cut into narrow lobes or segments.
Lanate. Woolly; densely clothed with long entangled hairs.
Lanulose. Short-woolly.
Lanceolate. Lance-shaped; much longer than broad, tapering from below the middle to the apex and more abruptly to the base.
Lateral. At or on the side.
Lax. Loose, distant.
Leaflet. A segment of a compound leaf.
Linear. Long and narrow, of a uniform width.
Lingulate. Tongue-shaped.
Lip. One of the two divisions of a bilabiate corolla or calyx.
Lobe. A division or segment of an organ, usually rounded or obtuse.
Membranaceous. Of the nature of a membrane; thin, soft and pliable.
-merous. Having parts, *i.e.*, 5-merous, having 5 parts.
Midrib. The central rib of a leaf or other organ.
Monoecious. Having staminate and pistillate flowers on the same plant.
Mucro. A small and short abrupt tip of an organ, as the projection of the midrib of a leaf.
Mucronate. With a mucro.
Nerve. A simple vein or slender rib of a leaf or bract.
Node. The joint of a stem; point of insertion of a leaf.
Nut. A hard-shelled and one-seeded indehiscent fruit.
Nutlet. Diminutive of nut.
Ob-. Prefix signifying the reverse.
Obconic. Inversely conical, point of attachment at the small end.
Obcordate. Inversely cordate, the notch at the apex.
Oblanceolate. Inversely lanceolate.
Oblong. Much longer than broad with nearly parallel sides.
Obovate. Inversely ovate.
Obovoid. Inversely ovoid.
Obsolete. Rudimentary or not evident.

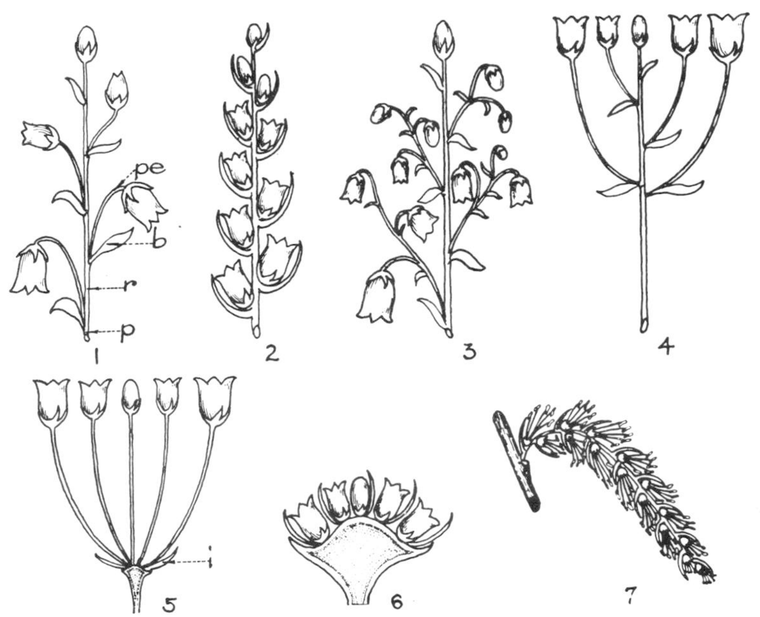

Fig. 88. Kinds of inflorescences: 1, raceme; p, peduncle; pe, pedicel; b, bract; r, rachis; 2, spike; 3, panicie; 4, corymb; 5, umbel; i, involucre; 6, head; 7, catkin. (McMinn & Maino, *Pacific Coast Trees,* 1937).

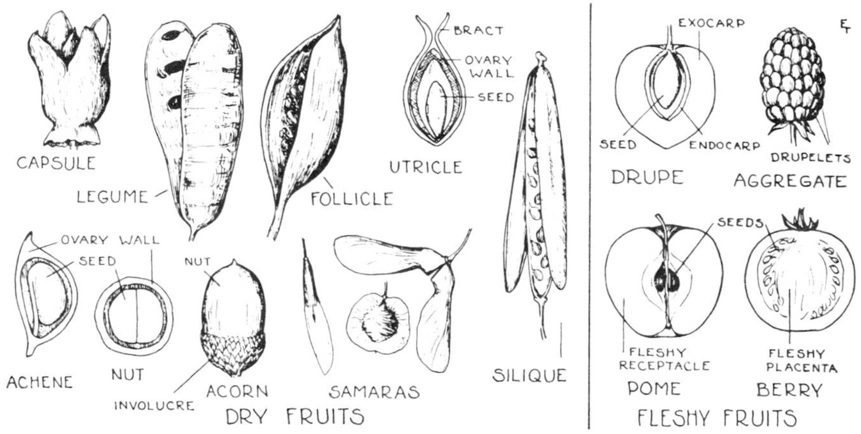

Fig. 89. Kinds of fruit. (McMinn & Maino, 1937).

Obtuse. Blunt or rounded at the end.

Offsets. Short lateral shoots from which new plants can develop.

Opposite. Set against, as leaves when two at one node.

Orbicular. Approximately circular.

Oval. Broadly elliptic.

Ovary. The part of the pistil that contains the ovules.

Ovate. With the outline of a hen's egg, the broader end downward.

Ovoid. Solid oval or solid ovate.

Palmate. Hand-shaped with fingers spread.

Panicle. A compound racemose inflorescence.

Paniculate. Borne in a panicle.

Papillate. Bearing minute conical processes or papillae.

Parted. Deeply cleft nearly to the base.

Pedicel. The stalk of a single flower.

Pedicellate. Having a pedicel.

Peduncle. The general term for the stalk of a flower or cluster of flowers.

Pedunculate. Having a peduncle.

Peltate. Shield-shaped; a flat body having the stalk attached to the lower surface instead of at the base or margin.

Pendent. Suspended or hanging, nodding.

Perfect. A flower having both stamens and pistils.

Perianth. The floral envelopes collectively; usually used when calyx and corolla are not clearly differentiated.

Pericarp. The ripened walls of the ovary, referring to a fruit.

Persistent. Remaining attached, as a calyx on the fruit.

Petal. One of the divisions of a corolla, usually colored.

Petaloid. Having the aspect of, or colored as petals.

Petiole. A leaf-stalk.

Petiolate. Having a petiole.

Pilose. Bearing soft and straight spreading hairs.

Pinna. A leaflet or primary division of a pinnate leaf.

Pinnate. A compound leaf, having the leaflets arranged on each side of a common petiole.

Pinnatifid. Pinnately cleft into narrow lobes not reaching to the midrib.

Pistil. The ovule-bearing organ of a flower.

Pistillate. Provided with pistils and without stamens; female.

Pitted. Having little depressions or pits; foveate.

Plane. Surface flat and even, not curved.

Plumose. Feathery; having fine hairs on each side of a plume.

Pod. Any dry dehiscent fruit; specifically a legume.

Pollen. The male fecundating spores found in the anther.

Polymorphous. Of several forms.

Pome. An apple-like fruit.

Prickle. Sharp outgrowth of the bark or epidermis.

Prickly. Armed with prickles.

Procumbent. Trailing on the ground, but not rooting.

Prostrate. Lying flat on the ground.

Puberulent. Minutely pubescent.

Pubescent. Covered with short soft hairs; downy.

Punctate. Dotted with punctures or with translucent pitted gland or with colored dots.

Pungent. Ending in a rigid, sharp point or prickle.

Raceme. A simple, elongated, indeterminate inflorescence with each flower subequally pedicelled.

Racemose. Of the nature of a raceme or in racemes.

Rachis. The axis of a spike or raceme, or of a compound leaf.

Ray. A primary branch of an umbel; the ligule of a ray-flower in the Asteraceae, the ray-flowers being marginal and differentiated from the disk-flowers.

Receptacle. That portion of the floral axis upon which the flower parts are borne.

Recurved. Bent backwards.

Reflexed. Abruptly bent downward.

Regular. Said of a flower having radial symmetry, with the parts in each series alike.

Reniform. Kidney-shaped.

Repand. With an undulating margin, less strongly wavy than sinuate.

Resiniferous. Producing resin.

Reticulate. With a network; not veined.

Revolute. Rolled backward from both margins, *i.e.*, toward the underside.

Rhombic. Somewhat diamond-shaped.

Rhomboidal. A solid with a rhombic outline.

Rootstock. An underground stem or rhizome, with scales at the nodes and producing leafy shoots on the upper side and roots on the lower side.

Rosette. A crowded cluster of radiating leaves appearing to rise from the ground.

Rotund. Rounded in outline.

Rugose. Wrinkled.

Sagittate. Shaped like an arrowhead with the basal lobes turned downward.

Salverform. A corolla with slender tube abruptly expanding into a flat limb.

Samara. An indehiscent winged fruit.

Scabrous. Rough to the touch, owing to the structure of the epidermis or to the presence of short stiff hairs.

Scale. Any thin scarious bract; usually a vestigial leaf.

Scandent. Climbing.

Scarious. Thin, dry and membranaceous, not green.

Scorpioid. Said of a unilateral inflorescence circinately coiled in the bud.

Scurfy. Clothed with small bran-like scales.

Seed. The ripened ovule.

Segment. A division or part of a leaf or other organ that is cleft or divided but not truly compound.

Sensu lato. In the broad sense.

Sepal. A leaf or segment of the calyx.

Sericeous. Silky; clothed with appressed fine and straight hairs.

Serrate. Saw-toothed, the sharp teeth pointing forward.

Serrulate. Finely serrate.

Sessile. Attached directly by the base, not stalked.

Setose. Clothed with bristles.

Setulose. Bearing minute bristles.

Shrub. A woody plant of smaller proportions than a tree, which usually produces several branches from the base.

Silky. See sericeous.

Simple. Unbranched, as a stem or hair.

Sinuate. With a strongly wavy margin.

Smooth. Not rough to the touch, without hairs.
Solitary. Borne singly.
Squamose. Covered with scales; scaly.
Squarrose. Spreading rigidly at right angles or more, as the tips of bracts.
Stamen. The male organ of the flower which bears the pollen.
Staminate. Having stamens but not pistils; said of a flower or plant that is male.
Stellate. Star-shaped.
Stigma. The receptive part of the pistil on which the pollen germinates.
Stipe. The stalk beneath an ovary.
Stipitate. Having a stipe or stalk.
Stipule. One of a pair of usually foliaceous appendages found at the base of the petiole in many plants.
Stomate. A breathing pore or aperture in the epidermis.

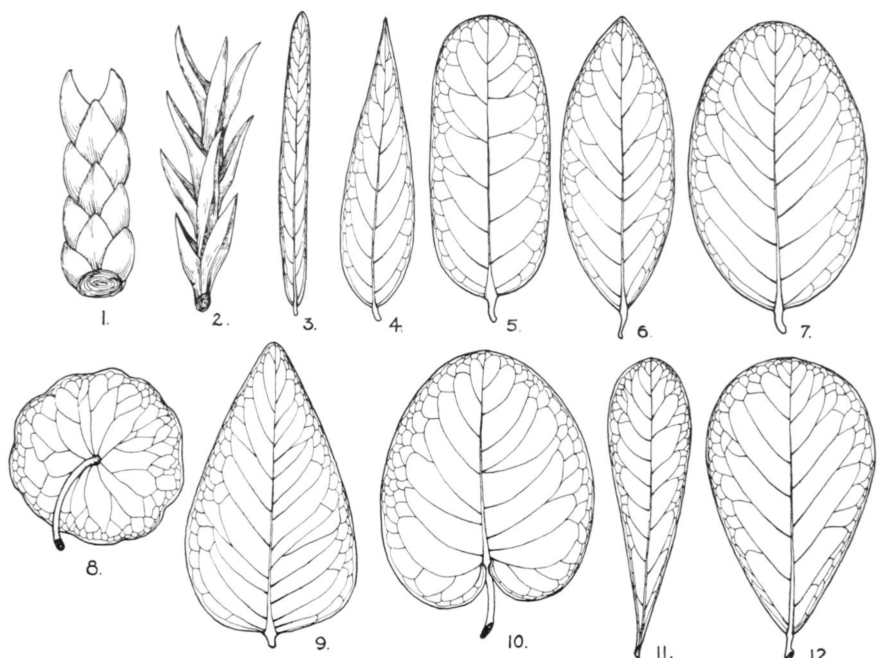

Fig. 90. Shapes of leaves: 1, scale-like; 2, awl-shaped; 3, linear; 4, lanceolate; 5, oblong; 6, elliptic; 7, oval; 8, orbicular; 9, ovate; 10, cordate; 11, oblanceolate; 12, obovate. (McMinn & Maino, *Pacific Coast Trees,* 1937).

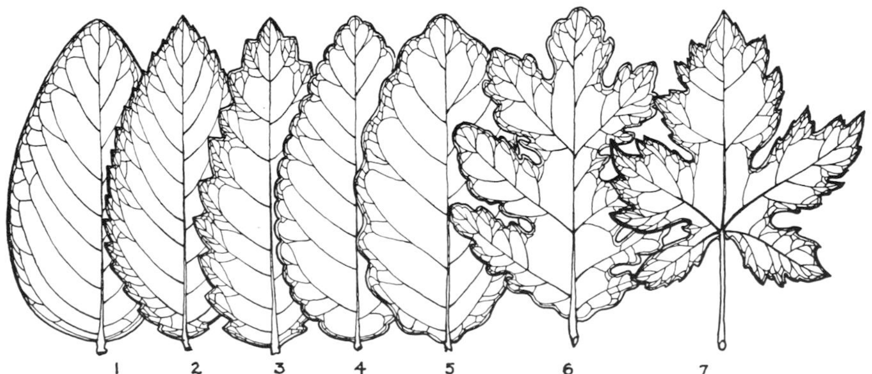

Fig. 91. Margins of leaves: 1, entire; 2, serrate; 3, dentate; 4, crenate; 5, sinuate; 6, pinnately-lobed; 7, palmately-lobed. (McMinn & Maino, *Pacific Coast Trees,* 1937).

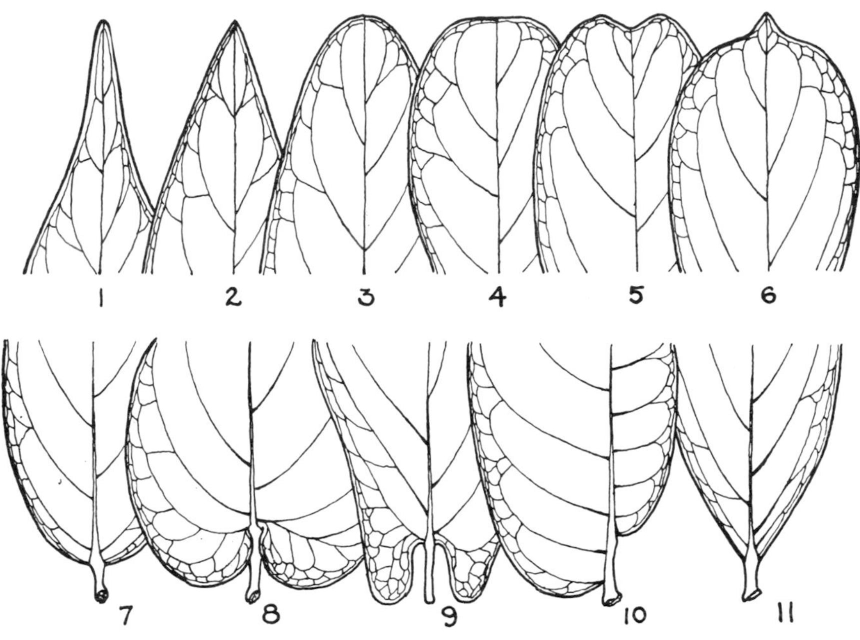

Fig. 92. Tips and bases of leaves: 1, acuminate; 2, acute; 3, obtuse; 4, truncate; 5, emarginate; 6, mucronate; 7, rounded; 8, cordate or heart-shaped; 9, auriculate; 10, oblique or unequal; 11, cuneate or wedge-shaped. (McMinn & Maino, *Pacific Coast Trees,* 1937).

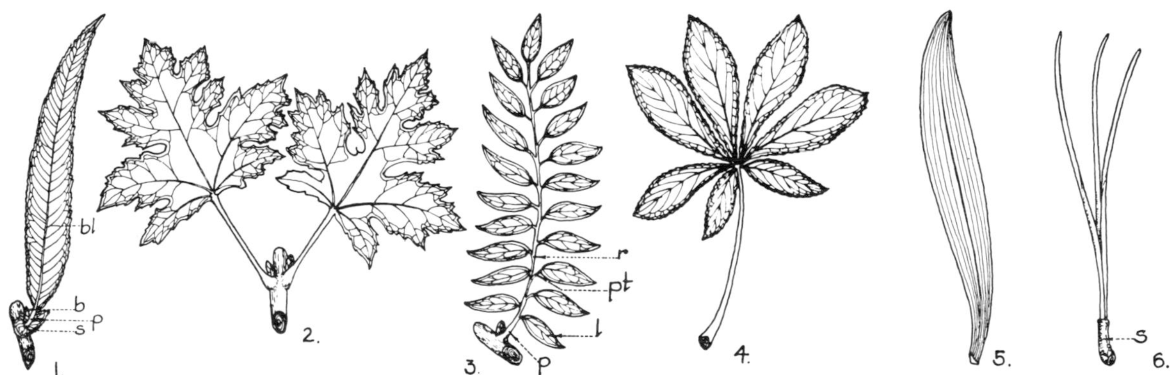

Fig. 93. Kinds, parts, arrangement and venation of leaves: 1, simple, alternate leaf with netted venation; b, bud; bl, blade; p, petiole; s, stipule; 2, simple, opposite, palmately veined and lobed leaves; 3, pinnately compound, alternately veined leaf; 4, palmately compound leaf; 5, simple leaf with parallel veins; 6, simple leaves in a fascicle; s, sheath. (McMinn & Maino, *Pacific Coast Trees,* 1937).

Stone. The bony endocarp of a drupe.

Strigose. Clothed with sharp and stiff appressed straight hairs.

Style. The contracted portion of the pistil between the ovary and the stigma.

Sub-. Prefix meaning somewhat, almost, of inferior rank, beneath.

Subtend. To be below and close to, as the leaf subtends the shoot borne in its axil.

Subulate. Awl-shaped.

Succulent. Juicy; fleshy and soft.

Suffrutescent. Obscurely shrubby; very little woody.

Symmetrical. Said of a flower having the same number of parts in each circle.

Sympatric. Growing together with, or having the same range as.

Syn-. Prefix meaning united.

Taxon. Any taxonomic unit, as an order, genus, variety, etc. (plural, *taxa*).

Tendril. A slender, coiling or twining organ by which a climbing plant grasps its support.

Terete. Cylindrical; round in cross section.

Ternate. In threes.

Thorn. A sharp-pointed stiff woody body derived from a modified branch.

Thyrse. A compact, ovate panicle; with main axis indeterminate, but with other axes cymose.

Tomentose. With tomentum; covered with rather short, densely matted, soft white wool.

Tomentulose. Diminutive of tomentose.

Tomentum. A covering of densely matted woolly hairs.

Tooth. A small marginal lobe.

Toothed. Dentate.

Tri-. Prefix signifying three, thrice or triply.

Truncate. As if cut off squarely at the end.

Tube. The narrow basal portion of a gamopetalous corolla, or gamosepalous calyx.

Tubercule. A small tuber-like prominence or nodule.

Tubular. Shaped like a hollow cylinder.

Umbel. A flat or convex flower-cluster in which the pedicels arise from a common point, like rays of an umbrella.

Umbellate. Borne in an umbel.

Undulate. Wavy, less pronounced than sinuate.

Unisexual. Flowers having only stamens or pistils; of one sex.

Valve. One of the segments into which a dehiscent capsule or legume separates.

Vein. A vascular bundle of a leaf or other flat organ.

Velutinous. Velvety; covered with a fine and dense silky pubescence.

Verrucose. Warty; covered with wart-like excrescences; tuberculate.

Villous. Bearing long and soft and not matted hairs; shaggy.

Viscid. Sticky, glutinous.

Whorl. A ring of similar organs radiating from a node.

Wing. A thin and usually dry extension bordering an organ.

Woolly. Having long, soft, entangled hairs; lanate.

Zygomorphic. Bilaterally symmetrical.

Index to Common Names

condalia: *Zizyphus parryi*
manzanita: *Arctostaphylos parryana*
nolina: *Nolina parryi*
pine: *Pinus quadrifolia*
Peach, desert: *Prunus andersonii*
Pecho manzanita: *Arctostaphylos pechoensis*
Penstemon, climbing: *Keckiella cordifolia*
snapdragon: *Keckiella antirrhinoides*
whorlleaf: *Keckiella ternata*
yellow: *Keckiella antirrhinoides*
Pepperwood: *Umbellularia californica*
Phlox, prickly: *Leptodactylon californicum*
Pine, bigcone: *Pinus coulteri*
Coulter: *Pinus coulteri*
digger: *Pinus sabiniana*
fourleaf pinyon: *Pinus quadrifolia*
Jeffrey: *Pinus jeffreyi*
oneleaf pinyon: *Pinus monophylla*
sugar: *Pinus lambertiana*
Torrey: *Pinus torreyana*
yellow: *Pinus ponderosa*
Pine Hill fremontia: *Fremontodendron californicum* ssp. *decumbens*
Pinkbract manzanita: *Arctostaphylos pringlei* var. *drupacea*
Pinyon pine, fourleaf: *Pinus quadrifolia*
oneleaf: *Pinus monophylla*
Piper barberry: *Mahonia piperiana*
Pipestem clematis: *Clematis lasiantha*
Plum, Sierra: *Prunus subcordata*
Poison oak: *Toxicodendron diversilobum*
Poppy, Catalina tree: *Dendromecon rigida* ssp. *rhamnoides*
island tree: *Dendromecon rigida* ssp. *harfordii*
matilija: *Romneya coulteri*
tree: *Dendromecon rigida* ssp. *rigida*
Prickly phlox: *Leptodactylon californicum*
Pt. Reyes ceanothus: *Ceanothus gloriosus* var. *gloriosus*
Purple sage: *Salvia leucophylla*
Quail bush: *Atriplex lentiformis*
Quixote plant: *Yucca whipplei*
Rabbitbrush: *Chrysothamnus nauseosus*
Red alder: *Alnus oregona*
Redberry: *Rhamnus crocea*
Redbud, western: *Cercis occidentalis*
Red bush monkeyflower: *Diplacus puniceus*
Red-flowered currant: *Ribes sanguineum*
Redheart ceanothus: *Ceanothus spinosus*
Redshanks: *Adenostoma sparsifolium*
Redwood, coast: *Sequoia sempervirens*
giant: *Sequoiadendron giganteum*
Reed, common: *Phragmites australis*
giant: *Arundo donax*
Refugio manzanita: *Arctostaphylos refugioensis*
Roble: *Quercus lobata*
Romero: *Trichostema lanatum*
Rose, California wild: *Rosa californica*
small-leaf: *Rosa minutifolia*
wood: *Rosa gymnocarpa*
Rowe ceanothus: *Ceanothus papillosus* var. *roweanus*
Saffron buckwheat: *Eriogonum crocatum*
Sage, black: *Salvia mellifera*
bladder: *Salazaria mexicana*
Cleveland: *Salvia clevelandii*
mountain desert: *Salvia pachyphylla*
purple: *Salvia leucophylla*
white: *Salvia apiana*
Sagebrush, basin: *Artemisia tridentata*
coastal: *Artemisia californica*
Saltgrass: *Distichlis spicata*
San Diego bursage: *Franseria chenopodiifolia*
bush monkeyflower: *Diplacus aridus*
ceanothus: *Ceanothus cyaneus*
mountain mahogany: *Cercocarpus minutiflorus*
tetracoccus: *Tetracoccus dioicus*
sunflower: *Viguiera laciniata*
Sandmat manzanita: *Arctostaphylos pumila*
Santa Barbara ceanothus: *Ceanothus impressus* var. *impressus*
Santa Cruz Island buckwheat: *Eriogonum arborescens*
manzanita: *Arctostaphylos subcordata*
Sawtooth goldenbush: *Haplopappus squarrosus*
Screwbean: *Prosopis pubescens*
mesquite: *Prosopis pubescens*
Scrub oak: *Quercus dumosa*
Sequoia, coast: *Sequoia sempervirens*
giant: *Sequoiadendron giganteum*
Serpentine manzanita: *Arctostaphylos obispoensis*
Shadscale: *Atriplex confertifolia*
Shaw agave: *Agave shawii*
Sierra bladdernut: *Staphylea bolanderi*
gooseberry: *Ribes roezlii*
plum: *Prunus subcordata*
Silver buffaloberry: *Shepherdia argentea*
Silverleaf manzanita: *Arctostaphylos silvicola*
Silver lupine: *Lupinus albifrons*
Singleleaf ash: *Fraxinus anomala*
Small-leaf mahogany: *Cercocarpus intricatus*
rose: *Rosa minutifolia*

Literature Cited

Abrams, L. R. 1934. The mahonias of the Pacific Coast. *Phytologia* 1: 89-94.

Ahrendt, L. W. 1961. *Berberis* and *Mahonia. J. Linn. Soc.* (London), Bot. Ser. 57: 1-408.

Anderson, D. L. 1971. The San Andreas fault. *Sci. Amer.* 225: 52-66.

Bailey, H. P. 1954. Climate, vegetation and land use in southern California, pp. 31-44. *In* R. H. Jahns (ed.) *Geology of southern California.* Bull. 170. Dept. of Nat. Resources, State of California, San Francisco.

————. 1966. *Weather in southern California.* Univ. Calif. Press, Berkeley, California. 87 pp.

Bailey, L. H. 1936. *Washingtonia. Gentes Herb.* 4: 53-82.

Barbour, M. G. & J. Major. 1977. Introduction, pp. 3-10. *In* M. G. Barbour & J. Major (eds.) *Terrestrial vegetation of California.* John Wiley & Sons, New York. 1002 pp.

Beeks, R. M. 1961. Variation and hybridization in southern California populations of *Diplacus* (Scrophulariaceae). M. S. thesis. Claremont Graduate School, Claremont, California. 103 pp.

Bergman, R. 1977. Wind, sand and tears: gardening in the low desert. *Pacif. Hort.* 38 (1): 3-5.

Clark, D. E. (ed.) 1979. *Western Garden Book.* Sunset Books, Palo Alto, California. 512 pp.

Clausen, J., D. D. Keck & W. M. Heisey. 1945. Experimental studies on the nature of species; II. Plant evolution through amphiploidy, and autoploidy with examples from the *Madiinae.* Carnegie Inst. Wash. *Pub.* 564. 174 pp.

Cooper, W. S. 1922. The broad-sclerophyll vegetation of California — an ecological study of the chaparral and its related communities. Carnegie Inst. Wash. *Pub.* 319. 122 pp.

Cornell, R. D. 1938. *Conspicuous California plants, with notes on their garden uses.* San Pasqual Press, Pasadena, California. 192 pp.

Cronemiller, F. P. 1941. Chaparral. *Madroño* 6: 199.

Daubenmire, R. 1968. *Plant communities; a textbook of plant synecology.* Harper & Row, New York. 300 pp.

Davis, P. H. & V. H. Heywood. 1963. *Principles of angiosperm taxonomy.* Van Nostrand, Princeton, New Jersey. 532 pp.

Dewey, J. F. 1972. Plate tectonics. *Sci. Amer.* 226: 56-68.

Dourley, J. 1977. Developing new cultivars, pp. 55-62. *In* D. R. Walters, et al. (ed.) Native plants: a viable option. California Native Plant Society. *Spec. Pub.* 3. 213 pp.

Durham, J. W. & E. C. Allison. 1960. The geological history of Baja California and its marine faunas. *Syst. Zool.* 9: 47-91.

Fritts, H. C. & G. A. Gordon. 1980. Annual precipitation for California reconstructed from western North American tree rings. Agreement No. B53367. California Dept. of Water Resources. Sacramento. 43 pp.

Gastel, R. G., R. P. Phillips & E. C. Allison. 1975. Reconnaissance geology of the state of Baja California. The Geol. Soc. Amer. Inc. *Memoir* 140. 170 pp.

Gentry, H. S. 1958. The natural history of jojoba (*Simmondsia chinensis*) and its cultural aspects. *Econ. Bot.* 12: 261-295.

Greene, E. L. 1889. *Illustrations of West American oaks from drawings of the late Albert Kellogg, M.D.* Privately printed. San Francisco, California. 84 pp.

Griffin, J. R. 1977. Oak woodlands, pp. 383-416. *In* M. G. Barbour & J. Major (eds.) *Terrestrial vegetation of California.* John Wiley & Sons, New York. 1002 pp.

Hallan, A. 1972. Continental drift and the fossil record. *Sci. Amer.* 227: 56-66.

Hanes, R. L. 1977. Chaparral, pp. 417-470. *In* M. G. Barbour & J. Major (eds.) *Terrestrial vegetation of California.* John Wiley & Sons, New York. 1002 pp.

Hellmers, H., J. S. Horton, G. Juhren & J. O'Keefe. 1955. Root systems of some chaparral plants in southern California. *Ecology* 36: 667-678.

Hildreth, W. R. 1977. Bringing natives into cultivation, pp. 42-47. *In* D. R. Walters, et al. (eds.) *Native plants: a viable option.* California Native Plant Society. *Spec. Pub.* 3. Berkeley. 213 pp.

Horton, J. S. 1960. Vegetation types of the San Bernardino Mountains. Forest Service, Pacific Southwest Forest and Range Experimental Station, United States Department of Agriculture. *Tech. Paper, PSW-44.* Berkeley.

Jahns, R. H. 1954. Investigations and problems of southern California geology, pp. 5-29. *In* R. H. Jahns (ed.) *Geology of southern California.* Bull. 170. Dept. of Nat. Resources, State of California. San Francisco.

Jepson, W. L. 1914. *A flora of California.* Vol. 1, pt. IV. pp. 439-441. Associated Students Store, Berkeley.

————. 1936. *A flora of California.* Vol. 2, pt. III, p. 217. Associated Students Store, Berkeley.

————. 1936a. *A flora of California.* Vol. 2, pt. III, p. 227. Associated Students Store, Berkeley.

————. 1925. *Manual of the flowering plants of California.* pp. 918-919. Associated Students Store, Berkeley.

Johnson, H. B. 1976. Vegetation and plant communities of southern California deserts — a functional view, pp. 125-164. *In* J. Latting (ed.) *Plant communities of southern California.* California Native Plant Society. *Spec. Pub.* 2. Berkeley. 164 pp.

Kimball, M. H. 1959. Plant climates of California. *Calif. Agric.* 13: 7-12.

Lathrop, E. W. & R. F. Thorne. 1978. A flora of the Santa Ana mountains. *Aliso* 9: 197-278.

Latting, J. (ed.) 1976. *Plant communities in southern California.* California Native Plant Society. *Spec. Pub.* 2. Berkeley. 164 pp.

Lenz, L. W. 1977. Rancho Santa Ana Botanic Garden — the first fifty years. *Aliso* 9: 1-156.

McMinn, H. E. 1939. *An illustrated manual of California shrubs.* J. W. Stacey, San Francisco. 689 pp.

————. 1949. Two new artificial hybrids suitable for garden ornamentals. *Leafl. Santa Barbara Botanic Garden* 1: 58-63.

Mirov, N. T. 1967. *The genus Pinus.* Ronald Press, New York, 602 pp.

Mooney, H. A. 1977. Southern coastal sage scrub, pp. 471-490. *In* M. G. Barbour & J. Major (eds.) *Terrestrial vegetation of California.* John Wiley & Sons, New York. 1002 pp.

Munz, P. A. 1974. *A flora of southern California.* Univ. Calif. Press, Berkeley. 1086 pp.

————. 1950. California plant communities — supplement. *El Aliso* 2: 199-202.

————. 1959. *A California flora.* Univ. Calif. Press, Berkeley. 1681 pp.

———— & D. D. Keck. 1949. California plant communities. *El Aliso* 2: 87-105.

Nobs, M. 1963. Experimental studies on species relationships in *Ceanothus.* Carnegie Inst. Wash. *Pub.* 624. 94 pp.

Nord, E. C. 1959. Bitterbrush ecology — some recent findings. Forest Service, Pacific Southwest Forest and Range Experiment Station, United States Department of Agriculture. *Research note* 148.

————. 1959. Bitterbrush plants can be propagated from stem cuttings. Forest Service, Pacific Southwest Forest and Range Experiment Station, United States Department of Agricultural. *Research note* 149.

Oosting, H. J. 1948. *The study of plant communities.* W. H. Freeman & Co., San Francisco, California. 389 pp.

Philbrick, R. N. & J. R. Haller. 1977. The southern California islands, pp. 893-908. *In* M. G. Barbour & J. Major (eds.) *Terrestrial vegetation of California.* John Wiley & Sons, New York. 1002 pp.

Rafols, E. S. 1977. Analog tree and shrub species, pp. 128-143. *In* N. J. Thrower & J. E. Bradley (eds.) *Chile-California mediterranean scrub atlas.* Dowden, Hutchinson & Ross, New York. 237 pp.

Raven, P. H. & D. I. Axelrod. 1978. Origin and relationships of the California flora. *Univ. California Pub. Bot.* 72: 1-134.

Roberts, W. G. 1979. California plants for the Central Valley. *Pacif. Hort.* 40 (2): 27-36.

Roderick, W. 1977. Natives introduced (or should be introduced) into the trade, pp. 63-67. *In* D. R. Walters, et al. (eds.) *Native plants: a viable option.* California Native Plant Soc. *Spec. Pub.* 3. Berkeley. 213 pp.

Roof, J. B. 1967. Editor's note. *Four Seasons* 2: 14.

Russell, R. J. 1926. Climates of California. Univ. Calif. *Publ. Geogr.* 2: 73-84.

Russo, R. A. 1979. *Plant galls of the California region.* Boxwood Press, Pacific Grove, California. 203 pp.

Sacamano, C. M. & W. D. Jones. 1975. Native trees and shrubs for landscape use in the desert southwest. Cooperative Ext. Serv. College of Agriculture Univ. Ariz. *Bull.* A82. 40 pp.

Schimper, A. F. 1903. *Plant geography on a physiological basis.* (Trans. by W. R. Fisher.) Clarenden Press, Oxford, U.K. 839 pp.

Schopmeyer, C. S. 1974. *Seeds of woody plants in the United States.* Forest Service, United States Department of Agriculture. *Agricultural handbook* 450. 883 pp.

Shaw, G. R. 1914. The genus *Pinus.* Arnold Arboretum. *Publ.* 5. 96 pp.

Shreve, F. & I. L. Wiggins. 1964. *Vegetation and flora of the Sonora Desert.* Vol. 1, Stanford Univ. Press, Palo Alto, California. 840 pp.

Skinner, H. T. 1960. A new plant hardiness map for the United States and southern Canada. *Am. Hort. Mag.* 39: 204-205.

Stebbins, G. L. 1974. A California botanist in Chile. *Fremontia* 2: 8-13.

Storie, R. E. & W. W. Weir, n.d. [1964] *Generalized soil map of California.* Division of Agricultural Sciences, Univ. Calif. Publ. 4028.

Sudworth, G. B. 1908. *Forest trees of the Pacific slope.* Forest Service, United States Department of Agriculture. Washington. 441 pp.

Thorne, R. F. 1976. The vascular plant communities of California, pp. 1-31. *In* J. Latting (ed.) *Plant communities of southern California.* California Native Plant Society. *Spec. Pub.* 2. Berkeley. 164 pp.

——————. 1977. Montane and subalpine forests of the Transverse and Peninsular ranges, pp. 537-558. *In* M. G. Barbour & J. Major (eds.) *Terrestrial vegetation of California.* John Wiley & Sons, New York. 1002 pp.

Thrower, N. J. & D. E. Bradbury (eds.) 1977. *Chile-California mediterranean scrub atlas.* Dowden, Hutchinson & Ross, New York. 237 pp.

Tilforth, C. W. 1977. Survival of the fittest, pp. 20-31. In *Director's Report, 1976.* Rancho Santa Ana Botanic Garden, Claremont. 31 pp.

Trewartha, G. T. 1954. *An introduction to climate.* McGraw-Hill Book Co., New York. 402 pp.

Van Rensselaer, M. & H. E. McMinn. 1942. *Ceanothus.* Santa Barbara Botanic Garden, Santa Barbara, California. 308 pp.

Vasek, F. C. 1980. Creosote bush rings in the Mojave Desert. *Fremontia* 7: 10-13.

—————— & R. F. Thorne. 1977. Transmontane coniferous vegetation, pp. 797-834. *In* M. G. Barbour & J. Major (eds.) *Terrestrial vegetation of California.* John Wiley & Sons, New York. 1002 pp.

Walheim, L. (ed.) 1977. *The world of trees* (Western edition). Ortho Books, San Francisco. 113 pp.

Wiggins, I. L. 1960. The origin and relationships of the land flora [of Baja California]. *Syst. Zool.* 9: 148-164.

——————. 1980. *Flora of Baja California.* Stanford Univ. Press, Stanford, California. 1025 pp.

Wilson, R. & R. J. Vogl. 1965. Manzanita chaparral in the Santa Ana mountains. *Madroño* 18: 47-62.

Wolf, C. W. 1942. *Diplacus longiflorus rutilis* — bush monkeyflower. *Leafl. Popular Information* 49. Rancho Santa Ana Botanic Garden. Santa Ana Cañon, Orange Co., California. 2 pp.

——————. 1943. *Comarostaphylis diversifolia planifolia* — summer holly. *Leafl. Popular Information* 61. Rancho Santa Ana Botanic Garden. Santa Ana Cañon, Orange Co., California. 2 pp.

——————. 1944. *Cornus glabrata* — brown dogwood. *Leafl. Popular Information* 72. Rancho Santa Ana Botanic Garden. Santa Ana Cañon, Orange Co., California. 2 pp.

Yermanos, D. M. 1974. An agronomic survey of jojoba in California. *Econ. Bot.* 28: 160-174.

Young, D. A. 1974. Introgressive hybridization in two southern California species of *Rhus* (Anacardiaceae). *Brittonia* 26: 241-255.

Index

Common names will be found in Index to Common Names. Main entries are italicized; **figures** are indicated by boldface.

A

B

C

D

E

California Native Trees & Shrubs was composed and lithographed by The Day Printing Corporation, Pomona, and bound by Weber-McCrea Co., Glendale. The body of the text is set in Garamond type. The paper is Haworth improved Quintessence. Color separations by Mueller Color Plate Co., Monterey Park. Design by Guy Moore & Associates, Beverly Hills. All photographs, unless otherwise noted, are by Lee W. Lenz.